AUDITORY
COGNITION
AND
HUMAN
PERFORMANCE
RESEARCH AND APPLICATIONS

AUDITORY COGNITION AND HUMAN PERFORMANCE
RESEARCH AND APPLICATIONS

CARRYL L. BALDWIN

CRC Press
Taylor & Francis Group
Boca Raton London New York

CRC Press is an imprint of the
Taylor & Francis Group, an **informa** business

CRC Press
Taylor & Francis Group
6000 Broken Sound Parkway NW, Suite 300
Boca Raton, FL 33487-2742

First issued in paperback 2019

© 2012 by Taylor & Francis Group, LLC
CRC Press is an imprint of Taylor & Francis Group, an Informa business

No claim to original U.S. Government works

ISBN-13: 978-0-415-32594-3 (hbk)
ISBN-13: 978-0-367-37647-5 (pbk)

Library of Congress Cataloging-in-Publication Data

Baldwin, Carryl L.
  Auditory cognition and human performance : research and applications / Carryl L. Baldwin.
    p. cm.
  ISBN 978-0-415-32594-3 (hardback)
  1. Listening. 2. Auditory perception. 3. Cognition. 4. Ability. I. Title.

BF323.L5B28 2012
152.1'5--dc23

2011031377

Visit the Taylor & Francis Web site at
http://www.taylorandfrancis.com

and the CRC Press Web site at
http://www.crcpress.com

AUDITORY COGNITION AND HUMAN PERFORMANCE

RESEARCH AND APPLICATIONS

CARRYL L. BALDWIN

CRC Press
Taylor & Francis Group
Boca Raton London New York

CRC Press is an imprint of the
Taylor & Francis Group, an **informa** business

CRC Press
Taylor & Francis Group
6000 Broken Sound Parkway NW, Suite 300
Boca Raton, FL 33487-2742

First issued in paperback 2019

© 2012 by Taylor & Francis Group, LLC
CRC Press is an imprint of Taylor & Francis Group, an Informa business

No claim to original U.S. Government works

ISBN-13: 978-0-415-32594-3 (hbk)
ISBN-13: 978-0-367-37647-5 (pbk)

Library of Congress Cataloging-in-Publication Data

Baldwin, Carryl L.
   Auditory cognition and human performance : research and applications / Carryl L. Baldwin.
      p. cm.
   ISBN 978-0-415-32594-3 (hardback)
   1. Listening. 2. Auditory perception. 3. Cognition. 4. Ability. I. Title.

BF323.L5B28 2012
152.1'5--dc23
                                                                    2011031377

Visit the Taylor & Francis Web site at
http://www.taylorandfrancis.com

and the CRC Press Web site at
http://www.crcpress.com

Contents

List of Figures and Tables

Preface

How people acquire and process information has been a fundamental question in psychology since its inception. Cognitive science has devoted much effort to addressing the question, but typically in the domain of vision. Auditory processing has generally received less-extensive inquiry, whether in basic perceptual or cognitive psychology or in applied areas such as human factors. Moreover, even within the field of audition, higher-order auditory processes have received less study than such lower-level processes as loudness and pitch perception (Bregman, 1990; Plomp, 2002). Yet, since the 1990s there has been a growing base of empirical research on auditory cognition and its role in human performance at work and in everyday life. I describe this work in this book.

The notion that listening requires attention and that it can at times be a difficult undertaking is well known to the elementary schoolteacher. Less well appreciated is the effort that we adults must put forth to comprehend auditory information in our everyday lives. Auditory processing relies on mechanisms of the brain as well as the ear. Describing the mental effort involved in these interacting mechanisms is the primary purpose of this book.

I first became interested in this interaction when I was a graduate student studying age differences in mental workload for complex tasks, driving in particular. I had just completed an investigation in which, as I had expected, the classic dissociation between the performance of younger and older participants was observed in the most challenging dual-task conditions, while no performance differences were seen in the single-task or simpler dual-task conditions. Closer inspection of the results, however, revealed that this age difference could be attributed to a small minority of the participants. The overwhelming majority of older participants had not only performed just as well, but in fact many had performed slightly better than their younger counterparts. Conventional wisdom at the time strongly suggested that the performance decrements observed by the small group of older participants were indicative of cognitive impairments. Perhaps this interpretation was true. However, I could not rule out that something else might have contributed to their poor performance. As was also conventional at the time, and sadly remains the dominant protocol, I had not collected basic measures of visual or auditory acuity that would allow me to rule out the possibility that perhaps sensory decrements were contributing to or masquerading as cognitive decrements. Convinced that this possibility must be ruled out before any conclusions could be drawn, I embarked on a series of investigations that has led to my current position on the matter and forms one of the primary themes of this book.

Auditory processing is a remarkable process demanding mental effort and relies heavily on the interaction between sensory and cognitive mechanisms.

There are many people to whom I am grateful for contributing to the development of this book, either directly or indirectly. Gratitude goes to my husband and colleague, Raja Parasuraman, for both encouraging me to write the book in the first place and for his support throughout the entire process. I am also grateful to the many graduate students who have worked in my lab who have discussed different aspects of the research described in this book and offered valuable comments. Preparation of this book was also made possible by grants that I have received for my research from the National Aerospace and Space Administration, the National Institutes of Health, the Office of Naval Research, and the National Highway Traffic Safety Administration, although the views expressed here are my own and not necessarily endorsed by these federal agencies. Finally, I dedicate this book to Paeton, to who she is now and all she will become. May she forever find solace and joy in the soundscape of life.

About the Author

Carryl L. Baldwin, PhD, is an associate professor at George Mason University, where she directs an auditory research group and the Mason Transportation Institute's Driving Simulation Laboratory. She received her graduate training in human factors at the University of South Dakota in Vermillion, where she began to conduct research leading to her current interests in applied auditory cognition, mental workload assessment, and transportation. She received her undergraduate training from the University of Nebraska–Lincoln, where she developed a strong interest in auditory processing. She has taught classes at Western Iowa Technical Community College in Sioux City, where she was also chair of the Social and Behavioral Sciences Department. She was tenured at Old Dominion University in Norfolk, Virginia, before moving to join the faculty in George Mason's ARCH Lab and Human Factors and Applied Cognition Program. She currently resides in Falls Church, Virginia. She spends her nonworking hours playing music—particularly mandolin and guitar—gathering with friends and family, taking long walks outside, and traveling.

1 Hearing
The Neglected Sense

INTRODUCTION

Being able to hear and understand sounds—auditory processing—greatly enriches our lives and enables us to accomplish many tasks essential to survival. Although we engage in this process continuously throughout our lives, many may fail to appreciate that the seemingly automatic task of auditory processing often involves considerable mental effort to accomplish. Consider the following examples:

> Lara is driving down the highway listening to a news program on the radio when she hears the name of her hometown. To better listen to and understand the subsequent story, she turns up the volume on the radio and quits eating popcorn from the sack on the seat next to her.

> Johan is really hoping to get a chance to interview for a new position and feels he must show his knowledge of the topics discussed. The restaurant is crowded and noisy, and he struggles to hear so that he can follow the conversation. He concentrates, turns his visual attention to each speaker, and then realizes at the end of the lunch that he has barely touched his food.

In each of the scenarios, the effort required for auditory processing became more evident because it occurred in a situation when the person was engaged in other tasks (i.e., driving and trying to eat lunch). Simultaneous visual demands from the driving task were placed on Lara, and she chose to temporarily shed the task of eating as well as turn up the volume on the radio to allow her to focus more intently on the listening task. For Johan, visual speech cues aided his ability to understand the verbal cues. Auditory processing requires effort, even under the best of listening circumstances, although this effort may go unnoticed until the situation becomes more challenging. Challenges to auditory processing can stem from noisy or degraded listening situations, faint signals, or the concurrent demands of other tasks that must be performed simultaneously. Understanding these relationships is the focus of this book.

People use their auditory capabilities to communicate with each other, to locate sirens, oncoming traffic, and a host of other potentially dangerous objects in the environment. Auditory processing also enriches our lives in countless ways: Consider the pleasure of listening to one's favorite music, the relaxing sounds of a babbling brook, or the heartwarming sound of a child's debut in the grade school choir. Although

FIGURE 1.1   Paul Broca.

most of us seem to accomplish such tasks with little conscious effort, our ability to process the auditory world around us is nothing less than remarkable.

The human brain, in conjunction with the ear, has evolved in such a way that it enables humans to organize and interpret the complex array of sounds heard in everyday life. Remarkably, we are able to simultaneously segregate multiple sources of sounds into their individual units while combining individual components of each sound stream into meaningful wholes. During the Baroque era (1600–1750), composers such as Johann Sebastian Bach made use of some of these remarkable principles of auditory processing that were evidenced, if not fully understood, to suggest to the listener something other than what was actually presented. For example, Bach used this auditory streaming illusion in his Partita no. 3 for solo violin in E major to suggest two melody lines. This technique came to be known as *virtual polyphony* (Bregman, 1990).

In the 19th century, localized brain structures that had evolved to carry out specific language-processing tasks were identified by the seminal work of physicians such as Paul Broca (Figure 1.1) and Carl Wernicke (Kaitaro, 2001). This early work relating language functions to hemispheric specialization and modular organization of the brain was a significant contributing factor to the development of modern neuroscience (Banich, 2004; Corballis, 2000). However, despite the significance of early contributions in the area of auditory processing, and language processing in particular, the auditory sense has often been less appreciated than its cousin sense, vision.

THE BATTLE OF THE SENSES: VISION VERSUS AUDITION

In his engaging essays, *The Five Senses*, Gonzalez-Crussi (1989) reminded us that Aristotle first noted that sight and hearing were what distinguished humans from the animals because it was these two senses, he argued, that allowed the unique human ability of aesthetic appreciation—of art and music—a quality that animals and robots lack. Although Aristotle's argument can be debated, most people seem to agree with the supremacy of vision and of the secondary role of audition.

If you could only retain one of your five senses, which one would you choose? Over the last several years, I have posed this question to hundreds of students in my sensation and perception classes, asking them to choose the sense (among sight, hearing, taste, smell, or touch) that they consider most valuable and cherished. Generally, about two thirds of the class will choose sight. A majority of the remaining third will choose hearing, and a few students will choose from the remaining three.

Helen Keller, who had neither sight nor hearing, is reported to have differed from this majority sentiment. She said that of the two, she missed hearing the most. For though an inability to see separated her from experience with objects, the inability to hear separated her from experience with people (Ackerman, 1990). Auditory processing, and speech in particular, play a critical role in communication and social interaction for most humans across the entire life span. Still, most people (unless they are accomplished musicians) seem to value their ability to see more than their ability to hear. As if to underscore this common sentiment, the study of vision has received considerably more attention from the scientific community than has audition over the course of the last century.

Vision has been the object of prolonged and more extensive scientific research than has auditory processing. This is true both of the basic sciences of physics, biology, and psychology and in applied areas such as engineering and human factors (Bregman, 1990). Moreover, even within the field of audition, higher-order auditory processes (auditory scene analysis, auditory streaming, auditory recalibration, and language perception) have received considerably less study than lower-level processes such as loudness and pitch perception (Bregman, 1990; Plomp, 2002). The relative neglect of higher-level audition relative to vision is evidenced in the applied areas of human performance research as well as in the related disciplines of sensory, perceptual, and cognitive psychology. As a result, many key issues regarding how humans process auditory information remain poorly understood, whereas the basic low-level mechanisms of hearing are well known.

Despite its secondary status (to vision) in the scientific mainstream, many landmark contributions to the science of auditory processing were made in the 20th century. A notable example was work at Bell Laboratories, particularly when it was under the directorship of Harvey Fletcher (1927–1949). Fletcher published a widely read book, *Speech and Hearing*, in 1929. For a more recent edition, see Fletcher's 1953 edition. By the late 1950s and early 1960s, a few prominent researchers had made major progress in areas that would lay the foundation for our current understanding of auditory cognition. Much of this early work involved the examination of differences between processing information in visual versus auditory modalities. A few researchers intentionally focused their attention on auditory processing. This focus is illustrated by the introductory comments of Robert Crowder at a conference entitled, The Relationships Between Speech and Learning to Read, held in the early 1970s. In the proceedings of the conference, Crowder (1972) stated:

Direct comparisons between the visual and auditory modes will be drawn where appropriate; however, since (as invariably seems to be the case) visual work is considerably farther advanced than auditory, most of my talk will be directed to the properties of auditory memory. (p. 252)

Other researchers who contributed significantly to our early understanding of auditory processing include Donald Broadbent, Colin Cherry, Anne Treisman, Reiner Plomp, Albert Bregman, and Neville Moray. Much of their work as it relates to information processing and mental workload, or auditory cognition, is examined throughout further chapters of this book. As a prelude to a discussion of auditory cognition, I turn first to a brief history of early work focusing on the role of sensory processing in applied human performance research.

EARLY HUMAN PERFORMANCE RESEARCH

Research in the area of human performance capabilities and limitations in applied settings gave rise to the field now known as human factors, or ergonomics. During World War I, prior to the growth of human factors as a discipline, much of the applied work of psychology focused on the area of personnel selection and training, or what has sometimes been called "fitting the human to the task or machine" (Wickens, 1992). Human factors as a field gained major impetus during and after World War II as it became clear that the contributions of psychology to personnel selection and training were inadequate for the successful implementation and use of emerging technologies (Sanders & McCormick, 1993). Experimental psychologists who became the pioneers of early human factors research, such as Norman Mackworth and Paul Fitts, were called in to analyze the operator-machine interface (Wickens, 1992). Mackworth's early work on factors affecting sustained attention or vigilance in radar monitors and the situations in which visual detection abilities become degraded (Mackworth, 1948, 1949) is illustrative of early human factors research. The proliferation of visual displays resulted in an emphasis on visual research in this early human factors effort.

In the postwar period, many psychologists turned from examining the immediate practical concerns of human performance in the operator-machine interface to more basic research issues. During the 1950s through the 1970s, the field of cognitive psychology witnessed a rebirth (Anderson, 2000). The new direction in cognitive psychology had at its core a focus on examining how humans process information, with particular emphasis on the role of attentional factors. The information-processing approach is now evidenced throughout the field of cognitive psychology, including the study of higher-level auditory processing (Anderson, 2000).

EARLY DEVELOPMENTS IN MODERN COGNITIVE PSYCHOLOGY

Donald Broadbent played a key role in the reemergence of cognitive psychology. His work revived interest in cognition following several decades during the 1920s to 1950s when the behaviorist approach dominated American psychology. More importantly, Broadbent was the principal architect of the information-processing approach to psychology, which was later to become the dominant model in cognitive psychology during the second half of the 20th century. Broadbent's (1958) seminal book, *Perception and Communication*, presented considerable research in the area of auditory selective attention as well as the first systematic model of information processing. Broadbent's filter theory of attentional processing is discussed in a subsequent chapter.

During this era after World War II, auditory tasks, and the dichotic listening paradigm in particular, were used extensively to explore attentional processes (Cherry, 1953a; Moray, 1969; Moray, Bates, & Barnett, 1965; Treisman, 1960, 1964b). The dichotic listening paradigm (listening to two messages simultaneously—usually a different message in each ear) provided an important means of early exploration in the new era of cognitive psychology. A wealth of information exists in this early literature, much of which has direct relevance to our current understanding of auditory cognition and therefore is explored in subsequent chapters. However, outside the arena of investigations of attentional processing, visual processing still maintained a primary role in cognitive psychology.

George Sperling, another key figure in the early stages of the "cognitive revolution," made major contributions to advancing knowledge and understanding of information processing. In particular, Sperling's seminal work presented evidence for the existence of a temporary sensory store capable of holding information until it could be attended to and processed by relatively more long-lasting memory systems (Sperling, 1960, 1967). Sperling's "partial report" paradigm was to become an extremely useful tool for many psychologists seeking to further examine the role of sensory characteristics in the initial stages of information processing.

While Sperling and numerous others of his day concentrated primarily on visual information processing, a few researchers, notably Neville Moray (Moray, 1969; Moray et al., 1965) as well as Crowder and colleagues (Darwin, Turvey, & Crowder, 1972), applied Sperling's paradigm to examinations of auditory information processing. Their work is also discussed in further chapters. However, again with a few notable exceptions, investigations of visual processing took precedence over auditory processing in the field of cognitive psychology. There was a similar relative neglect of auditory processing in the field of sensation and perception.

SENSATION AND PERCEPTION RESEARCH

Visual processing dominated the early days of sensory and perceptual research. Textbooks on sensation and perception prior to 1965 limited their coverage of auditory processing to basic psychophysical auditory qualities, such as loudness and pitch perception, the physiology of the auditory system, and at most perhaps coverage of the perceptual aspects of auditory localization (Bregman, 1990). Due to an emphasis on visual processing, our current understanding of the psychophysics and neural mechanisms involved in vision is far advanced relative to that of audition.

The neural mechanisms and structural mechanisms relevant to visual processing received considerable early attention. In the late 1950s and early 1960s, Hubel and Wiesel conducted pioneering work examining information coding and feature detection of individual cells in the lateral geniculate nucleus and the visual striate cortex of the cat brain (Hubel, 1960; Hubel & Wiesel, 1959, 1962; Moray, 1969). Analogous work on the auditory neural pathways was sparse and received little attention at the time, relative to the impact of Hubel and Wiesel's Nobel Prize-winning work.

In fact, in many areas ranging from contextual effects stemming from the neural processes of lateral inhibition (Holt & Lotto, 2002) to the identification of parallel processing streams for visual object identification and object location (Courtney & Ungerleider, 1997; Ungerleider & Mishkin, 1982), our current understanding of visual phenomena greatly exceeds our knowledge of analogous auditory phenomena. However, this research gap is narrowing.

Recent empirical research indicated that the central auditory system consists of numerous neural mechanisms and modules that carry out specific auditory processing tasks that in many ways are analogous to the specialized mechanisms observed in the visual processing system. For instance, recent evidence indicated that, much like the parallel processing streams found in the visual system, the auditory system is characterized by separate processing streams—one stream involved in processing speech stimuli and another utilized in processing auditory spatial information (Belin & Zatorre, 2000; Rauschecker & Tian, 2000; Zatorre, Bouffard, Ahad, & Belin, 2002). Further, the auditory system appears to have a specific voice-selective region that responds primarily to vocal stimuli rather than nonvocal stimuli in ways similar to the face-selective areas of the visual cortex (Belin, Zatorre, Lafaille, & Ahad, 2000). Contemporary findings such as these are discussed in detail if they are relevant to an understanding of auditory cognition, particularly the mental workload requirements of auditory processing tasks.

SCOPE OF THIS BOOK

Plomp (2002) pointed out that the aim of hearing research is to understand how sounds presented to the ear are translated by the hearing process into perception. The main premise of this book is that this process requires mental effort. This book presents for the first time a comprehensive examination of the mental effort involved in several different aspects of auditory processing. In addition to theory, numerous recent empirical investigations and everyday examples are presented to illustrate the interaction of sensory and cognitive processes. The effect of acoustic degradation on task performance and the impact of combining tasks that require auditory processing in addition to sensory processing in other modalities are emphasized.

Auditory processing involves extensive coordination between peripheral sensory detectors and central processing mechanisms located in both hemispheres of the brain. The more degraded the peripheral input, the harder the brain has to work, leaving fewer resources available for remembering what has been heard. For example, when listening to spoken material that is "hard to hear," fewer cognitive resources will be left for interpreting the semantic and emotional content of the communication and remembering what was heard, relative to when the speech is clearly audible (Baldwin & Ash, 2010; Wingfield, Tun, & McCoy, 2005). At the level of the brain, the left and right hemispheres work together to segregate the acoustic information into meaningful units, using context to interpret both *what* is heard and *how* it is presented (Scott et al., 1997; Tervaniemi & Hugdahl, 2003). For spoken words, how the message is presented often underlies the practical and emotional significance of the communication. At the same time, the ventral (lower) and dorsal (upper) auditory pathways work together to process both what is being heard and where it is coming from (Romanski et al., 2000; Zatorre, Bouffard, et al., 2002b).

Disruptions at any stage or process in this complex interaction, which can become more common as we age, can result in devastating personal and economical costs. To give an everyday household example, wives may begin to feel that their middle-aged husbands just do not listen anymore; conversely, husbands may feel that their postmenopausal wives misinterpret the intention of nearly every conversation. On an economical level, communication failures and misinterpretation of sound information can result in considerable tragedy, as seen in numerous aviation accidents, medical room errors, and industrial catastrophes. For example, Ballas (1999) described how the misinterpretation of sounds in the cockpit contributed to the crash of Delta Flight 1141 in August, 1988. One of the pilots apparently mistook compressor stalls for engine failures, and this likely resulted in the engines not being immediately fully engaged, which might have prevented the crash. In another example, miscommunication between an air traffic controller and the captain and first officer of flight Dan Air B-727 at Tenerife in 1980 resulted in a crash that killed everyone on board, the worst aviation accident in history (Beaty, 1995).

The systems utilized in everyday life are becoming more technologically advanced, and the role of the human operator increasingly becomes that of a supervisory monitor of displays rather than an active manipulator of machinery (Parasuraman, 2003; Parasuraman, Sheridan, & Wickens, 2000). With these changes, consideration of the way people extract and process information increases in importance. Mental workload often increases in these technologically advanced environments even as physical workload decreases. Heavy reliance on sensory detection and information processing is required by the supervisory monitoring role. Performance limits today are more likely to be based on the functioning of sensory mechanisms to extract information and adequate mental resources to process the extracted information rather than on the physical capabilities that limited performance in the past.

In the technologically sophisticated environments present in today's leisure and work environments, the potential for the demands on mental resources to exceed the human operator's available attentional capacity are great. For example, since 2000 we have witnessed the introduction of a host of new devices in our automobiles. These range from cellular phones to sophisticated routing and navigational systems, collision warning systems, and "infotainment" centers. The introduction of these devices threatens to compromise safety as drivers struggle to divide their attention between the task of driving and the operation of other in-vehicle systems. We examine many of these new sources of sound in the next chapter and return to research aimed at improving the design of these systems in subsequent chapters, most notably in the chapter on auditory display design. Interface design based on good human factors principles has the potential to reduce mental resource demands, thus facilitating performance and reducing error. Consideration of the mental effort required for auditory processing is imperative for today's operators, who face challenges associated with an increasingly complex array of informational displays.

Even if auditory devices are designed to facilitate safe operation, not all users will be able to use them effectively. Population aging, a global phenomenon, is resulting in an increasing proportion of these emerging devices being used by older adults. Age-related sensory and cognitive changes present in a large segment of today's population call for reanalysis and possibly a redesign of many existing human-machine interfaces (Baldwin, 2002). The importance of issues related to aging warrants that they be discussed in a separate chapter. Chapter 10 is devoted to this discussion.

An understanding of the mental workload involved in auditory processing is relevant to a wide range of operational environments, including aviation, surface transportation, and medical facilities, as well as in classrooms of all types. The material presented in this book will benefit a wide range of audiences, including the classroom educator, students, and practitioners of audiology, cognitive science, and the applied psychological fields of human factors and industrial organizational psychology, as well as those who are affected either directly or indirectly by hearing loss or other auditory processing disorders. Several introductory chapters are included so that this wide readership can be accommodated. The latter portion of the second chapter is designed to promote a basic understanding of auditory processing at the perceptual level, and Chapter 3 is designed to introduce the novice reader to current theories of attentional processing and mental workload. Readers already well grounded in these areas may wish to skim or skip these sections and move on to the materials covered in subsequent chapters.

The primary goal of this book is to promote an understanding of the relationship between auditory cognition and human performance, particularly to highlight the nature of and situations in which the mental resource requirements of auditory processing may be compromised. A second goal of this book is to bring to the forefront the importance of increasing our understanding of auditory cognition and its relationship to human performance. Despite the relative neglect that auditory processing has received in earlier years, the auditory modality remains a potent source of information with several advantageous and unique characteristics.

This book is not intended to replace the many works on physiological aspects of hearing or acoustics. For further readings in that area, see the work of Moore (1995) and Yost (2006). Rather, this book is intended to extend the existing literature by focusing specifically on the mental workload or attentional processing requirements of auditory processing and its application in complex real-world tasks. Extensive consideration is given to such everyday tasks as language processing, extracting information from auditory displays, and the impact of auditory processing in conjunction with performing tasks in other modalities (i.e., visual, tactile, and olfactory).

CHARACTERISTICS OF AUDITORY PROCESSING

Humans are biologically adapted to process complex auditory information, most notably speech. One need only point to the fact that speech is a universal characteristic of all community-dwelling human beings to begin to understand the importance of speech processing. While across time and civilizations humans have always developed a spoken language, reading and writing have been relatively rare (Liberman, 1995). The universal characteristic of auditory processing in the form of speech was pointed out by Liberman in the following six primary observations.

First, all communities of humans have developed a spoken language, while many languages do not have a written form. When a written form does exist, it is typically used much less frequently than the spoken form.

Second, speech developed much earlier than writing in the history of the human species. The development of speech was perhaps the single most salient development in distinguishing humans from other species.

Third, speech occurs earlier in the development of the individual. Humans begin to comprehend and produce speech in infancy. It is several years later, if at all, before they are capable of utilizing written language forms (Liberman, 1995).

In addition, speech must be learned but need not be taught. Individuals of normal intelligence and functional capacity will learn to understand and produce speech with mere exposure. It is in this sense therefore that Liberman (1995) referred to speech as a precognitive process, much like learning to localize sound. Reading and writing, on the other hand, must be taught and therefore represent an intellectual achievement rather than a precognitive process.

Specific brain mechanisms have evolved to process spoken language. Reading and writing presumably utilize these mechanisms to some extent. However, the processing of written language must also rely on brain mechanisms that have not evolved for that specific purpose.

Last, spoken language is extremely flexible, adaptable, and capable of conveying a nearly infinite number of expressions. Written language has no independent existence without its spoken language base and therefore can only have utility to the extent that it transcribes its spoken counterpart (Liberman, 1995). The natural ease with which humans learn and process speech makes it an important component to be utilized in the design of human-machine interfaces. Also, the auditory modality has several unique and important characteristics as an information-processing channel.

THE AUDITORY CHANNEL

The auditory channel has at least two characteristics distinct from the visual channel that have important implications for our understanding of human information processing and mental workload (Wickens, 1992). First, auditory information can be perceived from any direction and therefore is said to be omnidirectional. This characteristic of the auditory channel means that the listener does not need to focus on a specific spatial location or even be oriented in any particular direction to perceive a sound. Provided the auditory signal is salient enough to be heard, the listener is free to move or direct visual attention to other tasks without signal loss. On the contrary, visual signals require that an observer be in the direct line of sight of a display and further that the observers' attention be directed to the visual display. As discussed in a subsequent chapter, the omnidirectional aspect of auditory processing allows drivers to keep their eyes on the road while receiving navigational instructions and allows pilots to maintain visual attention to flight displays while communicating with air traffic control.

The second distinct characteristic is that auditory information is typically transient. This characteristic of a limited temporal duration translates to the imperative that the operator must have sufficient mental resources and time to process an auditory signal in real time. Unlike visual signals, which typically remain in view and can often be reexamined, the listener is not free to repeatedly refer back to check the status or clarify the information presented in the auditory signal. Fortunately, the human auditory processing system has evolved in such a way that the auditory sensory store (referred to as *echoic memory*) is of much longer duration, relative to the visual sensory store (or *iconic memory*). The benefits of a more persistent echoic sensory trace and the factors that affect its duration are discussed in more detail in further chapters because of their important implications for auditory processing in real-world settings.

The unique characteristics of auditory processing and the auditory channel have important implications for human performance. As previously stated, it is the premise of this book that all auditory processing requires mental effort. We turn now to the topic of mental effort and its quantification in terms of mental workload.

MENTAL WORKLOAD

Mental workload is a multidimensional construct that in essence describes the level of attentional engagement and mental effort that a person must expend to perform a given task (Wickens, 1984). Measurement of mental workload essentially represents the quantification of this level of engagement in mental activity resulting from performance of a task or set of tasks. Specific techniques for accomplishing this quantification procedure are discussed in a subsequent chapter. However, regardless of the technique utilized, some common assumptions are as follows: Mental workload theory entails the assumption that (a) people have limited mental and attentional capacity with which to perform tasks; (b) different tasks will require different amounts (and perhaps different types) of processing resources from the same individual; and (c) two individuals might be able to perform a given task equally well, but one person may require more attentional resources than the other (Baldwin, 2003). Quantifying the mental workload of a given task or task set is critical to understanding, designing, implementing, and improving the systems humans use in their everyday lives. A key element of this design process (characterized by ergonomic/human factors design principles) is to develop systems that make efficient use of human mental processing capabilities to allow people to perform several simultaneous tasks without exceeding their mental processing capacity (Baldwin, 2003; Gopher & Donchin, 1986; Ogden, Levine, & Eisner, 1979).

The study of mental workload has played an important role in human factors research since the 1970s (Casali & Wierwille, 1983b; Gopher & Donchin, 1986; Hancock & Desmond, 2001; Kramer, Sirevaag, & Braune, 1987; O'Donnell & Eggemeier, 1986; Wierwille & Connor, 1983; Wierwille, Rahimi, & Casali, 1985). In fact, mental workload took a "central theme in laboratory-based empirical work," as pointed out by Flach and Kuperman (2001, p. 433), during the 1970s and 1980s. Auditory tasks have frequently been used as secondary task indices of mental workload (Backs, 1997; Fowler, 1994; Harms, 1986, 1991; Kramer et al., 1987). This area of research is explored in Chapter 6, which focuses on the use of auditory tasks in cognitive research and in research aimed at assessing mental workload. Emphasis is placed on the intersection of peripheral auditory and central information-processing mechanisms and the critical role that this interaction has in influencing the mental workload requirements of auditory processing. The implications of this sensory-cognitive interaction for human performance are a unique contribution of this book.

CONCLUDING REMARKS

Relative to visual processing, auditory processing has received little attention in the early research on human performance, including human factors research, much of early cognitive psychology, and early sensation and perception research. Some

major exceptions to this relative neglect include investigations of attentional processing and the study of language, which has been a focus of considerable research for
decades (Bagley, 1900–1901; Broadbent, 1958; James, 1890/1918). Over a century
ago, Bagley (1900–1901) realized that speech understanding involves the interaction
of sensory and cognitive processes. Bagley observed the essential role that context
played in speech processing. This book expands on that early realization, looking at
the many factors that affect the mental workload associated with auditory processing. It also may lay the groundwork for new growth in a burgeoning area of human
performance research: applied auditory cognition.

.

2 The Auditory World

INTRODUCTION

Sound is around us at all times. Close your eyes for just a few moments, and you can better appreciate the many different sounds present in the environment. Many of these sounds typically go unnoticed in our day-to-day existence. But, when we close our eyes for just a few minutes we begin to notice the barrage of electronic buzzing sounds from lights, ventilation systems, and other home or office appliances that incessantly bombard our senses. If you are fortunate enough to escape these modern-day devices for a period of time, such as on a hike in the woods or a camping trip, then you may notice a new orchestra of the sounds of nature, which are often masked by the technological sounds of modern-day life.

Before one can investigate how we process sounds, whether natural or artificial, we need a sound understanding (pun intended) of their physical characteristics. This chapter highlights some of the many sources of sound present in everyday and work environments and describes their characteristics in relation to the human auditory processing system. This will provide a foundation for understanding the subsequent material. However, the reader who is already an expert on these topics may wish to skip to the next chapter. To begin, a discussion of the many different sources of sound in the home and in various work settings is presented.

SOURCES OF SOUND

Music and speech are the two most valued, as well as cognitively complex, sources of sound (Zatorre, Belin, & Penhume, 2002). However, a plethora of additional sources of sound (i.e., auditory displays, alerts, and warnings) has proliferated in modern societies. It is common to have conversations and other activities interrupted by one or more myriad sounds stemming from modern technologies, such as mobile phones, pagers, and home/office electronics. Both music and speech are discussed in further sections. First, the multitude of nonverbal sounds is considered.

NONVERBAL SOUNDS

The majority of sounds we hear are nonverbal, although many of these receive little or no direct attention. As I sit at my desk, I can hear the hum of a ventilation system, the whir of a fan in a nearby computer, the shuffling of papers, doors closing in nearby hallways, and the sounds made by my keyboard as I type. By definition, any of the many sounds that form the constant auditory canvas of our environment that are unwanted or irrelevant to the current task at hand can be broadly classified as noise. Irrelevant or distracting sounds can increase the mental workload of processing information from auditory as well as other sensory channels. Because of their impact on mental workload, noise and particularly irrelevant speech are important topics discussed in more detail. A discussion of nonverbal noise can be found in Chapter 7, "Nonverbal Sounds and Workload," and the particular case of irrelevant speech is discussed in Chapter 8, pertaining to speech processing.

Setting the topic of noise aside, nonverbal sounds can be used to provide a wealth of information and have powerful alerting capabilities. Therefore, nonverbal sounds are used to provide information in a wide variety of home and occupational environments.

The following personal example illustrates the challenge that can be posed by the cornucopia of sounds in the everyday environment.

After just moving into a new home, I was awakened one night by an intermittent, although seemingly persistent, faint auditory tone. After noticing the sound and not immediately being able to identify its source, my curiosity was piqued, not to mention my annoyance from having been awakened. My first thought was that it might be coming from one of the appliances, perhaps a signal that the washer, dryer, or dishwasher had finished its cycles. A quick check indicated this was not the case. The next logical guess was that perhaps a smoke detector battery was low. However, remembering that the smoke detectors were wired into the electrical system, this possibility was quickly ruled out. After more consideration, it occurred to me that the sound was a familiar alert. In fact, it sounded just like the new message alert on my mobile phone. However, after checking the phone several times, I was satisfied that the sound was not coming from my mobile phone. So, the question remained, where could it be coming from? Tired and a bit annoyed at this point, I put a pillow over my head and went back to sleep. When I awoke in the morning the sound was still present, intermittent, but nevertheless present. Again, I asked myself, was it something that actually needed my attention, or was it just a source of annoyance at this point? After many more failed attempts at identification, I finally discovered the source of the sound. It seemed to have been coming from a box, among a stack of boxes yet to be unpacked. Inside the box, I discovered that my old mobile phone had inadvertently been switched on, and the sound was indeed a new message alert signaling me of the arrival of a message from my mobile phone provider. The phone that I had had disconnected some months ago had no service plan, and the former service provider had been kind enough to send a message stating this. How thoughtful of them!

This trivial little story points to just one of the many sounds present in our homes that have been ushered in with the digital age. Homes, offices, vehicles—all are alive with sound. Microwaves, washers, dryers, dishwashers, toaster ovens, and cappuccino makers, to name a few, provide us with auditory displays. Many of these sounds are in the form of alerts and buzzers that have been intentionally designed by the manufacturers for a particular purpose. Others, such as the sound of water swirling in the washer or dishwasher, are merely sounds inherent to the operation of the machine. But, even such intrinsic sounds can provide information. A particular favorite of mine is the gurgling sound produced by my coffee pot in the morning signaling that a fresh pot of java is nearly ready.

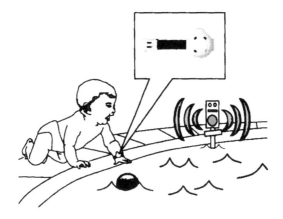

FIGURE 2.1   Water-activated wristband safety alarm. (From Bridget Lewis.)

The modern home environment contains many sources of sound, including alerts and alarms. Doorbells and chimes, appliance signals, smoke and carbon monoxide detectors, and intruder alarms are examples of common household alarms. More recently, specialized alarm systems have been designed. As illustrated in Figure 2.1, one example is a system designed to sound an alarm if someone wearing a special wristband (e.g., a small child) enters a swimming pool.

It is clear that the home environment is replete with sound. We turn the focus now on the many sources of sound in the work environment. It is worth pointing out, however, that as many of us take more of our work home and engage in telecommuting and various other work-related activities outside the confines of the conventional work environment, there may be no clear distinction between sources of sound in the workplace and in the home.

SOURCES OF SOUND IN WORK ENVIRONMENTS

As with the home, there are a multitude of sources of sound in the modern work environment. We begin by discussing auditory alerts, warnings, and displays in general and then move on to discuss sources of sound in three work environments that have received considerable attention in human factors research. These three work settings are aviation, surface transportation, and health care. All three have witnessed a proliferation of auditory alarms, including bells, such as the simple one pictured in Figure 2.2, to whistles, buzzers, and gongs, to name but a few.

In recent years, considerable human factors research has led to improved understanding of the human performance capabilities and performance issues associated with warnings. Critical issues in current warnings theory include not only signal characteristics associated with detection and comprehension but also such issues as urgency mapping, compliance, and preferred display format (i.e., visual, verbal vs. nonverbal, haptic, or multisensory).

FIGURE 2.2   Fire bell.

AUDITORY WARNINGS

Auditory alerts and warnings are most abundant in high-workload, high-stress environments where the consequences of performance failure are dramatic (Edworthy & Adams, 1996). Auditory warnings come in a wide variety of formats, including bells, whistles, buzzers, and gongs (Stanton & Edworthy, 1999b), and have proliferated in recent times in both number and type. Identification and rapid processing of these warnings is frequently critical to both human and system safety. Numerous investigations have established empirical support for human factors design guidelines for auditory warnings. Acoustic characteristics such as intensity, frequency, and onset-offset time affect important warning parameters such as noise penetration, alerting effectiveness, perceived urgency, and annoyance (Baldwin, 2011; Baldwin & May, 2011; Edworthy & Stanton, 1995; Hellier, Edworthy, & Dennis, 1993; Patterson, 1990a). For example, auditory warnings of high amplitude and frequency are generally perceived as more urgent than warnings of low amplitude and frequency (Hellier & Edworthy, 1999b; Hellier, Edworthy, Weedon, Walters, & Adams, 2002). A wide variety of auditory alerts, along with a host of auditory displays and interfaces, is common in the modern workplace.

Auditory warnings can be nonverbal tones and signals, verbal alerts or messages, and the more recently developed categories consisting of auditory icons and earcons (Belz, Robinson, & Casali, 1999), hearcons (Sodnik, Dicke, Tomazic, & Billinghurst, 2008), and spearcons (Walker, Nance, & Lindsay, 2006). Traditional nonverbal auditory alerts are defined by their acoustic parameters, while auditory icons are representational sounds, such as breaking glass or screeching brakes (Belz et al., 1999). Earcons are made of structured musical tones that listeners learn to identify as representing a specific system state or condition. As one example of the use of auditory icons, Belz and colleagues found that, compared to conventional nonverbal auditory alerts, auditory icons can improve brake response time and driver performance in simulated collision situations. As will be discussed further, various auditory alerts are increasingly being installed in modern vehicles to aid in collision avoidance. Common examples of nonverbal auditory alerts include the familiar backup alerts on forklifts and many commercial trucks, emergency vehicle sirens, fire alarms, telephone rings, and the auditory bell sound signifying receipt of a new e-mail message.

Nonverbal Alarms

Nonverbal alarms are frequently used in many home and occupational environments. They come in a variety of forms, ranging from simple tones and melodic patterns to complex representational sounds. We begin with a discussion of sirens.

Sirens

Stationary sirens are used to indicate various critical states, such as emergency weather conditions (e.g., tornados and hurricanes) and fires. Stationary sirens must be easily recognizable since the action responses may be quite different. For example, if the siren signifies a tornado, the action response is to seek shelter inside a permanent structure; however, if the siren indicates a fire, the action response is to vacate the building as quickly as possible. For these reasons, siren signals must be designed to consist of distinct acoustic patterns, and people must be trained to recognize these patterns and practice appropriate action responses. Sirens on moving vehicles require additional constraints.

Sirens on emergency response vehicles not only must be easily detected and recognized but also localized. Since humans cannot easily localize pure tones or sounds consisting of limited-frequency bandwidths, moving sirens should consist of a broad range of frequencies. However, many sirens currently use a frequency range of 500 Hz to 1.8 kHz, too narrow a band to be easily localized (see discussion by Withington, 1999).

Patterson (1982) and Hass and Edworthy (1996) identified new sounds for use in sirens and other alerts that have better detection and localization rates. In one on-road investigation, compared to conventional sounds, the new siren sounds improved the accuracy of determining the direction of an ambulance siren by as much as 25% (Withington, 1999).

Auditory Icons

Auditory icons are nonverbal representational sounds (Belz et al., 1999). Since they represent familiar sounds, auditory icons do not require overt learning (Keller & Stevens, 2004). In addition to the screeching brake example described, auditory icons have been incorporated in a number of computer applications (Gaver, 1986), including educational software programs (Jacko, 1996). A taxonomy developed by Keller and Stevens for classifying auditory sounds in terms of their association with the object or referent they represent is described in detail in Chapter 11, "Auditory Display Design."

Earcons

Earcons are musical patterns that operators learn to associate with system states or events. Ronkainen (2001) gave the example that all Nokia mobile phones come with a default auditory pattern of two short notes followed by a pause and then two short notes again to signify the presence of a new text message. This "musical" pattern or earcon is quickly learned by users, although without learning it would have no inherent meaning in and of itself. Earcons can be used to alert trained listeners to critical system states and therefore can be considered a class of nonverbal warning.

FIGURE 2.3   Fire truck.

However, they can also be considered more broadly as auditory displays and share similarities with data sonification. The difference is that with earcons listeners learn to associate a musical pattern with a certain state, whereas in sonification, changing data or values (i.e., temperature or velocity) are represented by sounds in a manner similar to an auditory graph (Walker, 2002).

Verbal Warnings

As Stanton and Baber (1999) pointed out, to be effective alarms must be recognized and comprehended in addition to being heard. Learning and experience aid in the recognition of common auditory alarms, such as the well-known siren sounds emitted from a fire truck, like the one illustrated in Figure 2.3. Verbal warnings can capture attention and convey information at the same time (Hawkins & Orlady, 1993). In this way, they are similar to auditory icons. However, verbal warnings can quickly convey a wide range of information that might not be associated with an easily identifiable sound. Auditory warnings are found in numerous environments, including the automobile and the airplane. For example, in the modern cockpit a verbal warning associated with the Ground Proximity Warning System (GPWS) and the Altitude Alerting System instructs pilots through a digitized or synthesized voice to "pull up" when too close to terrain. Verbal warnings have also been examined for use in collision warning systems (CWSs) in automobiles (Baldwin, 2011; Baldwin & May, 2011).

AUDITORY DISPLAYS

When sounds, whether verbal, nonverbal, or both, are added to a device or computer-based system to provide information to the human operator, the result is an auditory display. Effective design of these displays is an important human factors issue and is examined in the concluding chapter of this book, after a detailed discussion of various aspects of auditory cognition and human performance. At this point, however, it is instructive to examine briefly some major categories of auditory displays.

Sonification

Sonification refers to representing multidimensional information that would normally be presented in a graphic format in an auditory format. Sonification has obvious implications for assisting visually impaired individuals but may also be used in the sighted population when visual information channels are either overloaded or temporarily unavailable (e.g., in dark caverns or tunnels). Sonification has been used to develop usable computer interfaces with many of the same benefits as graphical user interfaces for blind individuals (Mynatt, 1997).

Empirical evidence indicates that information presented in visual and auditory scatterplots is used in much the same way (Flowers, Buhman, & Turnage, 1997). People are able to estimate the direction and magnitude of data points as well as detect the presence and magnitude of outliers in auditory scatterplots in efficient ways comparable to visual displays.

More complex data, such as statistical distributions and time series data, can also be conveyed with equal clarity through auditory relative to visual displays (Flowers & Hauer, 1995). Flowers and colleagues pointed out that in addition to developing interfaces to assist blind individuals, data sonification can be used to develop efficient data representation tools for scientists and engineers, particularly when visual attention must be directed somewhere else (Kramer et al., 1999). We turn now to one more general format for using sound to convey information, the verbal display.

Verbal Displays

Verbal displays have increasingly been added to a number of human-machine interfaces. Probably the most familiar verbal display is the automated voice messaging systems that are utilized by computerized call centers and automated phone messaging systems that are the default settings on voice mailboxes and telephone answering systems (the synthesized voice that comes prerecorded telling callers to "Please leave a message"). Verbal displays are also used in a variety of other settings. For example, mall parking garages may utilize automated verbal displays to remind shoppers where they have parked their cars, and moving walkways caution travelers that they are "approaching the end of the moving walkway." These verbal displays are often presented in synthesized voice (Hawkins, 1987).

Synthesized Voice

Considerable effort beginning in the 1960s was expended toward developing synthesized voices that were intelligible and acceptable to the operator (Hawkins & Orlady, 1993; Simpson & Marchionda-Frost, 1984; Simpson & Williams, 1980). Acoustic factors such as speaking rate and fundamental frequency affect speech intelligibility as well as listeners' perceptions of the personality characteristics of the synthesized voice. For example, synthesized voices presented at fast speaking rates are perceived as less benevolent and more competent, while voices with high fundamental frequencies are perceived as not only less benevolent but also less competent (Brown, Strong, & Rencher, 1974). Synthesized voices are currently used in a wide variety of settings, ranging from automated voice messaging systems to synthesized verbal alerts on aircraft and in other operational environments. As discussed in more depth in further chapters, synthesized voice requires more effort to process relative to natural or digitized natural speech. But, because synthetic speech (particularly when generated in real time from text-to-speech [TTS] synthesizers) requires considerably less data storage capacity relative to digitized voice recordings, it is preferable for many mobile applications.

Spearcons

A less-common variety of verbal display that is increasingly being examined for use in auditory menus is the spearcon. Spearcons are created by speeding up a spoken phrase to the point that it is nearly, if not completely, unintelligible without altering its frequency (Walker & Kogan, 2009; Walker et al., 2006). Although spearcons may not be comprehensible as the actual word they represent in isolation, they resemble the word closely enough that they are learned more effectively than arbitrary sounds and show promise as a method of providing a means of navigating complex menus. Spearcons may be particularly beneficial when used in small portable devices without enough space to provide an adequate visual display and yet complex menus are needed and storage capacity.

SUMMARY

Operational sources of sound range from simple auditory alerts and alarms in the form of bells, beeps, and buzzers to sonographic and verbal displays capable of presenting detailed information in an auditory format. As previously mentioned, sources of sound in three particular work environments (the cockpit, the modern vehicle, and medical care facilities) are discussed next. The multitude of sounds used to convey information in these environments will help to underscore the importance of understanding the mental workload of auditory processing.

SOURCES OF SOUND IN THE COCKPIT

The modern cockpit is replete with sound. Heavy demands are placed on visual attention, as pilots are continually required to shift attention across numerous flight deck displays and to maintain awareness of views outside the cockpit window. Therefore, the auditory modality is a good channel for presenting time-critical information. Sources of auditory workload include the simultaneous processing of radio communications, copilot communications, and auditory flight deck alerts, warnings, and displays. Hamilton (1999) pointed out that the multiple sources of simultaneous auditory information require pilots to choose to ignore some information to attend to prioritized sources of information. Ideally, flight deck instruments should assist in minimizing external workload influences by presenting required information in an efficient, integrated manner. In practice, this goal has proven difficult to accomplish, and a vast array of flight deck displays implemented in piecemeal fashion has great potential for increasing total pilot workload.

PROLIFERATION OF AUDITORY AVIONICS DISPLAYS

Dramatic changes in avionics displays took place between 1981 and 1991, including the introduction of the new generation of "glass cockpit" airliners (Learmount, 1995). Concomitant developments in aircraft warning systems have resulted in centralized alerting and monitoring systems that present information regarding flight status, system states, automation mode changes, and traffic management assistance in integrated multifunction displays directly to pilots (Noyes, Starr, Frankish, & Rankin, 1995).

Auditory Avionics Alarms

Auditory alarms are used on the flight deck to alert crew to dangerous conditions, potentially dangerous conditions, the arrival of new information on visual displays (Patterson, 1982) and to signal changes in system states. According to Patterson (1982), existing auditory warning systems perform their alerting function well, too well in some cases, presenting unnecessarily loud sounds. Loud tones capture attention but can also startle, annoy, and block crew communications at critical points (Patterson, 1982; Wickens & Hollands, 2000). Further, loud noises can increase arousal. During time-critical situations arousal may already be too high; thus, further increases may exacerbate the risk of performance impairment.

According to Patterson's estimate in 1982, there were as many as 16 different auditory warnings and alerts on some aircraft. Modern aircraft have the same number of alerts, and the trend has been for further increases as new avionics systems are added (Hawkins & Orlady, 1993; Noyes, Cresswell, & Rankin, 1999). For example, Noyes et al. (1999) pointed out that for visual alerts alone, there were 172 different warnings on the DC 8, and this number increased to 418 on the DC 10. Similarly, the number of visual warnings on the Boeing 707 was 188, and this increased to 455 on the Boeing 747 (Hawkins & Orlady, 1993). According to Hawkins and Orlady (1993), during the jet era alone a number of auditory alerts were also incorporated, including a bell, clacker, buzzer, wailer, tone, horn, intermittent horn, chime, intermittent chimes, and later a synthesized voice.

This number is well above the recommended guidelines for the number of auditory dimensions that can be identified on an absolute basis (see discussion in Sanders & McCormick, 1993). For example, Deatherage (1972) suggested that for absolute judgments, four or five levels could be identified on the basis of intensity, four to seven on the basis of frequency, two or three on the basis of duration, and a maximum of nine on the basis of both intensity and frequency. This means that a pilot must develop an extensive repertoire of stored knowledge associations between individual warnings and the events they represent.

The extensive number of alerts in modern aircraft has led to the recognition that performance issues will most likely be related to cognitive factors (attention, learning, memory, and understanding) rather than the perceptual factors such as detection (Hawkins & Orlady, 1993). For example, in one investigation of auditory warnings in military aircraft, Doll and Folds (1986) found high numbers of confusable auditory warnings and alerts (11–12 of similar type on some aircraft). More problematic, however, was the lack of consistency in audio signaling across aircraft. That is, different sounds were used to signify the same event in different aircraft. These factors increase the mental workload of processing auditory warnings. In addition, Doll and Folds found that warning sounds were typically not matched appropriately with the level of hazard or urgency for the event they signified. Inappropriately urgent auditory warnings can cause distraction and increase the mental workload of communicating with flight crew members and performing other flight-related tasks.

To improve perceptual factors associated with auditory warnings, Patterson (1982) suggested auditory guidelines to minimize the adverse effects of loud noises without compromising reliability. Edworthy and colleagues proposed guidelines for improving auditory warning urgency or hazard mapping (Edworthy, Loxley, & Dennis, 1991; Edworthy & Stanton, 1995; Hass & Edworthy, 1996). And, to address cognitive issues, recommendations have been made for attensons (attention-getting sounds) coupled with voice alerts and visual displays within an integrated warning system (Hawkins & Orlady, 1993). All of these issues can potentially have an impact on mental workload and are therefore discussed more thoroughly in subsequent chapters.

In addition to warnings and alerts, aviation operations rely on extensive communication between personnel on the flight deck, flight attendants, and air traffic controllers.

Radio Communications (Radiotelephony)

Communication between air traffic control (ATC) and pilots primarily occurs over voice-radio channels. Recent implementations of a text-based system called datalink in some airports/aircraft aside, the majority of pilot and ATC communications take place via speech. Miscommunication has been a major contributor to aviation accidents, possibly accounting for over 50% of all major incidents (Nagel, 1988).

In an effort to reduce communication-related incidents, the Federal Aviation Administration (FAA) implemented the use of a standard set of phrases and terminologies for ATC and pilot communication (Mitsutomi & O'Brien, 2004; Roske-Hofstrand & Murphy, 1998). This standardized set of communications, referred to as ATC terminology (Mitsutomi & O'Brien, 2004), requires that both pilots and controllers learn what essentially amounts to a new vocabulary or what Mitsutomi referred to as an example of English for special purposes (ESP). The use of ATC terminology along with standardized procedures for communication have reduced but not eliminated the potential for errors due to miscommunication.

Current ATC terminology is designed to be brief and concise, with an emphasis on promoting accuracy (Mitsutomi & O'Brien, 2004). Grammatical markers and conventions are often eliminated, and a set of "carrier" syllables may be used to promote accurate communication of alphanumeric characters. For instance, the communications code alphabet (alpha, bravo, Charlie, etc.) increases the correct identification of alphabetic characters (Wickens & Hollands, 2000). Procedural deviations and the trade-off between keeping communications brief yet maintaining accuracy remain a challenge, and controller-pilot miscommunications are still cited as a contributing factor to a substantial number of aviation accidents and incidents.

Controllers and pilots must learn the ATC terminology as well as the accepted procedures for communication. Once learned, this communication frequently takes place in an atmosphere of congested frequencies from multiple aircraft attempting to maintain communication with a limited set of controllers, thus adding to the information-processing requirements of all parties. In addition, controller-pilot communications occur amid resource competition stemming from a host of visual displays. The impact of resource competition stemming from multiple sensory input channels on auditory processing is further discussed in Chapter 9. We turn now to another arena in which sounds are increasingly being used as an important source of information.

IN-VEHICLE AUDITORY TELEMATICS

In-vehicle telematics have proliferated in recent years, increasing the complexity and potentially the attentional demands of the driving task. Many of these advanced systems utilize auditory interfaces. Examples of advanced in-vehicle auditory interfaces include CWSs, lane departure warning systems, route guidance systems (RGSs), infotainment systems (such as speech-based e-mail, Web surfing, and satellite radio that allow users to make MP3 recordings), and advanced traveler information systems (ATISs). Further, despite considerable recent controversy, the use of personal cellular phones while driving is also on the rise. Assessing the mental workload required by processing information from these devices is essential to transportation safety (Harms & Patten, 2003; Horrey & Wickens, 2002; Verwey, 2000). Much like the task of piloting, driving requires heavy visual demands. Empirical research indicates that even small changes in the direction of gaze result in lateral steering deviations (Readinger, Chatziastros, Cunningham, Bulthoff, & Cutting, 2002), making the auditory modality well suited for many auxiliary in-vehicle systems. However, research clearly indicated that the auditory processing requirements of even hands-free devices have the potential to exceed drivers' attentional capabilities (Lee, Caven, Haake, & Brown, 2001). We begin with a discussion of the proliferation of cellular phone use, some implications for driving, and the different types of cellular phones commonly found in modern vehicles.

CELLULAR PHONES

One need only take a quick look around anywhere that people congregate to notice the exponential growth in the use of mobile cellular communications. Recent figures from the Cellular Telecommunications Industry Association (CTIA) reports verify this proliferation in cellular phone usage, citing an increase from roughly 28 million users in 1995 to well over 276 million users in 2009 (CTIA, 2009) and more recently over 302 million users as of 2010, with over a quarter of all households in the United States having only wireless telecommunications (CTIA, 2010). Correspondingly, there has been an increase in the number of individuals who choose to use their mobile phones while driving (Glassbrenner, 2005).

In a 1997 National Highway Traffic Safety Administration (NHTSA) survey, an overwhelming majority of drivers reported talking on a cellular phone on at least a few average trips (Goodman et al., 1997). Ten percent of the respondents indicated that they talked on cell phones on roughly 50% of their trips, and 16% reported use on most of their trips. The mental workload imposed by talking on a cellular phone (regardless of whether it is handheld or hands free) has been estimated to increase the probability of a crash by four times (Redelmeier & Tibshirani, 1997).

Many recent investigations have examined the impact of cellular phone use on driving. Cellular phone use while driving increases mental workload (Haigney, Taylor, & Westerman, 2000; Ranney et al., 2005), resulting in poorer driving performance. When using a mobile phone, responses to traffic hazards are delayed, and drivers exhibit increased steering error. Drivers use a number of compensatory strategies in an attempt to maintain driving within safe margins; however, empirical evidence suggested that these strategies may in fact not actually decrease risk (Haigney et al., 2000). For example, drivers tend to decrease speed when using a mobile phone. However, at least in simulation studies, decreased speed did not result in fewer lane deviations or off-road instances. These findings suggest that drivers may be more confident in their compensation strategies than actual performance would indicate.

Hands Free and Voice Activated

Although it may be easy to understand how tasks such as dialing could disrupt driving performance, Goodman and colleagues' (1997) review of the existing literature regarding crash data indicated that it is the conversation, not reaching for the phone or dialing, that increases crash risk. So, while hands-free, voice-activated cellular phones eliminate risks associated with tasks such as dialing a number (Ranney et al., 2005), performance degradations are still consistently observed during phone conversations (Caird, Willness, Steel, & Scialfa, 2008). For example Strayer, Drews, and Johnston (2003) observed delayed brake response times and impaired recognition of roadway signs in a driving simulator when drivers were engaged in simulated hands-free cellular phone conversations. In a similar investigation comparing the effects of hands-free phone conversations on young and older drivers, delayed reaction times and increased speed variability were observed when drivers were engaged in phone conversations (Strayer & Drews, 2004). Drivers in this investigation were also more likely to be involved in a rear-end collision when engaged in a phone conversation. In addition to cellular phones, there are a number of other sources of sound in the modern vehicle.

COLLISION WARNING SYSTEMS

Technological advances coupled with continued efforts to improve transportation safety have led to the development and implementation of in-vehicle CWSs. CWSs designed to alert drivers to potential hazardous situations are being installed in many new cars, public transportation buses, and commercial trucks (Hancock, Parasuraman, & Byrne, 1996; Siuru, 2001). These CWSs have been demonstrated to improve collision avoidance behaviors (Brown, Lee, & Hoffman, 2001), but they also have the potential to distract drivers, particularly when the systems are unreliable or overly sensitive.

As with auditory warnings in general, conveying the appropriate hazard level (perceived urgency) is a central component of effective CWSs (Hellier et al., 2002). Collisions are rare events. If a CWS alert is only provided when a collision is imminent, then the alert event would be so rare that it would possibly be unrecognizable to the driver (Brown, Lee, & McGehee, 2001). Conversely, a CWS that provides alerts too frequently, when the probability of a collision is low, is likely to be perceived as extremely annoying, and drivers may ignore or disable the system.

Verbal CWS messages show promise for appropriately matching the hazard level of a potential collision avoidance system with the urgency of the alert (Baldwin, 2011). Both presentation level and choice of signal word have been shown to affect ratings of the perceived urgency of spoken words (Baldwin, 2011; Baldwin & May,

2011; Hellier et al., 2002). Within a simulated driving context, increasing the amplitude of verbal CWS messages to moderately high levels (+4 dB signal to noise ratio or S/N) increases alerting effectiveness and perceived urgency without having a significant impact on annoyance levels (Baldwin, 2011). Further, verbal CWS alerts have been shown to decrease crashes in driving simulations (Baldwin & May, 2005), particularly among older drivers (Baldwin, May, & Reagan, 2006). Empirical evidence indicated that using auditory alerts in combination with visual warnings significantly improved driver performance and brake response time in potential collision situations (Belz et al., 1999; Brown, Lee, & Hoffman, 2001a).

DISPLAYS FOR MAINTENANCE OF SEPARATION

Auditory displays that are functionally similar to CWSs have been developed to alert drivers when their following distance is less than an acceptable range. These systems of headway monitoring, including a visual display coupled with an auditory alert, have demonstrated success in increasing the percentage of time drivers maintain a safe following distance in actual on-the-road investigations (Ben-Yaacov, Maltz, & Shinar, 2002; Shinar & Schechtman, 2002).

Some vehicles come with additional auditory displays to aid in parking. For example, Saab's optional park-assist system uses sensors mounted on the rear bumper to judge the distance between the rear of an operator's vehicle and another object (i.e., vehicle or wall). The system provides an auditory alert that changes frequency and pulse rate as the distance between the vehicle and the object changes. BMW has a similar auditory alert system they call the parking guidance system; it allows drivers to indicate the distance at which they would like the alert to come on. Another alert that is becoming more common in the modern automobile warns a driver when he or she makes an unplanned lane departure.

LANE DEPARTURE WARNINGS

Auditory warnings designed to alert drivers to unplanned lane departures have been the focus of considerable research. Several vehicle manufacturers have incorporated auditory lane departure warning systems in at least some of their recent models (i.e., Volvo, Infiniti, and Nissan). These safety systems provide an audible alert to drivers when the camera sensor system of the vehicle detects that the vehicle has drifted out of the appropriate pavement markings without the driver's intent. The audible alert may be presented alone or in conjunction with visual or tactile warnings (Navarro, Mars, Forzy, El-Jaafari, & Hoc, 2010). Simulated rumble strip vibrations to the steering wheel in combination with an auditory alert have been particularly effective at reducing the effects of unplanned lane departures (Navarro, Mars, & Hoc, 2007). Based on recent crash reports from the Fatal Analysis Reporting System (FARS) and the National Automotive Sampling System General Estimates System for reported crashes occurring from 2004 to 2008, Jermakian (2010) estimated that these systems have the potential to reduce up to 7,500 fatalities and 179,000 crashes per year.

Of course, for fatalities and crash rates to decrease, the systems must effectively alert drivers, and drivers must use them. In a survey of drivers who had lane departure systems in their cars, a substantial proportion indicated that they only sometimes (23%) or never (7%) use them (Braitman, McCartt, Zuby, & Singer, 2010). The most frequent (40%) reason given for not using the system was that they found the warning sound annoying, with another 24% indicating the reason to be because the system provided too many false alarms.

Navarro et al. (2010) examined different tactile vibrations as a means of decreasing lane departures. They used one vibration in the seat, two different types on the steering wheel (one designed to elicit motor priming or MP), and the MP vibration in conjunction with an auditory rumble strip sound. They concluded that the MP asymmetric oscillations of the steering wheel led to better driver response (i.e., shorter duration of lateral excursions, faster acceleration of appropriate steering wheel response, etc.). But, the MP system had the lowest subjective driver acceptance rating, while the auditory rumble strips had the highest. Combining the auditory rumble strips with the MP increased driver acceptance of the MP alert. Clearly, there is room for additional improvements in the design of lane departure warning sounds, a topic that is addressed further in Chapter 11.

These examples illustrate the many new ways that sound is being utilized in the modern automobile to improve safety. Auditory displays are also increasingly being used to provide drivers navigational information in a format that allows them to keep their eyes on the road.

AUDITORY ROUTE GUIDANCE SYSTEMS

Due to the heavy visual demands placed on drivers, auditory RGSs appear to have a number of safety advantages relative to their visual-only counterparts. Drivers using auditory RGSs have been shown to respond faster to targets in a visual scanning task (Srinivasan & Jovanis, 1997b) and to make significantly fewer safety errors (Dingus, Hulse, Mollenhauer, & Fleischman, 1997) relative to using a visual RGS. Auditory RGSs are now available in several formats. An issue of concern with auditory RGSs is the level of complexity that can be conveyed and the need to ensure that these systems do not exceed the information-processing demands of the driver.

In-vehicle routing and navigational systems (IRANSs) and RGSs have proliferated in recent years. RGSs have many potential economic and safety advantages, including reduced traffic congestion, enabling drivers to find destinations more easily while avoiding traffic congestion and delays, decreasing travel time and distance, and resulting in fewer instances of disorientation or getting lost, greater confidence, and less-stressful driving experiences (Eby & Kostyniuk, 1999). However, RGSs also have the potential to distract drivers by increasing the attentional processing requirements (i.e., mental workload) of the driving task. Therefore, the most effective system is one that assists the driver in navigating through and developing a cognitive map of the area without disrupting driving performance or significantly increasing mental workload. An RGS that assists the driver in establishing a cognitive map of an unfamiliar area may decrease the information-processing requirements of navigation and ultimately decreases reliance on the system in the shortest amount of time (Furukawa, Baldwin, & Carpenter, 2004).

Previous research (Dingus, Hulse, & Barfield, 1998; Noy, 1997; Tijerina et al., 2000) suggested that a combination of visual and auditory displays should be used in a comprehensive in-vehicle RGS. Complex routing information is facilitated by visual guidance information relative to a message of comparable complexity presented in the auditory modality (Dingus et al., 1998; Srinivasan & Jovanis, 1997a, 1997b). However, driving performance is degraded less by auditory guidance information relative to visual guidance information while the vehicle is in motion (Dingus et al., 1998; Noy, 1997; Streeter, Vitello, & Wonsiewicz, 1985).

Srinivasan and Jovanis (1997b) found improved driving performance and reduced workload, in complex driving situations in particular, when drivers used a system incorporating auditory guidance in conjunction with an electronic map, relative to using an electronic map alone. Specifically, when provided auditory guidance directions drivers spent more time with their eyes on the road and were able to better maintain their speed at or near the speed limit for the roadway traveled. In addition, drivers reported significantly lower workload levels while using the auditory system relative to the electronic map alone.

Although research supported using both auditory and visual guidance versus one modality alone, it could be argued that due to heavy visual demands of driving, the auditory modality should be used in lieu of visual formats when possible.

Auditory guidance is also frequently a preferred format relative to visual formats (Streeter et al., 1985). Streeter et al. (1985) reported that the auditory guidance format used in their experiment was rated easier to use than other formats. However, as mentioned previously, complex information such as spatial layouts is processed less efficiently when presented through the auditory channel relative to the same information conveyed visually. In fact, Walker, Alicandri, Sedney, and Roberts (1990) found that complex auditory instructions interfered with participants processing the instructions relative to less-complex instructions. The work of Walker and colleagues is frequently cited as a basis for current design guidelines, and at present auditory instructions are limited to terse commands. For example, a standard auditory instruction might be "turn left in two blocks." This type of command corresponds to what is termed a "route" style of navigation (Jackson, 1998; Lawton, 1994), a style characterized by a serial progression of instructions. Route instructions get the individuals to their destination without necessarily facilitating the individuals' formation of a thorough understanding of the area traveled.

Voice Guidance Formats

Many commercially available RGSs include a speech interface to provide turn-by-turn guidance to drivers. The speech interface typically consists of digitized prerecorded instructions, although some systems use synthesized speech (Burnett, 2000b). As discussed in a previous section, digitized speech is generally more acceptable to users than synthesized speech (Simpson & Marchionda-Frost, 1984); however, the trade-off is that digitized speech requires greater system memory; therefore, the number of instructions that can be issued tends to be fairly limited (Burnett, 2000b).

Standard voice messages most commonly include the direction of turn in an ego-centered or driver forward field of view reference and the distance to the turn. A few systems also include the street name (Burnett, 2000b). For example, systems such as the Garmin StreetPilot 2720 and Magellan RoadMate 800 announce the direction and street name for each turn. Some systems even allow voice queries and provide answers in the form of speech output. The new Earthmate GPS LT-20 by DeLorme, which works in conjunction with a portable personal computer (PC), allows the driver to ask questions such as, "How far to my destination?" and receive a verbal reply from the system.

Recent research indicated that although not commercially available yet, enhanced auditory RGSs providing drivers with salient landmark information in addition to conventional route guidance instructions can significantly decrease the time required to learn novel routes and form cognitive maps in unfamiliar locations without disrupting driving performance (Reagan & Baldwin, 2006), as well as reduce navigational errors during the drive (Burnett, 2000a, 2000b).

INFOTAINMENT SYSTEMS

In addition to the many emerging in-vehicle technologies previously discussed, a host of entertainment or infotainment systems is rapidly emerging on the market. Infotainment systems extend beyond the conventional audio devices for playing music (i.e., radios, CDs, and DVDs) and provide means of watching videos, checking e-mail, and surfing the Web. Bluetooth®-enabled systems allow motorists to sync their home computers and cellular phones with the car system. This means sharing contact lists, files, and other data between home, business, and mobile platforms.

Satellite or digital radio, also increasing in popularity, provides motorists with an extensive list of channels to choose from, including commercial-free music stations, talk shows, and weather reports. Recent advances allow users to record and play back songs on one radio (i.e., at home) for later use on another compatible system (i.e., in the car). The proliferation of auditory displays and devices that is found in modern automobiles is also present in medical care facilities.

SOUNDS IN MEDICAL CARE ENVIRONMENTS

Advances in technology have greatly improved medical care and have simultaneously ushered in a multitude of new sounds (Donchin & Seagull, 2002; Loeb, 1993; Loeb & Fitch, 2002; Wallace, Ashman, & Matjasko, 1994). A number of sophisticated technologies, including numerous patient-monitoring devices, are available to provide health care workers with continuous information. Donchin and Seagull (2002) described recent trends in medical environments:

> Each year, new devices were added—automatic syringe pumps, pulse oximetry and capnography, pressure transducers, and monitors—that were bigger and occupied more space. To increase safety, alarms were added to almost every device. Around the patient bed there are, at minimum, a respirator, a monitor, and an intravenous pool with two to ten automatic infusion pumps. (p. 316)

The omnidirectional characteristic of auditory displays and alarms is particularly useful to health care professionals. Previous research indicated that health care workers frequently failed to detect changes in visual displays (Loeb, 1993). Auditory alarms, in conjunction with visual displays, increased both detection and recognition of physiologic changes in patient state (Loeb & Fitch, 2002). However, auditory alarms in medical environments are known to have a number of critical problematic issues. False alarms are extremely common (Tsien & Fackler, 1997), which can add to stress in both health care workers and their patients (Donchin & Seagull, 2002).

Before leaving our discussion of the sources of sound, we turn now to one more important use of sound to convey information: auditory assistive devices for the visually impaired. Many of the devices and auditory displays previously discussed can be used by all normal-hearing users. However, increased attention has been devoted in recent years to the development of auditory displays aimed at increasing the mobility and safety of visually impaired persons.

AUDITORY DEVICES FOR THE VISUALLY IMPAIRED

A range of auditory assistive devices has been developed or is well under way in development to help visually impaired individuals. For example, auditory interfaces are being developed to assist blind people in navigating through hypermedia material, such as the auditory World Wide Web (Morley, Petrie, O'Neill, & McNally, 1999), and to learn algebraic problems (Stevens, Edwards, & Harling, 1997). For blind people interpreting tables and graphs, nonverbal sound graphs have been shown to result in significantly less workload, less time to complete tasks, and fewer errors than speech interfaces and less workload than haptic graphs used alone (Brewster, 2002).

Navigation systems for the blind have been developed for both obstacle avoidance and way finding. These systems are portable, wearable GPS-based navigation systems utilizing auditory interfaces such as spatialized speech from a virtual acoustic display (Loomis, Golledge, & Klatzky, 2001). Stationary acoustic beacons have also been developed that assist sighted individuals in conditions of low visibility but also have particular applications to assisting visually impaired individuals in auditory navigation (Tran, Letowski, & Abouchacra, 2000; Walker & Lindsay, 2006). Acoustic beacons are currently installed at many intersections with high traffic density in cities, such as Brisbane, Australia, to assist blind individuals at intersection pedestrian crossings. Although still needing further development, navigation systems for the blind and visually impaired show promise for facilitating mobility egress in emergency situations.

SUMMARY

We have highlighted some of the myriad sounds present in our home and work environments. With the digital age, the number of auditory displays in the form of alerts and warnings as well as radio, telephone, and sonography has proliferated. The modern home and the modern flight deck, automobile, and medical environment are replete with sound.

CONCLUDING REMARKS

To process the multitude of sounds present in the everyday world, humans must be able not only to hear but also to interpret the incoming stimulus. In the next chapter, we distinguish between these two processes and discuss how they are integrated, also describing the mechanisms involved. Readers well acquainted with these sensory-cognitive issues may wish to skim or skip the next chapter, moving directly to the issues of the mental workload assessment in Chapter 4 and then the mental workload of auditory processing beginning in Chapter 5 and continuing on through the remainder of the book.

This book is concerned with the mental effort (workload) humans expend to identify, select, and extract meaningful information from the acoustic milieu constantly bombarding the airwaves. What characteristics aid or deter from this process? Under what circumstances does auditory processing succeed, and when does it fail? How does the amount of effort extended toward processing auditory information affect performance of other tasks? These are just some of the many questions addressed in the chapters to come. As Plomp (2002) so eloquently put it, the ability of human listeners to

> seek out which sound components belong together and to capture each individual sound with its associated characteristics ... are so sophisticated that they have to be seen as an active process. They operate so perfectly, interpreting every new sound against the background of earlier experiences, that it is fully justified to qualify them as *intelligent* processes. (p. 1)

In the next chapter, we begin to examine some of the characteristics of this intelligent process.

3 Auditory Pattern Perception
The Auditory Processing System

INTRODUCTION

To extract meaning from the many sources of sound in the environment, we must be able to sense, encode, and interpret auditory patterns. Fortunately, humans have a number of sophisticated physical structures in the auditory processing system that are specially designed to receive, organize, and encode acoustic information. These structures are what give humans the remarkable ability to recognize meaningful patterns from among the cacophony of sounds present in the acoustic environment: auditory pattern recognition. The auditory structures and the associated processes that make possible this ability are the focus of this chapter.

Auditory processing requires both hearing (a sensory process) and its interpretation (a perceptual-cognitive process). These processes can be closely tied to the concepts of bottom-up and top-down processing. Stimulus-driven sensory processes are often referred to as data-driven or bottom-up processes, while the interpretation of sensations can be referred to as contextually driven or top-down processing. Both bottom-up and top-down processes are essential to auditory processing, interacting and taking on various degrees of importance in different situations. A discussion of top-down processing is deferred to the next chapter on auditory cognition, while the current chapter focuses on the essentials of stimulus-driven, bottom-up processes. The physical characteristics of sounds, namely, the stimulus features that constitute bottom-up processing of the acoustic input to the ear, are discussed first, followed by a description of basic sound characteristics as well as the peripheral and central mechanisms involved in their perception. Readers already familiar with these sensory aspects of auditory processing may wish to skip ahead to the next chapter.

Auditory pattern perception must begin with the intake of the sound stimulus, the physical energy from the environment. The following section presents some of the important physical characteristics of sounds and how are they received and processed by the ear and brain.

CHARACTERISTICS OF SOUND PERCEPTION

In general, perception of sound involves three steps: generation of sound by a source, transmission of the sound through an elastic medium, and reception of the sound by a receiver (Hartmann, 1995). This chapter deals primarily with the first and last steps. First, a brief overview of the basic characteristics of sounds and the auditory sensory system is presented. It is beyond the scope of this chapter to provide a detailed account of this system. Readers interested in more detail may wish to consult one of the several excellent texts on the physical structures of the mechanisms of hearing (e.g., see Katz, Stecker, & Henderson, 1992; Moore, 1995; Yost, 2006). For the present purposes, some key fundamentals of the physical characteristics of sound are discussed in the next section.

THE SOUND STIMULUS

Sound is a mechanical pressure caused by the displacement of an elastic medium such as air or water (Coren, Ward, & Enns, 1999; Moore, 1995). The displacement results in changes in atmospheric pressure that are funneled into the auditory system by the pinna, through the auditory canal to the eardrum. For an in-depth description of the physical aspects of sound signals, see the work of Hartmann (1995). The physical structures involved in the process of collecting sound and converting sound into neural signals are discussed briefly in the next section. But first, a few key characteristics of sounds are described.

Frequency

Humans are capable of hearing sounds from roughly 20 to 20,000 hertz (Hz) or cycles per second. As illustrated in Figure 3.1, the peak-to-peak period represents the amount of time taken to complete one full cycle. The number of cycles per second is referred to as the frequency of the tone. Frequency is subjectively perceived as the pitch of a sound. As the sound wave moves faster (completes more cycles per second), the frequency and its perceived pitch typically both increase. In other words, sounds with higher frequency are perceived as higher in pitch.

The frequency of everyday speech has a range of roughly 10 kHz. However, humans are able to understand speech when a much narrower range is presented, on the order of 500–3,500 Hz, such as the limited bandwidth or range of analog telephone transmissions (Kent & Read, 1992). Humans are most sensitive to the frequency range between 2,000 and 4,000 Hz. Fortunately, this range encompasses

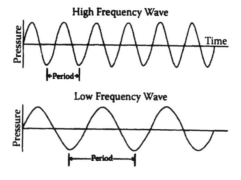

FIGURE 3.1   Sound waves and frequency.

most of the frequencies critical to speech processing (Kryter, Williams, & Green, 1962). However, the fundamental frequency of the human voice is much lower, with males having a fundamental vocal frequency (FVF) approximating 100 Hz, females an FVF of approximately 200 Hz, and children's FVF ranging from approximately 250 to 400 Hz (Hudson & Holbrook, 1982; Kent & Read, 1992). There are considerable individual differences in FVF; in general FVF decreases with advancing age. People who smoke also tend to have lower FVF. Our ability to perceive vocal characteristics such as FVF provide important speaker cues in normal everyday conversation (Andrianopoulos, Darrow, & Chen, 2001; Gregory, Green, Carrothers, Dagan, & Webster, 2001), a topic addressed further in Chapter 8 on speech processing.

Humans are sensitive to frequency changes. Specifically, the human ear can detect minute changes in atmospheric pressure caused by the vibratory displacements of an elastic medium. This ability explains why so many of the courageous (or perhaps narcissistic) individuals who sing karaoke tend to sound so bad. Their pitch is often painfully off. How hard can it be to sing a familiar melody? Does the average karaoke singer really think he or she is in tune? In reality, the individual may be close to the intended familiar pitch. However, even slight differences can be easily detected by most listeners. Our sensitivity to minute changes in frequency also assists in a number of practical everyday tasks, such as distinguishing between various consonant sounds and in interpreting the emotional content of a spoken message. Theories of how we accomplish this task, commonly referred to as pitch perception, are discussed in Chapter 7 in reference to musical pitch perception; also, see reviews by de Boer and Dreschler (1987), Moller (1999), and Tramo, Cariani, Koh, Makris, & Braida (2005). Another primary characteristic of sound is its amplitude.

Amplitude

The amplitude of a sound wave is determined by the change in pressure between the medium in which it travels (i.e., air or water) relative to the undisturbed pressure. Thus, it is common to speak of pressure amplitude. The height of the wave, as illustrated in Figure 3.2, represents the pressure amplitude of a sound stimulus. The larger the displacement of the wave (in other words, the higher it is), the greater its amplitude is. For a simple sound wave in air, the pressure amplitude is the maximum amount of change relative to normal atmospheric pressure measured in the force per unit area expressed as dynes per square centimeter (Coren, Ward, & Enns, 1999).

Amplitude is subjectively perceived as loudness. In general, as amplitude increases, a sound is perceived as louder. However, a number of factors can affect perceived loudness (such as frequency and context); therefore, there is a far-from-simple, one-to-one relationship between amplitude and perceived loudness. As illustrated in Figure 3.2, sound waves also have what is referred to as a phase. That is, relative to a reference wave, they are rising or falling. Presenting a sound of equal amplitude in opposing phase cancels out the sound. This principle is the basis behind active noise cancellation headphones, as illustrated in the right side of Figure 3.3: destructive interference.

Humans can hear a range of sound amplitudes on the order of several thousand billion to one units of intensity. To represent this enormous range of sound levels more easily, sound intensity is conventionally measured on a base 10 logarithmic

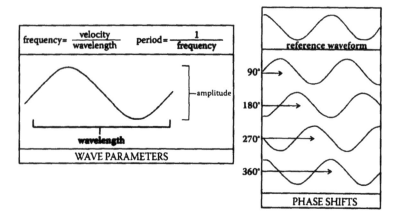

FIGURE 3.2   Sound waves and amplitude.

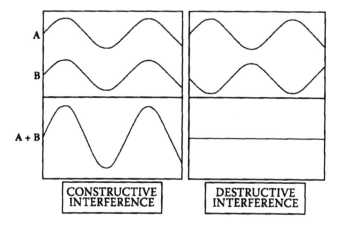

FIGURE 3.3   Phase-based noise cancellation.

scale using a unit called the decibel (dB). The decibel scale represents the intensity of a sound relative to a reference level, which is usually close to the absolute threshold.

The amplitude of everyday speech spans a range of roughly 60 dB (Kent & Read, 1992). Vowels are generally of higher amplitude, while consonants, and fricatives in particular (such as the "f" in fish), have the lowest amplitude. The amplitude range thus varies for a given speaker from utterance to utterance. Sound amplitude is also an important factor in determining the mental workload involved in processing speech. This topic is discussed in considerably more detail in the chapter on speech processing. For the present purposes, only the interaction between frequency and amplitude is examined.

Frequency-Amplitude Interactions

Frequency and amplitude interact in some interesting ways. For instance, in general, lower tones (e.g., 200 Hz) seem slightly lower as amplitude is increased, while higher tones (6,000 Hz) are perceived as slightly higher as amplitude increases (Rossing,

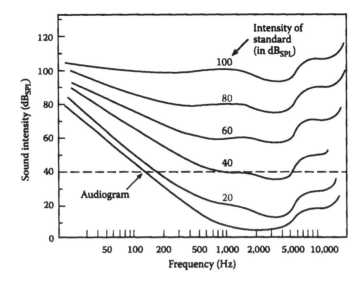

FIGURE 3.4   Equal loudness contours.

1990, as cited by Matlin & Foley, 1997, p. 312). In addition, humans are differentially sensitive to different frequencies. That is, sounds presented in certain frequency ranges are audible at lower intensity levels than are sounds presented at other frequencies. Figure 3.4 illustrates the audibility threshold at different frequencies. In particular, notice the lower threshold (meaning greater sensitivity) across the range of human speech. In higher-order, central auditory processes, we see that context also affects the perception of loudness. This is illustrated by a phenomenon referred to as recalibration (Marks, 1994), which is described in a further section.

Because the human ear is differentially sensitive across frequencies, sound-level meters (SLMs), which are used to measure intensity levels, typically use a weighting system (designated A, B, C, and D). The two most commonly used weighting systems are the A-weighted network and the C-weighted network. The A-weighted network most closely simulates the response of the human ear by discriminating against (showing less sensitivity to) low frequencies. Specifically, the A scale assigns weights according to equal loudness contours at 40 dB (Loeb, 1986). The dBA scale is therefore frequently used to assess noise levels in various environments. The C-weighted scale presents equal weighting across frequencies or unweighted sound pressure level (SPL) and therefore is noted as dB (SPL) (Loeb, 1986). In addition to the physical properties of individual sounds, the acoustical context can affect perceptions of loudness. Recalibration is one such contextual effect.

Masking

Auditory masking refers to the phenomenon that occurs when one sound prevents or blocks the perception of another sound. Sounds have their greatest potential to mask other sounds that are close in frequency. For example, a loud 1,000-Hz tone will tend to mask other sounds with frequencies of 1,010, 1,100, and 990 Hz. Sounds tend to have more of a masking effect on sounds that are higher than they are relative to those that are lower in frequency, a phenomenon that gives rise to the term *upward spread of masking* (Scharf, 1971). However, at high loudness levels, masking effects are more symmetrical (Scharf, 1971).

Recalibration

Perceived loudness is subject to context effects through an adaptation-like change referred to as recalibration (Arieh & Marks, 2003; Marks, 1994). Essentially, if one is listening to a relatively intense tone at one frequency and a weak tone is presented at another frequency, the latter tone will be perceived as louder than it actually is (Arieh & Marks, 2003, see p. 382) Adaptation can result from either relatively brief continuous presentation of a single tone or from prolonged exposure to intense tones, which results in temporary threshold shift. Marks (1994) discovered an interesting phenomenon he termed loudness recalibration. Specifically, Marks noticed that for tones above threshold, if one tone frequency is presented at a high SPL and a tone of another frequency is presented at a lower SPL, the latter lower-SPL tone will sound louder than it really is. Marks distinguished loudness recalibration as distinct from sensory adaptation, "suprathreshold sensitivity changes in response to transient stimuli, under conditions in which sensory adaptation is not expected" (Arieh & Marks, 2003, p. 382).

Frequency and amplitude can be thought of as the two most primary characteristics of a sound stimulus. Up to this point, we have discussed these characteristics for pure tones. Pure tones consisting of only a single frequency rarely occur in the natural environment. Tuning forks are one example. Among musical instruments, the flute comes closer than most to being capable of producing a pure tone.

Pure tones can be created easily in laboratory settings using one of a variety of software packages or electronic devices, such as pure-tone generators. Pure tones can be described simply by discussing the frequency and amplitude pressure changes occurring in the basic pattern illustrated in Figure 3.1 and referred to as a simple sine wave.

Most naturally occurring sounds are complex, being composed of the interaction of a number of waves of different frequencies and phases. The human ear separates the components of complex sounds (up to six or seven components with training) in a crude version of Fourier analysis known as Ohm's acoustical law. The individual components making up a complex tone are not heard as individual units but rather as a single tonal event (Plomp, 2002). Figure 3.5 illustrates a complex tone. This complex blend, referred to as timbre, is another key characteristic of most everyday sounds.

Timbre

Two tones of the same frequency and amplitude may still sound very different (such as the same note played by a piano vs. a clarinet). This difference in sound quality is due to a difference in timbre (pronounced like *timber* or *TAM-ber*; either is correct). Timbre is sometimes referred to as the *color* of a musical sound. Differences in timbre play an integral role in central auditory processing and in auditory stream segregation (a topic discussed in more detail in this chapter). Since most sounds present in the everyday auditory scene are complex, they are made up of a rich blend of frequencies referred to as harmonics. This complex blending of harmonics is what

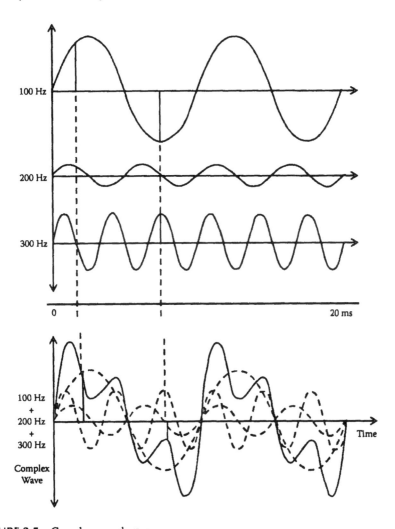

FIGURE 3.5 Complex sound wave.

gives a sound its unique timbre. As the name implies, we refer to these blended sounds as complex and distinguish them from simple or pure tones. The objectives of the current text preclude further discussion of other, more minor characteristics of the sound stimulus. Instead, we turn now to a look at the auditory mechanisms, both peripheral and central, that are involved in processing the sound stimulus.

PERIPHERAL AND CENTRAL AUDITORY PATHWAYS

The auditory system can be divided into two primary components, the peripheral hearing structures, encompassing the outer, middle, and inner ear, and the central auditory pathways located primarily within the cerebral cortex. Both peripheral and central auditory mechanisms are essential to auditory processing, and as we shall

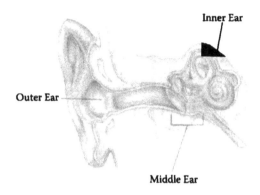

FIGURE 3.6   The three ears. (Drawn by Melody Boyleston.)

see, disruption of either mechanism can greatly affect the processing of auditory stimuli in everyday life.

Peripheral Auditory System

The human ear is thought to have evolved from the sense of touch and in fact shares several similarities with that sense modality. A structure in the inner ear called the cochlea contains specialized groups of cells with protruding hairs that resemble the hairs on our skin. These hairs respond to mechanical stimulation (Coren, 1999; Siemens et al., 2004). The human ear consists of three major divisions: the outer, middle, and inner ear (Figure 3.6).

The Three Ears

The three divisions of the human ear each have key structures that aid in the reception of sound. The outer ear consists of the pinna and the auditory canal. The pinna, the fleshy visible portion of the ear, is designed to funnel the sound wave into the auditory canal. The folds that make up its shape assist with sound localization. The tube-like auditory canal is roughly 3 cm long, and its size results in the greater sensitivity observed for critical speech frequencies (2,000–4,000 Hz) (see discussion in Goldstein, 2002; Hong et al., 2001). The increased sensitivity to this range is due to resonance, the reflection of a sound wave in a closed tube that serves to increase intensity (Coren et al., 1999; Goldstein, 2002). Sound waves travel through the auditory canal, causing vibrations in the tympanic membrane, otherwise known as the eardrum.

The eardrum separates the outer ear and middle ear. Vibrations of the eardrum cause movement of the ossicles, three tiny bones located in the middle ear. These three bones—the malleus (hammer), incus (anvil), and stapes (stirrup)—are the smallest bones in the human body. Their primary function is to amplify sound. Recall that sound is the displacement of an elastic medium such as air or water. If you have ever tried to listen to sound while swimming, you will understand that it is much harder to hear when your ears are under water. This is because sound travels much easier through low-density air than through the much-higher-density medium of water. The ossicles function like a fulcrum, transmitting the vibration from the relatively large eardrum to the much smaller stapes, which causes vibrations of the oval window, the gateway to the inner ear. This concentration of pressure from the eardrum to the much smaller oval window is much like the process of concentrating the weight of a human being onto a pair of stiletto heels. The tiny point of the high heel concentrates the weight, which can cause much greater damage to soft surfaces, such as gym floors, than the more evenly distributed surface of a flat shoe (Matlin et al., 1997). The middle ear muscles are the final middle ear structure of note. These muscles are the smallest skeletal muscles of the human body (Coren et al., 1999). The muscles serve a protective role; when the ear is exposed to high-intensity sound, they contract to dampen the movement of the ossicles.

The inner ear contains one of the most important structures in the human auditory system, the cochlea. The fluid-filled cochlea is a coiled bony structure resembling a snail in appearance. It is divided into two sections by a structure called the cochlear partition. The base of the cochlear partition is located near the stapes, and the apex is at the far end, deep inside the cochlea. The cochlear partition includes the organ of Corti, which contains the receptors for hearing. These receptors are called hair cells since tiny protrusions called cilia at the end of each cell resemble the tiny hairs found on the surface of the skin. Human hairs in the inner ear are specialized mechanosensors that transduce or convert the mechanical forces stemming from sound waves into neuroelectrical signals, providing us with our sense of hearing and balance (Siemens et al., 2004).

The organ of Corti rests on top the basilar membrane and is covered by the tectorial membrane. Movement of the stapes causes pressure changes in the fluid-filled cochlea, causing movement of the basilar membrane, which in turn results in an up-and-down movement of the organ of Corti and a back-and-forth movement of the tectorial membrane relative to the hair cells. The basilar membrane is about 3 cm long in humans and is stiffer at the base near the oval window, becoming wider (0.5 mm) and narrower (0.08 mm) at the apex (Coren et al., 1999).

The approximately 15,000 hair cells are divided into two groups, including a single row of about 3,000 inner hair cells and three to five rows of outer hair cells arranged in V- or W-shaped rows. Each of the inner and outer hair cells has a set of smaller hair-like protrusions called cilia of differing sizes protruding from a single hair cell (Coren et al., 1999). The tallest cilia of the outer hair cells are embedded in the tectorial membrane, while the shorter cilia do not reach. The bending of these hair cells results in transduction (conversion of the mechanical motions of the ear into electrical activity in the auditory nerve). The hair cells nearest to the ossicles respond preferentially to high-frequency sounds, while those farthest away respond best to low-frequency sounds. These tiny hair cells can be damaged by exposure to intense sounds, although recent evidence suggests that some regeneration in mammals can occur (Lefebvre, Malgrange, Staecker, Moonen, & Van De Water, 1993; Shinohara et al., 2002). The hair cells synapse on the spiral ganglion nerves, which form the auditory nerve projecting toward the primary auditory cortex and the central auditory pathways of the brain. It is in these central auditory pathways that higher-order processing occurs.

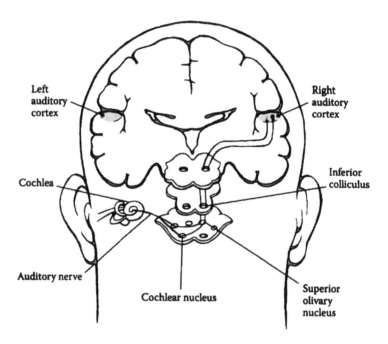

FIGURE 3.7   Auditory pathways. (Drawn by Melody Boyleston.)

Central Auditory Pathways

The auditory nerve carries the signals beyond the ear to structures of the central nervous system, which lead to the primary auditory receiving area in the cortex. The first structure encountered is the cochlear nucleus (Figure 3.7). The cochlear nucleus located in either hemisphere receives signals from the ipsilateral (same-side) ear and then projects the information to both the ipsilateral and contralateral (opposite side) superior olivary nucleus (Banich, 2004; Goldstein, 2002). This dual-hemisphere organization is a unique characteristic* of the auditory sensory system. In most other sensory systems (with the exception of olfaction), stimuli from one side of the body, once transduced into neural signals, project solely to the contralateral hemisphere of the brain (Banich, 2004, p. 29).

From the superior olivary nucleus, the pathway proceeds to the inferior colliculus and then on to the medial geniculate bodies in the thalamus. The medial geniculate bodies then project to several cortical areas, including the primary auditory cortex and several adjacent areas. The primary auditory receiving area (A1) is located in the superior portion of the posterior temporal lobe in an area referred to as Heschl's gyrus, located in Brodmann area 41 (Figure 3.8). Several adjacent areas, including Brodmann area 42, referred to as the second auditory cortex or A2, and Brodmann area 22 (Wernicke's area for speech), also receive signals from the medial geniculate

---

* This unique characteristic of the auditory system has important implications. For example, sound stimuli from both ears can still be processed if damage occurs to the primary auditory cortex in one hemisphere.

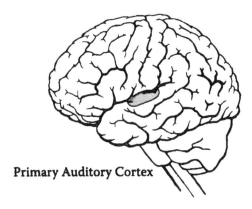

Primary Auditory Cortex

FIGURE 3.8   Auditory cortex. (Drawn by Melody Boyleston.)

nucleus. The primary auditory cortex is arranged tonotopically, or organized in such a way that particular areas are more sensitive to certain acoustic frequencies, with neighboring areas demonstrating sensitivity to neighboring frequencies (Fishbach, Yeshurun, & Nelken, 2003; Hall, Hart, & Johnsrude, 2003). Low-frequency tones are processed in areas closer to the scalp (rostrally and laterally), and higher-frequency tones are processed in areas deeper in the cortex (caudally and medially) (Cansino, Williamson, & Karron, 1994).

Evidence also indicated that pure tones are processed in the primary auditory cortex or core areas, while more complex tones are processed in the surrounding belt areas (Wessinger et al., 2001). After initial processing in the core and surrounding belt areas, signals travel to several areas throughout the cortex. Neurophysiological evidence indicates that physical characteristics of the sound stimulus such as intensity, frequency, and duration are differentially represented in unique cortical areas within the auditory cortex (Giard, Lavikainen, Reinikainen, Perrin, & Naatanen, 1995; Takegata, Huotilainen, Rinne, Naatanen, & Winkler, 2001).

Giard and colleagues (1995) utilized the mismatch negativity (MMN) paradigm to investigate the neural mechanisms involved in processing frequency, intensity, and duration. The MMN is an event-related potential (ERP) of the brain and can be recorded noninvasively from the scalp. The MMN is elicited in response to any discriminable change in a repetitive sound (Giard et al., 1995) and is thought to reflect an automatic comparison process between neural traces for standard repeating stimuli and deviant stimuli (Naatanen, Pakarinen, Rinne, & Takegata, 2004; Sams, Kaukoranta, Hamalainen, & Naatanen, 1991). Giard and colleagues, using dipole model analysis to identify the cortical source of the scalp-derived MMN, observed that the cortical location of the MMN responses varied as a function of whether a stimulus deviated from the standard in frequency, intensity, or duration.

Corroborating evidence for the modularity of these processes came from case studies showing that some auditory disorders involve differential or selective disability in one function while other aspects of auditory discrimination remain intact. For example, a small percentage (roughly 4%) of the population has a condition known as amusia or tone deafness. These individuals demonstrate consistent difficulty in differentiating fine-grained pitch changes, while their ability to distinguish time changes related to the duration of tones remains intact (Hyde & Peretz, 2004). One well-known individual, Che Guevara, the highly educated revolutionary figure involved in the Cuban revolution and portrayed in the movie *Motorcycle Diaries*, probably suffered from amusia (Friedman, 1998, as cited in Hyde & Peretz, 2004).

That different aspects of pitch perception are processed in separable neural locations (Griffiths, Buechel, Frackowiak, & Patterson, 1998) provides additional evidence for modularity. Griffiths and colleagues used positron emission tomography (PET) imaging to identify the brain regions involved in processing short-term (pitch computations) and longer-term (melodic pattern) temporal information (see discussion by Zatorre, 1998). They observed that short-term temporal integration was associated with increased activity in and around the primary auditory pathway (namely, the inferior colliculus and medial geniculate body). However, long-term tone sequences resulted in greater activation in two bilateral cortical areas, the right and left posterior superior temporal gyri and anterior temporal lobes. It is interesting to note that the tone-sequence patterns were associated with greater bilateral activity rather than the more typical right-hemispheric asymmetric pattern frequently cited for music perception (Koelsch, Maess, Grossmann, & Friederici, 2003; Tervaniemi & Hugdahl, 2003). Bilateral activation patterns for both music and speech are more common among females than males (Koelsch et al., 2003); however, Griffiths and colleagues did not report the gender of their participants. Specific cases of speech and music perception are discussed in more detail in further chapters. For now, we return to the topic of auditory processing in the central auditory pathways.

Auditory Processing Streams

Two central auditory pathways with separate functional significance have been identified (Clarke & Thiran, 2004; Rauschecker & Tian, 2000; Read, Winer, & Schreiner, 2002; Zatorre, Bouffard, et al., 2002). These dual streams originate in separate nonprimary auditory cortical areas and terminate in distinct locations with the frontal lobes (Romanski et al., 1999). Rauschecker (1998) referred to the dorsal stream as the auditory spatial information pathway. This pathway appears critical for sound localization. Conversely, the ventral stream is essential for identification of sounds and is the pathway for processing auditory patterns, communications, and speech. Note the similarity to the parallel "what" and "where" pathways that are associated with processing visual stimuli (see Figure 3.9; Ungerleider & Haxby, 1994; Ungerleider & Mishkin, 1982).

Also similar to a debate in research on visual processing, there is ongoing discussion over whether the auditory where pathway might actually be more correctly referred to as the auditory "how" pathway. Belin and Zatorre (2000) suggested that both streams are necessary for speech perception, with the dorsal pathway critical to understanding the verbal content of the message and the ventral pathway critical for identifying the speaker. They suggested that characterizing the dual streams as the what and how pathways may more accurately reflect their functional segregation.

The ventral stream can be further separated into specialized hemispheric pathways. For example, Zatorre, Belin, et al. (2002) observed that temporal resolution,

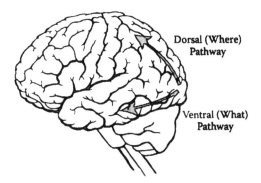

Dorsal (Where)
Pathway

Ventral (What)
Pathway

FIGURE 3.9   Dorsal and ventral pathways. (Drawn by Melody Boyleston.)

essential to speech, is considerably better in the left hemisphere, while spectral resolution, essential to music perception, is considerably better in the right hemisphere. In fact, a nearly complete dissociation can be demonstrated between intact speech processing and highly disturbed tonal processing (see Marin & Perry, 1999).

As already indicated, the neural mechanisms involved in speech and music perception have received considerable attention. Both have been utilized extensively in explorations of cerebral lateralization and a wide variety of other psychological phenomena. Due to their significant role in contributing to our understanding of auditory processing, we examine the particular cases of speech and music processing and the neural mechanisms associated with each. We begin with a look at speech processing. Some researchers suggested that speech is processed differently from other sounds (Belin et al., 2000).

Is Speech Special?

The debate over whether "speech is special," that is, whether speech is processed in distinctly different ways or utilizes different neural pathways from those of nonspeech sounds, is an ongoing one in psychology and neuroscience. Feature detectors for simple visual stimuli were discovered in the visual cortex of the cat nearly half a century ago (Hubel & Wiesel, 1962). However, despite considerable research, similar identification of neurons for processing auditory features or specific speech stimuli have remained elusive (Syka, Popelar, Kvasnak, & Suta, 1998). Throughout the 1970s, numerous research groups sought to find "call detector" neurons in the auditory cortex of several species (primarily awake monkeys), only to conclude that such pattern recognition must occur through an interconnected group of neurons rather than in feature-specific neurons (Syka et al., 1998).

In contrast, voice-selective areas have been found in regions of the superior temporal sulcus (STS) in humans. These regions are particularly sensitive to the speaker-related characteristics of the human voice (Belin et al., 2000). Belin and colleagues demonstrated that areas on the STS of the right hemisphere were selectively sensitive to speech stimuli. A more in-depth discussion of these issues is provided in Chapter 8 in reference to speech processing. However, for now an important method for examining the neural mechanisms of auditory processing that will be important to understanding the subsequent section on cerebral lateralization and music processing in particular is briefly presented.

Auditory Event-Related Potentials

The auditory ERP, including the auditory brain stem response (ABR) and early ERP components, can reveal important information regarding the function of the auditory processing system. ERPs are averaged electrical responses to discrete stimuli or "events" obtained from a continuous electroencephalographic (EEG) recording from the scalp. Responses to individual signals are generally too small to be distinguished from the background noise of neural events; therefore, the response to a number of discrete signals must be averaged together to observe ERP patterns. Depending on the size of the ERP component studied, signal averaging must typically be done over about 30–60 trials (for large components like P300) to several hundred trials (for small components like the brain stem potentials) (see Luck, 2005). Components or averaged waveforms are generally named according to the direction of their deflection (P for positive deflections and N for negative deflections) and then either their average expected latency or ordinal position (Luck & Girelli, 1998). For example, the first large negative wave deflection occurring about 100 ms poststimulus is referred to interchangeably as either the N1 or N100 component. See Figure 3.10 for a graphical depiction of functional ERP components. Early components, such as the N1, are often referred to as exogenous components since they are thought to reflect sensory processes that are relatively independent of cognitive control (see Coles & Rugg, 1995, for a review). Later components, such as P300 and N400, are thought to reflect cognitive processes and are therefore often referred to as endogenous components. N1 in response to auditory stimuli is modulated by physical aspects of the stimulus, such as intensity (see Coles & Rugg, 1995), but also shows reliable changes as a function of attention, generally being of greater amplitude when listeners attend to the stimuli and ignore competing sources (Hillyard, Hink, Schwent, & Picton, 1973). P300, an endogenous component, is modulated by context, expectancy, and task difficulty (Coles & Rugg, 1995), and the N400 component has been shown to reflect semantic processing (Federmeier, van Petten, Schwartz, & Kutas, 2003; Kutas & Hillyard, 1980). At the same time, these later endogenous components are not immune to the effects of changes in the physical aspects of a stimulus.

Components appearing around 100 ms poststimulus reflect sensory processes and are therefore said to be exogenous components; however, they can also be modulated by attention (Coles & Rugg, 1995; Federmeier et al., 2003; Hillyard et al., 1973; Naatanen, 1992). Later components (beyond 100 ms) generally reflect endogenous cognitive processes (Coles & Rugg, 1995). The P300 component in response to auditory stimuli has received considerable attention and has been used in a number of examinations as an index of mental workload. The P300 has most commonly been used in an oddball paradigm in which two or more stimuli of varying probability are presented. The low-probability target (or oddball) elicits a P300 response that typically decreases in amplitude as the demands of a secondary task increase. The auditory oddball paradigm is discussed in greater detail along with a number of examples of its application in Chapter 6, "Auditory Tasks in Cognitive Research." The ABR

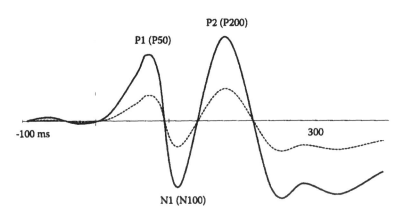

FIGURE 3.10   ERP components.

deserves further attention here as it is commonly used to assess the integrity of auditory structures that lie between the inner ear and the cortex.

Auditory Brainstem Response

The ABR occurring during the first few milliseconds after presentation of a stimulus provides an indication of the functioning of the specific relay structures in the pathway from the sensory receptors to the cortex (Buchwald, 1990). Figure 3.10 provides a characterization of a typical ERP graph. The ABR is a common clinical assessment tool. Benefits of ABR testing include the ability to test infants and others with no or diminished verbal capabilities. For example, ABR responses can be used to examine auditory functioning in persons unable to manually or vocally respond. The ABR can also be used to test the validity of claims made by adults that they suffer from partial hearing loss as a result of exposure to high-intensity noise in the workplace.

We now return to our discussion of central auditory processing. Considerable evidence exists for the hemispheric specialization of various auditory processes including both speech and music (Mathiak et al., 2003). A few of these investigations are discussed in the next section before going on to discuss some more general aspects of auditory pattern perception.

CEREBRAL LATERALIZATION

Cerebral lateralization or hemispheric specialization, particularly for language functions, has been investigated for well over a century, dating at least as far back as the work of Paul Broca and Carl Wernicke in the late 1800s (see Banich, 2004; Tervaniemi & Hugdahl, 2003). The importance of left hemisphere regions for language processing can be traced to Broca's work nearly a century and a half ago (Grodzinsky & Friederici, 2006; Powell et al., 2006). Left hemisphere language regions, namely, Broca's area for language production and Wernicke's area for language reception, were some of the earliest discoveries in brain physiology. Early work in modern cognitive neuroscience resulting from Sperry and Gazzaniga's (see Gazzaniga, 2000; Gazzaniga & Sperry, 1967) investigations of split brain patients further supported the important role of the left hemisphere for language processing, and production in particular. Sperry and Gazzaniga observed the behavior of patients who had undergone operations in which the two halves of their brains were separated by severing the connections at the corpus callosum and hippocampus, so-called a "split brain" individuals. When visual stimuli were presented to the left visual field (right hemisphere) of split brain individuals, the individuals would be unable to name the objects seen. Conversely, if presented to the right visual field (left hemisphere), the split brain individuals were able to identify and name the objects.

Since the early days of modern cognitive neuroscience, numerous experiments have been carried out examining the role of the left hemisphere in language processing. A dominant left hemispheric pattern of structural asymmetry for speech processing has been extensively documented (Just, Carpenter, Keller, Eddy, et al., 1996; Kasai et al., 2001; Pardo, Makela, & Sams, 1999; Tervaniemi & Hugdahl, 2003). Left hemisphere specialization for language has been observed using behavioral techniques (see, e.g., Hartley, Speer, Jonides, Reuter-Lorenz, & Smith, 2001), electrophysiological indices such as the EEG (Davidson, Chapman, Chapman, & Henriques, 1990), and brain imaging studies, such as PET scans and functional magnetic resonance imaging (fMRI; Drager et al., 2004; Peelle, McMillan, Moore, Grossman, & Wingfield, 2004; Posner, Petersen, Fox, & Raichle, 1988; Zatorre, Evans, Meyer, & Gjedde, 1992). Even nonauditory forms of language, such as American Sign Language, demonstrate greater activation of left hemisphere processing in individuals with congenital deafness (Bavelier et al., 1998). However, modern researchers pointed out that language processing is not solely a left hemisphere task. A number of important processes are carried out predominantly by the right hemisphere. For example, understanding prosodic cues relies more heavily on right hemispheric mechanisms than left (Friederici & Alter, 2004), and some syntactic processes have been found to activate right hemispheric regions (Grodzinsky & Friederici, 2006). In addition, there appear to be gender differences in cerebral asymmetry patterns for both speech and music.

A number of recent investigations have indicated that hemispheric asymmetry is stronger in males than females. This appears to be the case for both language processing (Nowicka & Fersten, 2001; Pujol et al., 2002) and music processing. For example, Pujol and colleagues (2002) found significantly more white matter content in left hemispheric regions relative to the right in areas associated with language processing in both males and females. But, males have significantly more left hemisphere white matter in these areas than females. Greater cerebral asymmetry among males relative to females is also found during music processing.

MUSIC PROCESSING

Humans possess an implicit ability to process both speech and music. In the case of speech, the supporting evidence comes from studies showing that neonates and infants are able to make phonetic discrimination, as measured by electrophysiological responses (Dehaene-Lambertz & Baillet, 1998; Dehaene-Lambertz & Pena, 2001). That music abilities are also implicit is supported by the observation that neonates detect changes in tonal patterns (Dehaene-Lambertz, 2000), and that adults with no formal musical training demonstrate both early (200 ms) and late (500 ms) ERP responses to unexpected changes in tonal patterns (Koelsch, Gunter, Friederici, & Schroeger, 2000). Koelsch and colleagues referred to these early ERP component patterns in response to unexpected sound patterns as ERAN (early right anterior negativity) responses. They suggested that the ERAN reflects a preattentive musical sound expectancy violation based on implicit knowledge of complex musical patterns (Koelsch et al., 2003). The neurophysiological exploration of musical processing is an area of growing interest, and a recent surge of investigations is beginning to shed light on how the human brain processes music. The perception of music is discussed in more detail in Chapter 7, "Nonverbal Sounds and Workload." Here, a brief discussion is presented of sex differences in speech and music.

SEX DIFFERENCES IN SPEECH AND MUSIC

Differences between the sexes have been observed for the cortical areas utilized when processing both speech and music (Koelsch et al., 2003). Illustrative of sex differences in language processing, Jaeger and colleagues (1998) used PET scans to measure cerebral blood flow (CBF) while men and women performed a task involving reading verbs and generating past-tense forms. Males and females had equivalent levels of accuracy and response time on the task but demonstrated different patterns of cortical activity. Males demonstrated left lateralized cortical activation patterns, while females demonstrated bilateral activation specific to the perisylvian region. A similar difference in lateralization is present for music processing.

Koelsch and his colleagues (2003) conducted what may be the first systematic examination of sex differences in music processing. They had participants listen to sequences of five chords while recording EEG and ERPs. In some of the chord sequences, an unexpected and harmonically inappropriate chord was presented at either the third or fifth position. The inappropriate harmonies elicited an ERAN response, reaching a peak amplitude around 200 ms following the presentation of the chord. Particularly germane to the current purpose was their observation that this ERAN pattern was different for males and females. Males demonstrated a strongly right-lateralized ERAN, while females exhibited a bilateral response. Note that for males this pattern is in direct opposition to the predominantly left-lateralized response to language processing, while females demonstrated bilateral activation for both speech and music tasks.

ESSENTIAL AUDITORY PROCESSES

A number of auditory processes can be considered essential even though they may be entirely unconscious. For example, as will be discussed, auditory stream segregation and auditory scene analysis help us understand the world around us in important ways; yet, most people never think about them. We begin with a discussion of one more recognized essential auditory process: sound localization.

Sound Localization

Humans and other animal species rely on sound localization for survival. It is essential that we be able not only to detect the presence of but also to localize the sounds of oncoming vehicles, falling objects, and so on, as well as find objects such as a ringing telephone or a distraught child. Since we have ears on both sides of our head, we can use a number of cues, such as the difference in arrival time at each of the ears (interaural time difference, ITD) and the difference in sound level at each ear (interaural intensity difference, IID) to aid in sound localization. ITD is best for sounds with low-frequency spectra, and the IID is utilized more effectively for high-frequency spectra.

Several neural areas are involved in sound localization. Neurons in the superior olivary nucleus respond maximally when a sound reaches one ear before the other, thus aiding in detection of ITD. ITD and IID appear to be carried out primarily in the primary auditory cortex as damage to this area significantly impairs these abilities (Banich, 2004; Middlebrooks, Furukawa, Stecker, & Mickey, 2005). Sound localization appears to utilize different neural pathways for approaching an object versus avoiding an object (Dean, Redgrave, Sahibzada, & Tsuji, 1986).

An area in the midbrain, called the superior colliculus (SC), aids in sound localization (Withington, 1999) and responds to light and touch. The SC assists in directing visual attention and guiding eye movements (Middlebrooks et al., 2005; Populin, Tollin, & Yin, 2004). The multimodal nature of the SC allows integration of alerts from multiple sensory channels. Other multimodal mechanisms facilitate our development of auditory spatial representations.

Auditory Space Perception

Auditory space perception refers to the ability to develop a spatial representation of the world (or the area one is experiencing) based on auditory information. Auditory information directs attention to distinct areas and objects. It helps us locate sound sources and avoid obstacles.

Although we may not consciously think about it, we can understand a considerable amount of information about the space around us solely though sound. For example, with eyes closed or in a completely dark room, most people can generally tell how big a room is, what type of materials are on the walls and floor, whether there are high or low ceilings, if there are many objects in the room, and so on. Sounds reverberate off walls and ceilings and provide the listener with a considerable amount of information. And, even if the information is not in conscious awareness, it can have an impact on mood and feelings of safety or security. Blesser and Salter (2007) referred to this as the aural architecture of the space. As they pointed out, the aural architecture of a space is a composite of the many surfaces, structures, and materials the space contains. We can tell by sound alone whether a space is crowded with objects or vacant. Perhaps you have had the experience of noticing how different a familiar room sounds after removing all of your furniture and personal items. Removing the contents of a room changes its aural architecture. Most of the time, auditory spatial information is integrated with visual spatial information. A number of similarities and some striking differences are observed between spatial perception in these two modalities. This topic is discussed further in Chapter 9.

The next stage of auditory pattern perception relies heavily on top-down processing mechanisms involving the integration of environmental context with the personal rich history of auditory experience, thus allowing us to make sense of the auditory scene. Acoustic signals must be analyzed into their respective sources as well as synthesized and reconstructed into meaningful units before an understanding of their informational content can be achieved. We turn now to an examination of the many processes that help us accomplish this no-less-than-astonishing feat.

AUDITORY SCENE ANALYSIS

Auditory scene analysis refers to the process of interpreting the auditory soundscape. It involves sorting out how many different sources of sound there are and which sounds go with what object or person. As discussed in this chapter, auditory perception stems from sound waves entering the ear and being channeled through the ear canal to the tympanic membrane. All of the various sound waves present in the environment must be processed through the same peripheral structures. Auditory scene analysis deals with the topic of how these various sounds are sorted out and interpreted. First, how are the various different sounds separated?

Auditory Stream Segregation

The everyday auditory scene is comprised of multiple competing acoustical stimuli generated from a variety of sources. Fortunately, humans possess a number of organizational processes that assist us in extracting useful information from these competing acoustical signals (Plomp, 2002). Two key organizational processes are auditory stream segregation and the continuity effect.

The everyday acoustic environment is filled with a host of sounds. To make sense of this auditory scene, the sounds must be separated into their respective sources. Some general principles of auditory stream segregation have been identified. The auditory system tends to group sounds that are similar rather than sounds that follow each other closely in time (Bregman, 1990; Plomp, 2002). These groupings or links are used to identify which sounds are being generated by the same source. This is the very essence of the process known as auditory stream segregation.

Stream segregation can be demonstrated in the lab by playing a series of high and low tones in quick succession. Within certain temporal restrictions, listeners will "hear" two streams, one made of the high tones and one made of the low tones, as if the two streams are being emitted from two different sources (Bregman &

Campbell, 1971). When the tones are played very slowly, they are heard as one continuous sequence; however, as the presentation rate increases, listeners begin to hear two separate streams. The phenomenon is particularly strong when notes are played 100 ms or less apart (Bregman, 1990; Bregman & Campbell, 1971). Bregman further noted that the greater the frequency separation between the notes the more compelling the segregation is.

Auditory stream segregation is a fundamental component of auditory scene analysis. Organizing information into coherent streams helps us process and retain the auditory objects and events in our environment. Bregman and Campbell's (1971) experiment demonstrated that organizing material into coherent streams not only is automatic but also aids in retention. Participants were presented a series of high notes (i.e., A, B, C) interspersed with a series of low notes (i.e., 1, 2, 3) in series, such as A12BC3. But when asked to recall the notes, they reported them as grouped by their perceived "stream" (i.e., ABC, 123). This grouping of the incoming sound information into auditory objects or wholes appears to occur automatically. In fact, Bregman (1990) argued that it is largely preattentive, and in fact that it takes additional effort to hear specific individual items outside their organized streams. This position remains controversial, with some arguing that it requires attention.

Attention and Auditory Stream Segregation

A recent surge of investigations has led to a debate over whether auditory stream segregation (separating sounds into their logical sources) requires attention. Little emphasis was placed on the role of attention during the early investigations of auditory streaming. However, several investigations have examined this issue (Carlyon, Cusack, Foxton, & Robertson, 2001; Macken, Tremblay, Houghton, Nicholls, & Jones, 2003). By examining apparent stream segregation with irrelevant sound effect (ISE) paradigms, Macken and colleagues presented compelling evidence that auditory stream segregation occurs outside focal attention.

By definition, the ISE paradigm investigates the disrupting effects of irrelevant auditory material on the serial recall performance of a concurrent task. Jones (1999) reviewed evidence of the situations and circumstances in which irrelevant sounds were detrimental to recall in a concurrent task. Macken and colleagues (2003) summarized this research and interpreted it in terms of the role of attention in auditory stream segregation, pointing out that segregation occurs outside the realm of focal attention.

However, despite the strong case made by Macken and colleagues (Jones, 1999; Macken et al., 2003) that focal attention is not required for auditory stream segregation to exist, it can still be argued that stream segregation, while clearly an obligatory process, nevertheless may still require mental resources. A case can be made for this interpretation by examining the situations in which ISE effects occur. Macken et al. (2003) pointed out that serial recall performance for an attended task is significantly disrupted when irrelevant babble speech (concurrent presentation of two or more voices) is presented from a single location. However, this disruption decreases as the number of simultaneous voices increases. Macken interpreted this as evidence that a cumulative masking effect results, which prevents segregation. When the simultaneous voices (up to six) are presented from separable locations, then the level of disruption is higher than when all voices are presented from a single location. This could be interpreted as evidence that when possible obligatory stream separation occurs even in channels outside focal attention this process requires some degree of mental resources and thus interferes with a concurrent task.

Continuity Effect

The continuity effect is another of the essential processes for organizing acoustical patterns. The continuity effect refers to our ability to perceive a continuous and coherent stream of auditory information in the face of disruption from simultaneously overlapping acoustical information. Our higher auditory processing mechanisms literally "reinstate" or reconstruct missing acoustical information. We accomplish this task by using the context of the signal to make predictions regarding the most likely acoustical stimulus to have been presented (Plomp, 2002). Miller and Licklider (1950) were among the first to document this remarkable process. More recently, numerous investigations have illustrated a special case of the continuity effect used to facilitate speech perception. This special case is called phonemic restoration, and it was illustrated in a now-classic experiment conducted by Warren and Warren (1970).

Warren and Warren (1970) demonstrated that when a phoneme within a sentence was replaced by a click or cough, listeners would "fill in" the missing phoneme, thus preserving the continuity of the sentence. Generally, listeners would not even be aware that the phoneme had not actually been presented. This process of phonemic restoration has been observed in several subsequent investigations (Samuel, 1996a, 1996b).

The continuity effect and the special case of phonemic restoration provide compelling evidence for the importance of context in auditory processing. The use of context relies on top-down cognitive processes. Additional organizational processes assist in auditory pattern recognition. As with all sensory experiences, there is a limit to how many sounds the human can process at any given time. Therefore, the gestalt principle of figure ground organization in which some sounds receive focal attention while others "fade" into the background can also be demonstrated during auditory processing. Further issues relevant to auditory stream segregation are discussed in Chapter 7 under the topic of auditory perceptual organization, which has particular relevance to the processing of nonverbal sounds. The role of attention and other information-processing topics are discussed in the next chapter, auditory cognition. But first, a summary of the physical characteristics of the acoustic stimulus and the mechanisms of the auditory processing system just discussed is provided.

SUMMARY

Auditory processing requires the interaction of hearing (a sensory process) and interpretation of the acoustic signal (a perceptual-cognitive process). The sensory process can be thought of as largely a bottom-up, data-driven process, while the interpretation can be said to be a top-down process. It is the interaction of these two processes that results in auditory cognition. The temporal pattern of a sound (the relationship of sound to silence across time) also provides an important source of information in the bottom-up sensory signal. Each of these physical characteristics and the location of the sound and other cues assist in the process of auditory stream segregation.

Together, these characteristics provide the soundscape open for interpretation, as discussed in the next chapter.

CONCLUDING REMARKS

The concepts presented in this chapter will reoccur through the remaining text. For example, we will discuss aspects of auditory scene analysis in greater detail, noting in Chapter 9 how information from the auditory channel is combined with information from other modalities. Chapter 11 discusses the importance of understanding how we perceive frequency and amplitude and the phenomenon of auditory masking in relation to the design of auditory displays and alarms. We also address in subsequent chapters the mental workload required by these intelligent processes. First, in the next chapter we address many of the key issues as well as methods and techniques for assessing mental workload.

4 Auditory Cognition
The Role of Attention and Cognition in Auditory Processing

INTRODUCTION

In this chapter, the discussion of auditory processing begun in the last chapter is continued by examining the role that attention plays in helping to identify and select an acoustic pattern of interest. Next, higher-order processes involved in audition, which can be viewed as the real heart of auditory cognition, are discussed. Some would go as far as to call these abilities *auditory intelligence* (de Beauport & Diaz, 1996), suggesting that as humans we vary in our ability to interpret and find meaning in the sounds we hear. Viewed in this way, auditory intelligence involves the degree to which we are able to go beyond merely hearing sounds to higher-order processing—involving not only taking in sounds, but also "words, tones of voice, and arriving at a sophisticated or comprehensive meaning ... and connecting inner meaning to a sound received from the outer environment" (de Beauport & Diaz, 1996, p. 45). Individuals may differ in their ability to construct meaning from sound because of talent or experience, but regardless, the same basic processes are used by all listeners.

Auditory cognition begins with our attending to an acoustic stimulus. Without such attention, further cognitive processing is unlikely. Therefore, we begin our discussion by examining attentional mechanisms in selective listening and the role that studies of selective listening have played in the development of theories of attention and information processing.

ATTENTION

It should be clear by now that listening requires more than simply passively receiving an auditory stimulus. A host of complex processes is involved in extracting and making sense of the acoustic environment. Selective attention plays one such important role in auditory pattern perception. Since we are constantly being bombarded by a wide variety of sounds, more stimuli than we could ever begin to process, it is fortunate that humans are capable of selectively attending to certain acoustic sources while ignoring others. Our ability to selectively attend to one source of sound in the midst of competing messages was examined extensively during the 1950s and 1960s using the selective listening paradigm. More recently, the role of attention in auditory processing has been examined using a variety of brain imaging techniques. This topic is discussed further in this chapter.

The selective listening paradigm, or dichotic listening task as it is more commonly referred to today, involves presenting two competing messages to the two ears. Most commonly, a different message is presented to each ear. The listener is generally required to attend to one of the two messages, often being asked to repeat aloud (shadow) the contents of the attended message. Investigations utilizing the dichotic listening paradigm made major contributions to the development of early theories of attention and information processing (Broadbent, 1958). The paradigm was used to systematically examine such issues as the aspects of the sound stimulus that help in auditory stream segregation and whether and to what extent one can attend to more than one input source at a time. The study of attention remains an integral part of modern experimental psychology. However, our discussion primarily focuses on what early research on auditory attention revealed about the processes involved in auditory pattern perception.

EARLY ATTENTION RESEARCH

The study of attention has a rich history. Dating at least as far back as William James (1890/1918), examinations of the role of attention in processing have been the focus of considerable theoretical and empirical research (Anderson, Craik, & Naveh-Benjamin, 1998; Ball, 1997; Barr & Giambra, 1990; Broadbent, 1982; C. Cherry, 1953a; Moray, 1969; Norman, 1976; Treisman, 1964c). It is well beyond the scope of the current discussion to review this long and fruitful history. The reader interested in more detailed coverage of research on attention is referred to one of the many excellent books on the topic (see, for example, Parasuraman, 1998; Parasuraman & Davies, 1984; Pashler, 1998b). However, a brief overview of the development of various theories of attention is relevant to the current aims. In a broad sense, this history can be divided into three primary phases: (a) filter theories, (b) capacity or resource theories, and (c) cognitive neuroscience models.

Early models of attention postulated the existence of a filter or bottleneck where parallel processing changed to serial processing. The primary distinction and point of debate between these early models was where in the processing chain the filter occurred. Early selection models (i.e., Broadbent, 1958) postulated that the filter occurred in preattentive sensory processing. Treisman's (1964c) attenuation theory suggested that rather than an early switch, unattended information was attenuated or processed to a limited extent. Late selection models (Deutsch & Deutsch, 1963) proposed that essentially all information reaches long-term memory (LTM), but we are unable to organize a response to all of the information. Use of the dichotic listening task was instrumental in the development of these early models of attention and shed light on a number of interesting features of auditory processing.

DICHOTIC LISTENING TASKS

Initial evidence for placing the filter early in the information-processing chain was demonstrated through the dichotic listening paradigm. As previously stated, this task involves simultaneous presentation of two distinct messages, usually a separate message to each ear. In the typical study, the listener was asked to answer questions about or shadow (repeat out loud) the message played to one of the two ears (Cherry, 1953; Moray, 1969). For example, Cherry (1953) implemented a dichotic listening paradigm and found that with relatively little practice participants could easily shadow one of the two messages. However, Cherry observed that participants generally had little recollection of the content of the unattended message. This finding indicated that we really cannot attend to and understand two messages at the same time. If you are listening to a lecture and the person sitting next to you begins a conversation, you must choose the information you want to understand. If you choose the neighbor's conversation, you will miss what is being said in the lecture and vice versa. Additional key generalizations arose from these early dichotic listening investigations.

In general, it was observed that the more two messages resembled each other, the more difficult they were to separate (Moray, 1969). Differences in the acoustical or physical characteristics of the messages (such as loudness, pitch, gender of speaker, or spatial location) assist listeners in distinguishing between and thus separating messages. When two simultaneous messages are presented and listeners are provided no instructions regarding which to pay attention to, louder messages are reported more accurately (Moray, 1969). However, if listeners are given instructions, they are able to attend selectively to one message at the expense of the other (Moray, 1969). These early investigations demonstrated that selective attention is a "psychological reality, not merely a subjective impression" (Moray, 1969, p. 19). The results of these studies were applied in settings such as aviation, in which it was determined that adding a call sign (i.e., the flight number or a person's name) could greatly assist listeners in selectively attending to one of two competing messages.

Dichotic listening investigations typically have used speech presentation rates approximating 150 words per minute (Moray, 1969). People can learn to shadow (verbally repeat) one of two competing messages rather easily, and performance improves with practice. Listeners are generally only able to report overall physical characteristics of the unattended message. For example, listeners are able to report whether the unattended message was speech or given by a male or female speaker, but generally they are not able to report the semantic content of the message or even the language of the spoken message.

Moray (1959) observed that even if a word was repeated 35 times in the unattended message, listeners did not report the word as having been present when given a retention test 20 s after the end of shadowing. Norman (1969) found that if the retention test was given immediately, listeners were able to report words presented within the last second or two. This finding provided initial evidence for storage of the unattended message in a brief sensory store.

Mowbray (1953) observed that when participants listened to one story while ignoring the other, comprehension of the unattended story was at chance level. Several researchers (C. Cherry, 1953b; Treisman, 1964c) investigated a listener's ability to separate two identical messages presented to each ear across different presentation lags; two identical messages were presented but one began at a temporal point different from the second. With temporal lags of approximately 5 s, listeners readily identified the two messages as separate. With shorter delays, an echo-like effect was produced. For the messages to be fused (or heard as stemming from a single source), the temporal lag could be no longer than 20 ms. These findings provided additional evidence for the existence of a brief sensory store capable of maintaining an echoic trace for a period of several seconds (Moray, 1969). Findings such as these led Donald Broadbent (1958) to propose that attention was subjected to a "filter"-type mechanism resembling a bottleneck.

FILTER THEORIES

The first formal theories of information processing in modern psychology suggested that some type of structural bottleneck keeps people from being able to process all available sensory information at any given time. That is, while we are constantly bombarded by a simultaneous cacophony of sounds, we can only process some limited set of these at any given time. A structural bottleneck or filter prevents all information from being processed. The early filter theories differed in terms of where in the information-processing stream this filter or bottleneck occurred.

Broadbent's Early Filter Model

Broadbent (1958) presented his classic bottleneck or filter theory of attentional processing in his seminal book *Perception and Communication*. The model essentially proposed that all sensory stimuli enter the sensory register and then are subjected to an attentional filter or bottleneck based on certain physical characteristics. Broadbent noted that in a dichotic listening paradigm, physical characteristics such as the ear of presentation or pitch could be used to allow one of many auditory messages selectively through the filter.

However, subsequent research provided evidence that under some circumstances more than one auditory message could be processed, at least to some extent. For instance, Moray (1959) found that sometimes when attending to one of two dichotically presented messages people were able to recognize their name in the unattended message. Moray reasoned that only information that was very important to the listener was able to "break through the attentional barrier" (p. 56).

Treisman (1960) found that when attending to one of two dichotically presented messages, subjects would sometimes switch their attention to the unattended ear to maintain the semantic coherence of the original message. That is, if the messages were switched to the opposite ear midsentence, listeners would sometimes repeat a word or two from the unattended ear, thus following the semantic context of the sentence for a short period of time. The semantic content of the "unattended" message appeared to influence the attended message for a period of time corresponding to the duration of the echoic trace (Treisman, 1964b).

Treisman's Attenuation Theory

Treisman (1964a) proposed that rather than a filter or all-or-none switch, attention functioned like an attenuation device. Sensory stimuli would pass through the sensory register, where only limited processing occurs, and then be passed on for further processing through a hierarchical progression of sets of gross characteristics.

The physical aspects of the stimulus were the most likely to be passed through the attenuation device and could be used to select the message to be attended. This would explain why listeners are able to report gross physical characteristics of an unattended message such as the speaker's gender. The semantic content of the unattended message would be attenuated but not completely lost.

Late Selection Models

Late selection models such as those of Deutsch and Deutsch (1963) and Norman (1968) proposed that filtering occurred much later in the information-processing stream, after information had passed on to LTM. According to late selection models, it is not so much that information is not processed but rather that we are limited in our ability to organize and select from the many representations available in LTM. So, in other words, all incoming sensory stimuli are matched with representations in LTM. The bottleneck occurs because we are incapable of choosing more than one representation at a time.

Although constructing experiments to conclusively decide between the different models has proven difficult, the investigations used to develop and examine these early theories of attention provided considerable knowledge regarding our ability to process multiple auditory messages. Perhaps as is often the case in psychological literature, the reality may be that some combination of the multiple theories explains the role of attention. Cognitive neuroscience data have recently added more to this debate; therefore, the early/late discussion is picked up in that subsequent section. Specifically, cognitive neuroscience data have provided converging evidence on when early selection is used and when late selection is possible. That is, perhaps the various filter-type (early or late) or attenuation-type mechanisms play a more dominant role in differing contexts and in different situations. Lavie's (1995) "perceptual load" theory may provide a reconciliation of the early/late debate. For example, late selection is typically used, and both relevant and irrelevant sources are processed through to LTM, except under high perceptual load (e.g., dichotic listening with fast rates, many messages), in which case early selection is used. But, it should be noted that Lavie's theory has mainly been validated with visual, not auditory, attention tasks (but see Alain & Izenberg, 2003).

Regardless of where the bottleneck occurs, if there is such a thing, it remains clear that humans can only respond to a limited amount of information. This notion became the cornerstone in the next stage of theories of attention.

CAPACITY MODELS

Evidence that humans have a limited capacity for processing multiple sources of information was discussed extensively in a seminal article by Moray (1967) and later elaborated in a now-classic book by Daniel Kahneman (1973), *Attention and Effort*. Kahneman's limited-capacity model suggested that attention was essentially the process of allocating the available resources from a limited pool. This pool of resources was not thought to have a fixed capacity but rather to change as a function of a person's overall arousal level, enduring disposition, momentary intentions, and evaluation of the demands of the given task. Since the concept of a limited supply of processing resources is a cornerstone of mental workload theory, we return to a discussion of capacity theories in the next chapter in our discussion of theories of mental workload and assessment techniques. However, for now we continue with a look at the third phase in the development of attention theories, which relies on cognitive neuroscience.

THE COGNITIVE NEUROSCIENCE OF ATTENTION

The perspective gained from application of the cognitive neuroscience approach to studies of attention, auditory processing in particular, has been quite remarkable. The techniques now available for recording the electrical activity of the brain and for brain imaging provide a window into the processes and structures involved in information processing that earlier theorists like William James could only dream about. That is not to say that much cannot still be learned from use of the earlier behavioral techniques. In fact, a powerful method of exploration involves combining established behavioral methods with measures of brain activity and brain imaging. For example, the dichotic listening task is frequently used in a number of different clinical assessments, including neurological assessment for central auditory processing disorder (Bellis, Nicol, & Kraus, 2000; Bellis & Wilber, 2001). However, the behavioral results can now be complemented and extended by techniques such as recording event-related potential (ERP) components of electroencephalographic (EEG) recordings to track the time course and level of brain activity. Brain imaging techniques such as functional magnetic resonance imaging (fMRI), positron emission tomographic (PET) scans, and magnetoencephalography (MEG) have also been powerful tools of investigation. The reader interested in detailed examination of the brain mechanisms of attention is referred to the work of Parasuraman (1998). Here, the discussion is limited to a few of the topics germane to auditory attention.

Event-Related Potential Indices of Auditory Attention

Auditory attention has been investigated with neurophysiological recordings. Evoked potentials (EPs) and other ERPs of brain activity generally show increased amplitude for attended auditory stimuli, relative to ignored auditory stimuli (Bellis et al., 2000; Bentin, Kutas, & Hillyard, 1995; Hillyard et al., 1973; Just, Carpenter, & Miyake, 2003; Parasuraman, 1978, 1980). Hillyard and colleagues (1973) conducted one of the first well-controlled examinations of the role of attention in affecting auditory EPs in a dichotic listening task. They had listeners perform a tone discrimination task while attending to stimuli presented to either the left or the right ear. Tones of low frequency (800 Hz) were presented to the left ear, while higher-frequency tones (1,500 Hz) were presented concurrently to the right ear. Tones were presented at a fast rate with a random pattern ranging from 250 to 1,250 ms between each tone. Roughly 3 of every 20 stimuli in each ear were presented at a slightly higher pitch (840 for the left and 1,560 for the right ear). Participants were instructed to count the deviant tones in the attended ear and to ignore all stimuli in the unattended ear. They observed a greater negative wave deflection occurring about 100 ms (N1) after each tone in the attended ear, relative to tones presented in the unattended ear. Specifically, N1 responses to tones presented to the right ear were higher than N1 responses to left ear tone when participants were attending to the right ear; conversely, N1 responses to left ear tones were higher than those to right ear tones when participants were attending to the left ear. Hillyard and colleagues noted that the fast event rate made the pitch discrimination task difficult, and that under these conditions, listeners were able to attend selectively to the relevant source while effectively ignoring the irrelevant source.

Extending the results of Hillyard et al. (1973), Parasuraman (1978) examined the influence of task load (i.e., mental workload) on auditory selective attention. Parasuraman presented a randomized series of tones to the left (1,000 Hz), right (500 Hz), or an apparent midway position between the left and right ears (2,000 Hz) at 50 dB above threshold. Tones were presented at either a fast rate (average of 120 tones per minute or less than 500 ms between each) or a slow rate (average of 60 tones per minute or less than 1,000 ms between each). Listeners were asked to monitor one, two, or all three auditory channels for the presence of targets (tones presented 3 dB greater in intensity). At the fast presentation rate, the amplitude of the N1 component was significantly higher in the attended channels relative to the unattended channels. Importantly, however, Parasuraman pointed out that this occurred only when the presentation rate was high, which can be attributed to a situation with high mental or informational workload (Parasuraman, 1978). It has been shown that auditory selective attention is enhanced by both learning to ignore distracting sounds and attending to sounds of interest (Melara, Rao, & Tong, 2002).

The results of these studies showed that selective listening paradigms played an integral role in the development of theories of attention. Further, from these early studies much was learned about the way people process information through the auditory channel. The reader interested in more discussion of the role of attention in information processing is referred to works by Moray (1969), Parasuraman (Parasuraman, 1998; Parasuraman & Davies, 1984), and Pashler (1998b). We turn our focus now to a general look at information processing beyond the level of attending to auditory stimuli. We see that information-processing models have evolved from largely a "boxes-in-the-head" modal model of memory proposed in the late 1960s and early 1970s (Atkinson & Shiffrin, 1968, 1971) to elaborate theories grounded in models of artificial neural networks (ANNs). The development of many of these information-processing models relied heavily on an examination of how people process language. Thus, as was the case with theories of attention we will see again that the study of auditory processing both contributed to and benefited from research on information processing.

THE INFORMATION-PROCESSING APPROACH

Numerous models of cognitive processing have evolved over the last half century in an effort to explain and predict how humans process information in different situations. It is beyond the scope of the current text to review each of these models in depth, and the interested reader is referred to several more detailed descriptions of the development and characteristics of these many models. (See the work of Matthews, Davies, Westerman, and Stammers, 2000, for an excellent review.) Some common properties are worthy of discussion before presenting several influential models with particular relevance to auditory processing. First, a distinction must be made between processing codes and operations.

PROCESSING CODES

Codes can be thought of simply as the format of the information that is presented to the observer. Information presented through any of the senses must be translated into an internal code or representation. On the surface, this process appears superficially simple. All five sensory systems have unique processing codes. However, closer inspection reveals that information presented in the same sensory modality can be coded in different ways depending on the nature of the task being performed or the stage of processing. Auditory information can be encoded in several different ways—acoustically, phonologically, lexically, semantically, or even spatially. Similarly, while visual information typically invokes visual codes, visual stimuli involving speech or alphanumeric stimuli are frequently coded in an acoustic or phonological code (Conrad, 1964).

Processing code is included as a major dimension in Wickens's multiple-resource theory (MRT) of information processing, which is discussed further in a subsequent section (Wickens, 1984; Wickens & Liu, 1988). In addition, the same information may be coded in several ways during different stages of processing. For example, in the case of speech, the sound stimulus will be initially coded in acoustic format and then will progress to a lexical and then semantic code as processing continues.

PROCESSING OPERATIONS

Processing operations are the actions or computations performed on the stimulus information. Several types of operations may be performed on stimulus information, and these operations are typically carried out by different processing components that are now thought to occur in separate neurophysiological structures. Encoding, storage, and retrieval of the information are each separate operations that can be performed on an internal code. Encoding can be further subdivided into different operational strategies depending on the type of task to be performed. Maintaining information in short-term memory (STM) or working memory is an operation (referred to as maintenance rehearsal in some models)* that differs markedly from elaborative rehearsal, which involves relating information to existing schemas or frameworks in an attempt to permanently store information.

A number of operations may be performed on auditory information. For example, both speech and music may evoke the operations of listening to melodic patterns, temporal segmentation, rhyming, or segmentation into processing streams either automatically or intentionally. Processing operations are influenced by factors that have been divided into two primary categories, bottom-up and top-down processing.

---

* See discussions on maintenance versus elaborative rehearsal by Anderson and colleagues (Anderson, 2000; Anderson & Bower, 1972).

BOTTOM-UP AND TOP-DOWN CONTROL OF PROCESSING

Most models of information processing recognize the influence of bottom-up and top-down processing, although they may differ on whether these two influences are thought to be sequential or interactive (Altmann, 1990; Norris, McQueen, & Cutler, 2000). As discussed in Chapter 3, bottom-up processing essentially refers to the influence of the direct stimulus input or sensory components of the stimulus. Bottom-up processing is therefore often referred to as data-driven or data-limited processing. Conversely, top-down processing refers to the influence of existing memories and knowledge structures (such as the use of context) and is therefore often referred to as conceptually driven or resource-limited processing (Norman & Bobrow, 1975). We turn now to a discussion of the information-processing model that was dominant throughout most of the latter part of the last century.

ATKINSON AND SHIFFRIN'S MODAL MODEL

Atkinson and Shiffrin's (1968) modal model of information processing has generated a vast amount of research since its publication. According to this model, information processing occurs in a series of stages consisting of sensory memory, STM, and LTM (Atkinson & Shiffrin, 1968). Sensory memory is thought to contain separate storage systems for each sensory channel. Acoustic information is temporarily held in an echoic sensory store, while visual information is held in an iconic sensory store. Information in sensory memory is thought to be coded as a veridical or as exact replication of the form in which it was received. As discussed in Chapter 1, considerably more attention has been focused on the characteristics of visual sensory memory, or iconic sensory memory. However, auditory sensory memory, or echoic sensory memory, is more germane to our current purpose. Echoic memory is thought to hold an exact replica (in the form of an auditory trace or echo) of information presented for a brief period of time. The capacity and duration of echoic memory were examined through several key experiments in the late 1960s (see a review of much of this early work in the work of Hawkins & Presson, 1986).

ECHOIC MEMORY CAPACITY

Moray, Bates, and Barnett (1965) utilized Sperling's (1960) partial report paradigm to investigate the capacity of echoic memory. Sperling developed his partial report paradigm to examine the capacity and duration of visual sensory traces. He realized that people were processing more information than they could recall, and that some of this information was lost during the short time it took them to report their memories. Therefore, his ingenious solution was a partial report paradigm in which only a small subset of the information had to be recalled. The critical aspect was that the subset that was to be recalled was not made known to the viewer until *after* the stimulus array had disappeared. This allowed researchers to get a better estimate of both the capacity and the persistence of this echoic trace.

Using an auditory analogue to Sperling's partial report paradigm, Moray et al. (1965) found that when participants recalled information from only one of four locations, eight items could be recalled fairly consistently. This corresponded to approximately 50% of any given list when four letters were presented from four different locations. Similar to the analogous process in visual sensory memory, Moray's findings suggested that two mechanisms were at play in the auditory recall paradigm. One mechanism involved the amount of information that could be perceived in a brief auditory glance, and the second involved the number of items that could be recalled immediately after presentation. Moray referred to this brief auditory storage as the "immediate memory span." It is now more commonly referred to as echoic memory. Moray concluded that the recall limitations were most likely due to loss at the time of recall rather than limitations in the amount of information encoded. That is, its capacity is thought to be greater than STM or working memory. Moray's findings lend support to the notion that we are able to encode, at least briefly, more auditory information than we are capable of attending to and storing for further processing. The temporal limits of this brief storage system were investigated subsequently and are discussed in the section on echoic persistence.

ECHOIC PERSISTENCE

Current opinion on the topic of auditory sensory memory tends to suggest that two forms of precategorical acoustic storage exist (Cowan, 1984; Giard et al., 1995; Winkler, Paavilainen, & Naatanen, 1992). The first form lasts only a few hundred milliseconds, while the longer form lasts several seconds and is generally most synonymous with what is meant when using the term *echoic memory*.

Short-Term Auditory Store

The first form of auditory sensory memory is a short auditory store capable of retaining acoustic information for 200 to 300 ms (Giard et al., 1995). It is thought to occur in the primary auditory cortex. It begins within 100 ms following the presentation of an acoustic stimulus and decays exponentially over time (Lu, Williamson, & Kaufman, 1992). Investigations of the persistence of this short-term auditory store have been explored by varying the interval between successive acoustic stimuli through masking paradigms (see Massaro, 1972, for a review) and more recently through neurophysiological indices of auditory ERPs or mismatch negativity (MMN) paradigms (see Atienza, Cantero, & Gomez, 2000).

Naatanen and colleagues have demonstrated that several cortical areas may be responsible for the short-term storage of different aspects of the acoustic signal as well as conjunctions between aspects (Giard et al., 1995; Takegata et al., 2001). They have used the MMN paradigm with MEG to examine the location of neural traces for physical characteristics such as frequency, intensity, and duration. They observed that the MMN activity patterns observed on the scalp for each of the different auditory parameters differed. Further evidence that short-term auditory memory for frequency, intensity, and duration are at least partially a result of different underlying neural structures was seen in the results of their dipole model analysis. Next, the duration of the longer form of auditory sensory storage is discussed.

Long-Term Auditory Store

The longer auditory store is generally the one referred to when the term *echoic memory* is used. This convention is maintained here; thus, the term *echoic memory* is reserved from this point in reference to the long-term auditory store. It is found in the association cortex and is thought to hold information for several seconds (Lu et al., 1992). The duration of echoic memory was first examined using an auditory version of Sperling's (1960) partial report paradigm discussed in the previous section.

Darwin et al. (1972) expanded on the investigations of Moray and colleagues (1965) by examining the effect of a 1-, 2-, and 4-s delay between presentation and poststimulus cueing. Significant differences were found in the amount of information participants could recall after each poststimulus delay condition. Darwin et al. concluded that the time limit for the auditory sensory store was greater than 2 s but less than 4 s. This temporal limit has important implications for communication. Because a veridical representation of the auditory information is available for 2–4 s, it is possible that if one is engaged in a concurrent task when auditory information is presented, for a brief period of time the information will still be accessible for attentive processing. That is, it will be stored for 2–4 s, which may allow sufficient time to change focus and access the information content postpresentation. An everyday example of this is when a person is reading a book or watching TV and someone walks up and begins a conversation. The distracted person may automatically ask, "What did you say?" but then begin replying before the speaker has had a chance to repeat. Although the speaker may be a bit surprised and wonder why he or she was asked to repeat the message if the listener actually heard it, it makes perfect sense when we consider that the information remains in the listener's echoic memory for a brief period of time, thus allowing the listener to make an appropriate response to the speaker's question or comment.

The potential for presentation intensity to affect either the strength (veridicality) or the duration of the echoic memory trace was not examined in these early investigations. However, recent evidence indicates that presentation intensity may affect both of these aspects (Baldwin, 2007).

EMERGING DEVELOPMENTS IN ECHOIC MEMORY RESEARCH

Although a considerable amount of early work on sensory memory in the visual realm examined the visual parameters that had an impact on its veridicality and duration, this research seems not to have extended to the auditory modality. For example, researchers examined the impact of stimulus intensity and contrast on the persistence of visual sensory traces. They often found equivocal results, with increased intensity sometimes increasing the duration of the iconic traces and sometimes decreasing it. Many of these equivocal results can now be rectified by considering the frequency of the visual stimulus, particularly whether it relies on photopic (cone receptor) or scotopic (rod receptor) vision (see review in the work of Di Lollo & Bischof, 1995).

Baldwin (2007) examined the impact of auditory intensity (i.e., subjectively perceived as loudness) on echoic persistence. She sought to determine if the echoic traces of sounds that were presented at higher intensities lasted longer. Specifically, Baldwin found that at the upper temporal limit of echoic memory (4 s), matching performance for auditory tonal patterns was directly affected by presentation amplitude. That is, after 4-s delays, louder presentation amplitudes resulted in greater accuracy in determining whether a second tonal pattern matched the one presented previously.

Implications of Persistence

The impact of intensity level on echoic persistence has important implications for numerous auditory processing tasks, the processing of speech in particular. As discussed in Chapter 8, most contemporary models of speech processing assume that speech is processed in a series of stages. The initial stage begins with translation of acoustic signals into a pattern of abstract representations, followed by phonemic identification and then word or lexical processing utilizing higher-level representations constructed from contextual cues and knowledge of prior subject matter (Cutler, 1995; Fischler, 1998; Massaro, 1982; Norris, McQueen, & Cutler, 1995; Stine, Soederberg, & Morrow, 1996). Auditory memory is essential to this progression. Corso (1981) noted that initial stages of speech perception rely on the ability to discriminate between small changes in frequency or pitch. Later stages rely on the ability to integrate successively heard words, phrases, and sentences with previously stored information (Pichora-Fuller, Scheider, & Daneman, 1995).

Presentation conditions that facilitate echoic persistence have the potential to both facilitate auditory processing and decrease the mental resource requirements for the lexical extraction process. Imagine, for example, dual-task situations that require a person to perform an auditory and a visual task simultaneously. If the person is engaged in the visual task when the auditory task is presented, a long-duration echoic trace would assist the person in retaining the auditory information until he or she could shift attention toward the auditory stimulus. Most of us have had an experience for which this was useful. Imagine that you are reading a book when a buzzer sounds unexpectedly. Being able to retain the sound long enough to finish reading a sentence and then shift attention toward the processing of the sound aids in identifying the sound and taking the appropriate action. The particular role that echoic memory plays in speech processing is discussed further in Chapter 8. We turn our attention now to the next stage in processing information, working memory.

WORKING MEMORY

Working memory is a term given to describe a limited-capacity system that is used to hold information temporarily while we perform cognitive operations on it (Baddeley, 2002). This section provides a discussion of the role of working memory in auditory processing as well as its close link to mental workload. The construct now referred to as working memory stemmed from earlier depictions of an intermediate stage of information processing termed *short-term memory* (Atkinson & Shiffrin, 1968). STM was thought to play an integral role in storing new information transferred from the sensory register (i.e., echoic memory) for possible storage in LTM. Baddeley and colleagues (Baddeley, 1992; Baddeley & Hitch, 1974, 1994) introduced the term *working memory* as an alternative to the construct of STM. Unlike STM, working memory is thought to be a multidimensional transient storage area where information can be held while we perform cognitive operations on it. Working memory places less emphasis on storing information until it can be transferred into LTM, instead placing an emphasis on holding information while we engage in such operations as problem solving, decision making, and comprehension (Baddeley, 1997). Working memory can also be linked closely with the limited-capacity processing resources embodied in what we refer to as attentional resources or mental workload. Therefore, we examine the multiple dimensions of working memory in some detail in this chapter. However, a rich history of research utilizing the framework of investigations of STM provides important information regarding the nature of auditory processing. Therefore, this begins our discussion.

SHORT-TERM MEMORY

Considerable research has been conducted on the role of STM in both auditory and visual information processing. Early models emphasizing the importance of both a temporary storage and processing system and a more permanent longer-term storage system were developed (Atkinson & Shiffrin, 1968, 1971; Phillips, Shiffrin, & Atkinson, 1967; Waugh & Norman, 1965). These early models typically emphasized that a primary role of STM was to control an executive system, which functioned to oversee the coordination and monitoring of a number of subprocesses. Examinations of the capacity, duration, and code of STM led to the establishment of several well-documented characteristics of information processing in STM.

Recall paradigms of visually and auditorily presented letters, words, and sentences were frequently used to investigate the characteristics of STM (Baddeley, 1968; Conrad, Baddeley, & Hull, 1966; Daneman & Carpenter, 1980; Engle, 1974; Yik, 1978) and in fact are still frequently used in cognitive research (Baddeley, Chincotta, Stafford, & Turk, 2002; Craik, Naveh-Benjamin, Ishaik, & Anderson, 2000; Risser, McNamara, Baldwin, Scerbo, & Barshi, 2002). Two well-established findings from this body of literature are the modality effect and the suffix effect.

Modality Effect

When recall for lists of words presented in visual and auditory formats are compared, recall is consistently higher for items presented in the auditory format (Conrad & Hull, 1968; Murray, 1966). The recall advantage for auditorily versus visually presented material has been termed the modality effect. The modality effect is most salient for items at the end of a presented list. That is, in serial recall paradigms recency effects (better recall for items at the end of the list rather than items in the middle of the list) are strongest for material that is heard versus read. The modality effect provides evidence that short-term retention of verbal material benefits from an acoustic or phonological code, a point we will return to in further discussions.

Suffix Effect

What has been called the suffix effect provides another consistent and characteristic finding in investigations of STM. In recall paradigms in which to-be-remembered information is presented in an auditory format, retention of items at the end of a list (recency effects) is disrupted if the end of the list is signified by a nonlist word or suffix. For example, if a spoken list of items to be remembered is followed by the word *recall*, recency effects are diminished (Crowder, 1978; Nicholls & Jones, 2002; Roediger & Crowder, 1976). The role of an irrelevant suffix in disrupting recency effects provides further evidence that temporary retention of information benefits from an acoustic or phonetic code. Even more dramatic evidence of the acoustic code is found in observations of the recall advantages provided by articulatory rehearsal and acoustic confusions.

Articulatory Rehearsal

The recall advantages of articulatory rehearsal are evidenced by the observation that recall is higher if participants are allowed to engage in auditory rehearsal, such as silent vocalization, whispering, or speaking out loud the to-be-remembered stimuli (Murray, 1965). This process, known as *articulatory rehearsal*, facilitates memory in serial recall tasks (Larsen & Baddeley, 2003).

Acoustic Confusions

Acoustic confusions are instances when an acoustically similar item is substituted for a presented item during recall. Letters, due to their greater similarity, are more prone to acoustic confusion than are digits. For example, the consonants B, V, D, and T are acoustically similar and therefore prone to substitution. Interestingly, Conrad (1964) observed that people make acoustic confusion errors even when lists of items are presented visually. That is, people are more prone to incorrectly recall an acoustically similar substitute (e.g., V for B) than they are a visually similar substitute (e.g., L for V) even when the lists are presented visually. Acoustic similarity between items in a serial recall list dramatically affects recall in general. Lists that are more similar result in poorer recall performance (Conrad, 1964; Conrad & Hull, 1964).

It is beyond the scope of this chapter to cover the extensive body of research leading up to current models of working memory. However, two key models that have developed and are still currently extensively applied to the understanding of information processing in general, and auditory processing in particular, are discussed. The first of these models was first presented by Baddeley and Hitch (1974) and later refined by Baddeley (Baddeley, 1992; Baddeley & Hitch, 1994). The second model, which has been applied extensively in human factors research, was developed by Wickens (1984) and is termed *multiple-resource theory*. First, we turn to a discussion of Baddeley's concept of working memory.

WORKING MEMORY COMPONENTS

In 1974, Baddeley and Hitch published a seminal article discussing a series of 10 investigations systematically designed to determine if verbal reasoning, comprehension, and learning shared a common working memory system. Their results strongly suggested that the three activities utilized a common system that they referred to as working memory. The working memory system was postulated to be a limited-capacity work space that could coordinate the demands of storage and control. Baddeley (1992) has presented compelling evidence for a three-component model of working memory.

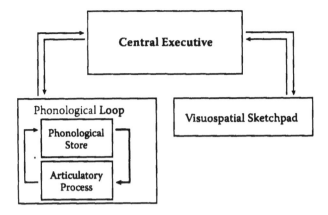

FIGURE 4.1 Baddeley's model of working memory.

Baddeley's (Baddeley, 1992; Baddeley & Hitch, 1974, 1994) three-component working memory system consists of a central executive, attentional controlling system and two slave systems (see Figure 4.1). The slave systems consist of a visuospatial sketch pad for processing and manipulating visual images and a phonological or articulatory loop for manipulation of speech-based information.

Baddeley (1998) further developed the model and later subdivided the phonological loop into two components. Considerable research has focused on revealing the characteristics and neural mechanisms of the phonological loop. This research is particularly germane to the current investigations.

Phonological Loop

Baddeley (1998) described the phonological loop as consisting of two components. The first is a phonological store capable of holding information for up to 2 s. The second component is an articulatory control process responsible for subvocal rehearsal, which feeds into the phonological store. The articulatory rehearsal process is thought to be capable of converting written material into a phonological code and then registering it in the phonological store.

Phonological Store Capacity

The phonological loop is of limited capacity, which has primarily been demonstrated by examining the so-called word length effect (Baddeley & Logie, 1999). The spoken length of words, rather than the number of phonemes (or syllables), appears to be a primary determinant in the number of to-be-recalled words that can be kept in the phonological store.

The phonological store capacity in Baddeley's model is postulated to vary considerably from individual to individual, with an average duration of 2 s (Baddeley, Gathercole, & Papagno, 1998). Baddeley et al. (1998) presented compelling evidence that the phonological store plays an integral role in language learning, particularly in children and for adults when learning new vocabulary or additional languages. Individuals with larger phonological stores acquire more extensive vocabularies as children and learn additional languages more easily as adults.

They cite evidence that the phonological store, however, plays a less-crucial role in verbal memory (i.e., word recall tasks) of familiar words. Evidence for this lesser role in familiar word recall stems in part from neurological patients with specific phonological memory deficits that demonstrate normal language comprehension and production capabilities but significantly lower recall in nonword memory tasks.

Experimental paradigms designed to disrupt the phonological store include (a) the word length effect, (b) phonological similarity, and (c) articulatory suppression (Baddeley et al., 1998). The word length effect can be demonstrated by impaired recall for words that take longer to say even though the number of syllables is equivalent. Phonological similarity is demonstrated by impaired recall for multiple words that sound alike, and articulatory suppression occurs when participants are required to say irrelevant words or syllables that block articulatory rehearsal.

Each of these processes is thought to activate distinct brain regions. Specifically, processes involving the phonological store have been shown to result in greater activation of the perisylvian region of the left hemisphere (Baddeley et al., 1998; Paulesu, Frith, & Frackowiak, 1993), while articulatory rehearsal processes are associated with greater activation of Broca's area and premotor areas (Paulesu et al., 1993; Smith & Jonides, 1999). The precise brain mechanisms associated with these processes are not particularly germane to the current discussion. Those interested in more in-depth coverage of this topic are referred to work by Banich (Banich, 2004; Banich & Mack, 2003); Zatorre (Zatorre et al., 1992; Zatorre & Peretz, 2003); and others (Hagoort, 2005; Logie, Venneri, Sala, Redpath, & Marshall, 2003; Petersson, Reis, Askeloef, Castro-Caldas, & Ingvar, 2000).

Speech is believed to gain automatic access to the phonological store, regardless of the mental workload demands imposed by a concurrent task (Baddeley, Lewis, & Vallar, 1984; Gathercole, 1994; Penny, 1989). Text, on the other hand, does not have this advantage. Verbal information presented visually (text) requires an extra step to convert it to a phonological form, typically using subvocal rehearsal. Therefore, according to Gathercole, a concurrent memory load can be expected to disrupt text processing to a greater degree than would be expected for speech processing.

Further, the phonological store is thought to hold information at a phonemic level rather than at a word level (Baddeley, 1997), as evidenced by the disrupting effects of nonsense syllables. Nonsense syllables that share the phonemic components of speech disrupt processing in the phonological store (Salame & Baddeley, 1987, 1990). However, speech is more disruptive than nonsense syllables, a phenomenon referred to as the irrelevant speech effect (Larsen & Baddeley, 2003; Salame & Baddeley, 1990).

In addition to these aspects of working memory, a number of other important concepts pertaining to the cognitive architecture involved in auditory processing are worthy of discussion. One such important concept involves distinguishing between serial and parallel processing.

SERIAL VERSUS PARALLEL PROCESSING

Much of the early work using the dichotic listening paradigm examined the notion of when information can no longer be processed in parallel but rather must be processed serially. This idea was the basis of the proposed bottleneck or filter concept in attentional processing. It was difficult to discern with any degree of certainty just where the filter was likely to occur, and the focus of research changed from looking for a physical filter to thinking of attention as a resource (see discussion in Hunt & Ellis, 2004; Kahneman, 1973).

Despite the difficulty in determining if a particular filter mechanism exists and if so, where, considerable insight was gained into our ability to process information in parallel versus serially. For example, in Treisman's feature integration theory (FIT; Treisman & Gelade, 1980), unique primitive features are thought to be processed without the need for conscious focused attention. Such sensory primitives are thought to be processed in parallel. Conversely, examining feature conjunctions (examining stimuli that may share one or more features with other stimuli in the environment) are thought to require focused attention and can only be processed serially. Although most of Treisman's work on FIT has been conducted with visual stimuli, it can be reasoned that similar phenomena are present with auditory stimuli. Therefore, sounds with unique characteristics varying from all other stimuli by one salient feature may stand out above the background noise, regardless of how many distracting sounds are present.

Parallel processing is a key component of more recent theories pertaining to the cognitive architecture of information processing. Parallel distributed processing (PDP) models, which are also referred to as ANN models, suggest that multiple streams of information may be processed simultaneously.

ARTIFICIAL NEURAL NETWORKS

It is remarkable that we understand language at all given the number of multiple simultaneous constraints. In language processing, for example, we must simultaneously consider the constraints of both syntax and semantics. As discussed further in a discussion of speech processing in Chapter 8, we simultaneously utilize the constraints imposed by the rules of syntax and semantics to arrive at an appropriate interpretation of a sentence (Rumelhart, McClelland, & Group, 1986). ANN or PDP models have made progress toward explaining how we might accomplish this significant feat.

ANNs have played a significant role in our understanding of speech processing (cf. Coltheart, 2004). Early network models developed by Collins and Quillian (1969) began to unravel the mysteries of our semantic knowledge structure. More recently, neural network models have been used to examine and explain how we accomplish lexical decisions by simultaneously considering syntactical and semantic constraints in everyday language-processing tasks (McClelland, Rumelhart, & Hinton, 1986). For example, syntactic information allows us to correctly interpret a sentence such as

The cat that the boy kicked chased the mouse.

However, as McClelland and colleagues (1986) pointed out, we need to consider the simultaneous constraints of semantic information to understand sentences such as the following:

I saw the dolphin speeding past on a catamaran.

I came across a flock of geese horseback riding near the beach.

ANN models have been used to demonstrate how we are able to simultaneously utilize information from several different sources interactively. These models
are based on the interaction of a number of simple processing elements (i.e., letter, phoneme, and word identification; syntactical rules; and semantic constraints).
Examples of ANN models of language processing are examined in greater depth in
the discussion of speech processing in Chapter 8. In particular, the TRACE model,
an interactive-activation model of word recognition developed by McClelland and
Elman (1986), and the dynamic-net model proposed by Norris (1990) are discussed.

SUMMARY

Humans have the capability of selecting and attending to a specific auditory pattern
in the midst of competing patterns. The chosen pattern can be located and held in
memory long enough to interpret it along several simultaneous dimensions. From
mere acoustic patterns, humans are able not only to recognize familiar acoustic patterns (i.e., Beethoven's *Fifth Symphony*) but also to gain considerable insight into the
nature of the sound source (i.e., whether a particular musical piece is played by one
type of instrument or another or whether it is comprised of a number of simultaneous
instruments in harmony). Humans not only can understand the words of a speaker
but also can gain considerable insight into the age, gender, and emotional state and
intent of the speaker.

CONCLUDING REMARKS

Bottom-up and top-down processing, auditory selective attention, auditory stream
segregation, and temporary storage of acoustic patterns while accessing the mental
lexicon are some of the many intelligent processes used to interpret auditory patterns. Each of these processes requires attentional resources; hence, each contrib-
utes in different ways to overall mental workload. We address the mental workload
required by these intelligent processes in the chapters to come. First, in the next
chapter we discuss many of the key issues as well as methods and techniques for
assessing mental workload.

5 Theories and Techniques of Mental Workload Assessment

INTRODUCTION

Mental workload refers to a psychological construct that has wide currency in areas of applied psychology and human factors. The concept has been found to be useful in understanding why and how the many different tasks that people perform at home, work, and leisure are carried out effectively at times and not so successfully at others. Mental workload assessment has played an integral role in designing and evaluating many human-machine systems found in working environments, as well as in many other facets of life. A primary aim of this book and the chapters to come is to discuss factors that affect the mental workload associated with processing auditory information.

In this chapter, the term *mental workload* is first illustrated by providing some everyday examples and then by discussing more formal definitions. Key theoretical issues pertaining to mental workload and workload assessment techniques form a central focus. This discussion includes an examination of the strengths and limitations of different ways in which workload is assessed, particularly as they apply to the processing of sound. In addition, the relationship between the concept of mental workload as used by human factors practitioners and that of working memory (WM) resources as used by cognitive scientists is explored.

An everyday example serves to illustrate the basic concept of mental workload and the complexity of its assessment. Most of us have had an experience of teaching someone a new task that we ourselves can do quite well. For example, perhaps you are teaching a child mental arithmetic or a teenager to drive. The task may be so well learned to you that you do not need to put much mental effort or thought into it. It has become nearly "automatic," to use Schneider and Shiffrin's (1977) term. However, to the novice, or person learning the task, considerable concentration and mental effort may be required to perform the task. The degree or quantity of this concentration and mental exertion is essentially what is meant by the term *mental workload*. Moreover, the amount of mental effort expended by the novice may or may not be reflected in the performance outcome: The child may sometimes correctly produce the right answer and sometimes not, and the teenager may successfully carry out a difficult driving maneuver or fail it. In both cases, the individuals will experience considerable mental workload. As this example illustrates, simply examining the outcome of a situation (i.e., whether the right sum is derived) often provides little insight into the amount of mental effort required or expended. Rather, the mental workload involved must itself be assessed, independent of performance outcome.

The term *mental workload* has been used extensively in the field of human factors and more generally the human performance literature for several decades. Mental workload is often used synonymously with terms such as mental effort, mental resources, and the attentional or information-processing requirements of a given task, situation, or human-machine interface. Precise definitions vary, and at present a single standard definition of mental workload has not been established. Mental workload is generally agreed to be a multidimensional, multifaceted construct (Baldwin, 2003; Gopher & Donchin, 1986; Hancock & Caird, 1993; Kramer, 1991), often referencing the relationship between the task structure and demands and the time available for performing the given task. Current theories of mental workload and its measurement stem from earlier work concerning theories of attention. Much of this early work examined the attentional demands of auditory processing and was discussed in the previous chapter. Chapter 5 examines theoretical perspectives pertaining to mental workload and the major assessment techniques used to measure it. Central to much of this work is the idea that humans are limited in the amount of information they can process or tasks they can perform at any given time, an idea clearly confirmed by early work in the cognitive psychology of attention (Broadbent, 1958; Moray, 1967; Kahneman, 1973). Establishing effective means to ensure that the mental workload required by a given task or set of tasks is within the limits of a human's processing capacity has played a critical role in several areas of application in human factors.

LIMITED-CAPACITY MODELS

The notion that humans can only process a limited amount of information at any given time can be traced to early filter models of attention, such as the one proposed by Broadbent (1958). Filter models, notably Broadbent's (1958) early selection model and Deutsch and Deutsch's (1963) late selection model, proposed that a structural mechanism acted like a bottleneck preventing more than a limited amount of information from being processed at any given moment. Later, the idea that information processing was regulated by a more general capacity limit rather than a specific structural mechanism was explored by Moray (1967) and further developed by Kahneman (1973).

The idea that there is an upper limit to the amount of mental effort or attention that one can devote to mental work became a central component in Kahneman's (1973) capacity model and remains a central tenet of mental workload theory today. Daniel Kahneman's (1973) influential book, *Attention and Effort*, described one of the earliest conceptual models of attention as limited (or limiting) processing capacity. In this model, mental effort is used synonymously with attention and is accompanied by the ensuing implication that humans are able to direct, exert, and invest attention among multiple stimuli.

As Kahneman (1973) pointed out, in everyday life people often simultaneously perform multiple tasks (i.e., driving a car while listening and engaging in conversation, taking notes while listening to a professor's lecture). Kahneman theorized that rather than examining the point at which stimuli are filtered, it might be more fruitful to consider cognitive processes in terms of resources, which much like physical resources have an upper limit. Because we are able to direct this mental effort, we are free to allocate our attention flexibly among the tasks, allotting more attention to one task than another at different times. So, for example, we might pay more attention (exert more mental effort) to the task of driving as we are entering a congested freeway and thus pay little attention to the conversation, or we might be paying so much attention to listening to a difficult or interesting point made by the professor that we neglect the task of writing notes. According to Kahneman's model, as long as the total amount of attention demanded by the concurrent tasks does not exceed our capacity, we are able to perform both successfully. However, when the concurrent demands exceed our capacity, performance on one or more tasks will degrade.

As discussed in Chapter 4, Kahneman's (1973) model proposed that attentional capacity was not a fixed quantity but rather varies as a function of an individual's arousal level, enduring disposition, and momentary intentions. So, for example, if an individual is sleep deprived or bored, he or she will have fewer attentional resources to devote to any given task or set of tasks. Illustrating this concept, nearly everyone can probably think of a time when he or she was tired and found it hard to pay attention to a conversation or lecture, particularly if the material discussed was complex, unfamiliar, or seemed uninteresting.

Kahneman's (1973) model focusing on a limited capacity of attentional resources remains an important concept in mental workload. It was also influential to later theories of time-sharing efficiency, such as the influential model of dual-task performance proposed by Wickens (1980, 1984) called multiple resource theory (MRT). Resource theories, in general, assume a limited amount of attentional resources that can be flexibly allocated from one task to another. As discussed in more detail in this chapter, MRT proposes that these resources are divided into separate "pools" of resources for different aspects of processing. As Moray, Dessouky, Kijowski, and Adapathya (1991) pointed out, of the many attentional models in existence (strict single-channel model, a limited-capacity channel model; a single-resource model, or a multiple-resource model), all share the concept that humans have a limited capacity of mental resources. Resource theories form a cornerstone of most theories of mental workload.

RESOURCE THEORIES

Rather than focusing on a structural filter mechanism, or attenuation device, resource theories build on Kahneman's proposal that human performance is a function of one's ability to allocate processing resources from a limited reserve capacity (Kahneman, 1973). Resource theory, as this position came to be called, and its focus on allocation of resources from a limited reserve is currently a widely held view of attentional processing (see discussions in Matthews et al., 2000; Pritchard & Hendrickson, 1985; Wickens, 1991, 2002). However, controversy still ensues over whether resources primarily stem from a single reserve or multiple pools. At present, the most commonly held view is that of multiple reserves or pools of resources.

MULTIPLE RESOURCE THEORY

MRT originated from examination of how people time-share two or more activities (Wickens, 1980, 1984). Numerous situations were observed for which the time-sharing efficiency of two tasks interacted in ways that were not easily predicted by examination of the performance on the individual tasks (Wickens, 1980; Wickens & Liu, 1988). Wickens's MRT stemmed primarily from observations that a single-pool model of attentional resources was inadequate to explain the results of many dual-task investigations in which visual-auditory task combinations could be time-shared more efficiently than either visual-visual or auditory-auditory combinations (Wickens, 1984). Wickens's MRT postulated that time-sharing efficiency was affected not only by the difficulty of each task but also by the extent to which each task competed for common mechanisms or structures (Wickens, 1984). According to MRT, tasks can compete for common mechanisms with functionally separate "reservoirs" or pools of attentional capacity in three ways. First, tasks may compete for the same modality of input (i.e., visual vs. auditory) or response (vocal vs. manual). Second, tasks can compete for the same stage of processing (perceptual, central, or response execution). And third, tasks may compete for the same code of perceptual or central processing (verbal vs. spatial). This version of MRT explained the results of a considerable number of dual-task investigations in which cross-modal (visual-auditory) tasks were time-shared more efficiently than intramodal (visual-visual or auditory-auditory) tasks.

Wickens (Wickens, 1991; Wickens & Liu, 1988) pointed out, however, that there are exceptions to the predictions of MRT with respect to the input modality resource pool dichotomy. One pattern of results in dual-task studies—when a continuous visual task is time-shared more efficiently with a discrete task that is visual rather than auditory—conflicts with the MRT prediction for input modality. For example, auditory air traffic control (ATC) communications (relative to visual [text-based] data-link-type communications) have been shown to be more disruptive in the visually demanding task of pilots performing approach scenarios during simulated flight (Latorella, 1998). Cross-modality performance is thought to suffer in this paradigm because the auditory task "preempts" the continuous visual task. That is, pilots disengaged from the visual task to perform the auditory task, resulting in efficient performance of the auditory task but degradation of the continuous visual task. Conversely, in this paradigm performance of a discrete visual task appears less disruptive to maintaining performance on the continuous visual task (Wickens & Liu, 1988). These findings suggest that, at least under certain circumstances, an aspect of processing other than the modality of input—in this case the power of auditory preemption—has a greater impact on task performance.

CRITICISMS OF RESOURCE THEORY

Resource theory is not without its critics. In famously bold words, Navon (1984) referred to mental resources as a theoretical "soup stone"—devoid of any true substance, as in the tale of the poor traveler with only a stone to make soup. Navon distinguished between two factors affecting information processing and task performance. Navon called these "alterants," which can be at several different states at any given time, thus affecting multiple levels of task performance (i.e., anxiety); and "commodities," akin to resources, which consist of units that can only be used by one process or user at any given point in time. Navon pointed out that with any given performance resource function (PRF; Norman & Bobrow, 1975) mapping dual-task decrement, it is impossible to distinguish between decrements stemming from commodities (or resources) and alterants. Therefore, according to Navon:

> Attempts to measure mental workload, to identify resource pools, to predict task interference by performance resource functions, or to incorporate resource allocation in process models of behavior *may prove as disappointing* as would attempts to isolate within the *human mind analogues of the functional components of the digital computer.* (p. 232)

An alternative idea is that, rather than separate resource pools, capacity limits may stem from constraints in the cognitive architecture of the processing mechanisms (see Matthews et al., 2000, for a review). A related and important construct that has received considerable attention in the cognitive science literature is the notion of working memory and its structures and resources. This topic is discussed in more detail in a further section.

Despite existing criticisms, MRT has been the single most influential theory in mental workload measurement to date (see the discussions in Colle & Reid, 1997; Sarno & Wickens, 1995; Tsang, Velazquez, & Vidulich, 1996). Here are just some of the many observations that can be explained within a MRT context: Numerous investigators have observed that performance on tracking tasks is disrupted more by concurrent spatial tasks than concurrent verbal tasks (Klapp & Netick, 1988; Payne, Peters, Birkmire, & Bonto, 1994; Sarno & Wickens, 1995; Xun, Guo, & Zhang, 1998). This implies that when the main task that an operator must perform is primarily visual or spatial (i.e., driving along a curvy road), then any additional information presented to the operator may be processed more efficiently if it requires a different type of processing (i.e., listening to verbal directions rather than reading a spatial map). In this situation, as in many others, the key to interference is the perceptual or processing code, rather than simply the modality in which information is received. In other words, it is more difficult to read text and listen to someone talking than it is to extract information from a spatial display while listening to someone talk. Both text and speech rely on a verbal processing code and therefore are more likely to interfere with another verbal task regardless of whether the concurrent verbal task is presented visually or auditorily (Risser, Scerbo, Baldwin, & McNamara, 2003, 2004). As Risser and colleagues pointed out, similar predictions for interference can be derived from Baddeley's (Baddeley & Hitch, 1974, 1994) model of working memory, in which the phonological loop and the visuospatial sketch pad are thought to process different types of information. Recall that these two mechanisms are thought to be relatively independent "slave systems" in working memory, but that each is controlled by the central executive system and therefore still subject to an overall processing limit. Parallels have been drawn between the concepts of working memory structure and MRT mental workload in previous literature (Klapp & Netick, 1988). More recently, similar parallels have been drawn between current concepts of working memory resources and mental workload. We therefore turn to a discussion of these new developments next.

WORKING MEMORY PROCESSES

An examination of the literature pertaining to mental workload and the literature pertaining to working memory indicates a striking overlap in terminology. Working memory has been defined as "a limited capacity system, which temporarily maintains and stores information, and supports human thought processes by providing an interface between perception, long-term memory and action" (Baddeley, 2003, p. 829). From this definition, we see reference to a limited-capacity system for performing mental activity, a theme central to most current theories of mental workload and its assessment. The association between the constructs of mental workload and working memory was discussed in an article by Parasuraman and Caggiano (2005). They pointed out that mental workload is influenced by external, bottom-up factors in the environment and top-down processes within the individual. These internal, top-down influences are regulated to a large extent by working memory capacity and individual differences in that capacity. Examining the performance impact of individual differences in working memory capacity has led to an extensive body of literature in recent years (Conway, Cowan, Bunting, Therriault, & Minkoff, 2002; Engle, 2001; Engle, Kane, & Tuholski, 1999). Much of this literature pertains to language processing and comprehension and is therefore discussed in subsequent sections and chapters. For now, we limit our discussion to more general concepts.

Daneman and Carpenter (1980) developed a means of assessing differences in working memory capacity. They called their test the *reading span test*. The test involves reading a series of sentences presented in sets of one, two or three, and so on and then recalling the last word of each sentence in the set. The number of sentence final words that can be remembered (the size of the sentence set that can be both processed and stored) is used to calculate an individual's reading span. Daneman and Carpenter found that individual reading spans ranged from two to five, and that individual scores correlated with several tests of reading comprehension. Several variations of Daneman and Carpenter's original paradigm have been developed and are widely used as measures of working memory capacity (for a review, see Conway et al., 2005).

Towse and colleagues reexplored the concept of working memory (Towse, Hitch, Hamilton, Peacock, & Hutton, 2005). As the authors pointed out, working memory is more than a capacity-limited temporary storage. They used an interesting analogy: comparing the capacity construct to the size of a traveler's suitcase. They suggested that individual differences in working memory might be compared to individual differences in the sizes of suitcases a traveler might choose. People travel with a variety of sizes. However, suitcase size alone does not determine how much one can bring along, let alone how useful the individual items packed might be. Efficient packing processes can result in tremendous variability in how many items and the appropriateness of the items included in the traveler's suitcase. Regardless of whether working memory performance relies primarily on the capacity, efficiency, or some combination of the two, working memory span scores show strong relationships with a wide variety of other cognitive tasks and therefore appear to represent some general processing capability. Current discussion surrounding working memory resources frequently centers on the extent to which working memory involves separate components.

MULTIPLE RESOURCES—MULTIPLE NETWORKS

Recall that MRT suggests that different modalities of input and different codes of processing will rely on separate pools or resources. This idea may be compared to theories coming from a separate line of reasoning that suggests that memories are processed and stored in the same neural pathways and networks initially used to encode the information. This idea can be traced to the writings of Carl Wernicke in the late 1800s (see Feinberg & Farah, 2006; Gage & Hickok, 2005). According to this perspective, retaining a memory of an auditory or visual event utilizes the same neural pathways that were activated when the event was first experienced. This theory has received support in more recent literature (see discussions in Jonides, Lacey, & Nee, 2005; Squire, 1986) and provides a potential neurophysiological explanation for the MRT concept of separate resource pools. Perhaps rather than thinking in terms of resource pools, the ability to time-share two tasks can be understood as more efficient if different neural mechanisms (i.e., visual vs. auditory pathways, object identification vs. location, occipital lobe vs. temporal lobe networks) are involved in the encoding, processing, storage, and response of each task.

Jonides et al. (2005) discussed the hypothesis that the brain mechanisms responsible for working memory might be the same as those responsible for perceptual encoding. They suggested that this would mean that there would be separate pathways for visual, spatial, and auditory stimuli and further that posterior regions would be involved in basic encoding and storage. However, if the storage occurs in the presence of distracting stimuli (interference), then frontal areas serving selective attentional processes would be involved to regulate and control the maintenance of information through rehearsal. Indeed, as Jonides and colleagues pointed out, there is evidence for their hypothesis from both brain imaging studies and from observations of persons with specific brain lesions.

Support for the shared neural mechanism account discussed by Jonides and colleagues (2005) comes from observations that sensory event-related potential (ERP) components are highly correlated with and thus good predictors of working memory span and fluid intelligence (Brumback, Low, Gratton, & Fabiani, 2004). Brumback and colleagues observed that N100 responses to stimuli in simple auditory tasks varied significantly between individuals with high and low scores on a working memory span measure thought to rely heavily on verbal working memory. The N100 responses of individuals with high working memory span were significantly larger than those with low working memory span. Similarly, a visual sensory ERP component (P150) differed between high and low performers on the Raven Progressive Matrices (RPM) test, a nonverbal assessment of general intelligence. Brumback and colleagues concluded that modality-specific sensory components are good predictors of working memory span and fluid intelligence measures presented in that same modality. In other words, visual sensory components are related to visual-spatial working memory processes, and auditory sensory components share a strong relationship with verbal working memory. This observation is in line with the hypothesis of Jonides et al. (2005) that sensory perceptual processes and working memory processes may share the same neural mechanisms. It also provides a possible neurophysiological explanation for numerous empirical investigations that demonstrated that cross-modality task pairings (visual and auditory) versus same-modality task pairings (auditory-auditory) often lead to more efficient performance. The extent to which two concurrent tasks rely on the same neural mechanisms or pathways would determine their degree of structural interference. A large amount of structural interference or competition between the two tasks would result in the need to process information serially (i.e., one task at a time in rapid alternation) to time-share the common neural mechanism. Conversely, if the two tasks relied primarily on separate mechanisms, more efficient parallel processing would be permitted.

The ability to store and manipulate information temporarily, referred to as working memory, plays an important role in our ability to process sounds, particularly speech. Numerous investigations conducted since the 1990s have been specifically designed to examine the nature of the working memory resources involved in language processing.

An ongoing debate regarding the nature of working memory resources in language comprehension is particularly germane to our current discussion. Two diverging perspectives exist regarding the degree to which deficits in working memory will automatically result in deficits in language comprehension. The debate centers around whether the working memory resources involved in language comprehension are domain general or domain specific: a single system for all or separate resources for different language processes (DeDe, Caplan, Kemtes, & Waters, 2004; Friedmann & Gvion, 2003; Just & Carpenter, 1992; Just, Carpenter, & Keller, 1996; Waters & Caplan, 1996). One perspective emphasizes capacity limits and a single general working memory system or resource serving all language functions (see, e.g., Daneman & Carpenter, 1980; Fedorenko, Gibson, & Rohde, 2006; Just & Carpenter, 1992; Just, Carpenter, & Keller, 1996). An opposing position argues instead that there are separate domain-specific resources serving different language processes (Caplan & Waters, 1999; DeDe et al., 2004; Friedmann & Gvion, 2003; Waters & Caplan, 1996). Support can be found for both positions, and major arguments for each perspective are briefly considered here due to their relevance for assessing mental workload from a resource capacity perspective. For more detailed examination and developments in this ongoing debate, see reviews in the work of Caplan and Waters (1999), Fedorenko et al. (2006), and MacDonald and Christiansen (2002).

The debate originated with a seminal article published by Just and Carpenter (1992) in which they proposed a capacity theory of language comprehension. According to their capacity theory, processing and storage of verbal material are both constrained by working memory activation. Variation in the ability to activate working memory processes (termed *capacity*) was thought to account for the substantial individual differences observed in language comprehension abilities. They reasoned that people with larger working memory capacities were able to hold multiple interpretations of initially ambiguous sentences in memory until subsequent information was presented that would disambiguate the message. Conversely, people with lower working memory capacity would be unable to hold these multiple interpretations in memory and would therefore be more likely to make sentence-processing errors if the later information supported the less-obvious interpretation of an ambiguous sentence. If these initially ambiguous sentences are later resolved in an unpreferred way (meaning the less-obvious form of the initial subject was used), they are called "garden path" sentences. The name refers to the idea that they initially support one interpretation (leading listeners or readers down a garden path) only to present conflicting information later in the sentence that requires reanalysis of the initial interpretation. An example of a garden path sentence used by King and Just (1991) is, "The reporter that the senator attacked admitted the error." As Just and Carpenter (1992) pointed out, this sentence is difficult for two reasons. First, it contains an embedded clause (a sentence part or clarifying idea in the middle of the sentence: "that the senator attacked") that interrupts the main part or clause of the sentence (which refers to someone doing the admitting). Second, the initial noun *reporter* plays a double role of both the subject of the main clause of the sentence (the person who did the admitting) and the object of the second clause (the person who was attacked), which is embedded in the middle of the sentence. So, this first noun, reporter, must be kept in memory while side information about it is obtained before the main point of the sentence is made clear.

In support of their capacity theory, Just and Carpenter (1992) pointed out that while these sentences are generally difficult for everyone, they are particularly difficult for people with low working memory spans. Note that working memory span can be assessed with a variation of Daneman and Carpenter's (1980) reading span paradigm discussed in this chapter. Working memory span, or simply "span," involves simultaneously processing and storing verbal material, and variations of Daneman and Carpenter's paradigm are in widespread use today (see Bayliss, Jarrold, Gunn, & Baddeley, 2003; Engle, 2001; Kane, Bleckley, Conway, & Engle, 2001).

Just and Carpenter (1992) proposed that the working memory capacity measured by span tasks is an essential factor constraining all language processing tasks. They suggested that people with high spans (greater working memory capacity) are better able to keep multiple meanings or roles of individual words within a sentence in memory while processing later components. They also pointed to the observation that older adults tend to have more difficulty comprehending difficult garden path sentences than young adults. Since older adults also tend to have smaller span scores than their younger counterparts, this is discussed as supporting evidence for a general working memory system that affects language comprehension in general.

Neuropsychological evidence for a general working memory resource capacity comes from the observation that people with a variety of different aphasias (impaired ability to produce or comprehend language) share a common deficit involving greater difficulty processing more syntactically complex sentences (Miyake, Carpenter, & Just, 1995). People with left hemisphere aphasia (regardless of the specific type) perform more poorly on tasks requiring the use of syntactic structure to extract meaning from sentences, like in the garden path sentences previously described. Performance impairment in these individuals relates more directly to the overall working memory resources that can be devoted to the task (as assessed by span scores) and the degree of overall global neurological impairment rather than the specific type of impairment (see review in Caplan & Waters, 1999). Support for a common deficit in aphasic sentence comprehension performance has also been found using a computational modeling approach (Haarmann, Just, & Carpenter, 1997).

Waters and Caplan (Caplan & Waters, 1999; Waters & Caplan, 1996) have argued instead that separate working memory resources support different types of language tasks. This theoretical position has been referred to as a domain-specific theory of working memory (Caplan & Waters, 1999; Fedorenko et al., 2006). They suggest that language tasks that require online processing (interpretation of immediate verbal information like processing syntactic complexity) rely on a verbal working memory that is at least partially independent of a more general controlled attention form of working memory resource involved in "off-line" language comprehension tasks. In their view, traditional working memory span tasks, such as those modeled after Daneman and Carpenter's (1980) reading span task, should be related to the controlled attention working memory resource system, but rather unrelated to the verbal working memory system involved in syntactic processing. Note this is contrary to Just and Carpenter's (1992) position that all language processes are related to a general working memory activation capability as assessed by reading span scores. As evidence for their position, Waters and Caplan (1996) discussed the observation that individuals with reading span scores of 0 to 1 (i.e., patients with Alzheimer's disease) maintain their ability to utilize syntactic structure to interpret the meaning of sentences, and that they are able to do so even when their working memory systems are busy processing concurrent verbal material (i.e., a digit memory load) (Waters, Caplan, & Rochon, 1995). Further support for their model is derived from the results of structural equation modeling (SEM) approaches examining the relationship between age, verbal working memory, and performance on online syntactic processing versus off-line sentence comprehension tasks (DeDe et al., 2004). Results of their modeling investigation indicated very different patterns of relationships for the different language processing tasks. A direct relationship was found between age and performance on the syntactic processing task, and this relationship was not mediated by verbal working memory. However, performance on the sentence comprehension task was directly related to verbal working memory, and this relationship was not mediated by age.

Neurophysiological and neuropsychological evidence can be found to support the position that there are distinct multiple pathways for different types of language processing tasks, with a distinction generally arising between syntactic and semantic aspects. For example, using positron emission tomography (PET) imaging, Caplan and colleagues (Caplan, Alpert, & Waters, 1998) observed that different cortical areas were associated with increased regional cerebral blood flow (rCBF) when using syntax to extract meaning from sentences versus when they were making semantic plausibility judgments in sentences with varying numbers of propositions. Specifically, rCBF increased in the dominant perisylvian association cortex, a portion of Broca's area, when participants were using syntax to extract the meaning of sentences irrespective of the number of propositions the sentence contained. Conversely, rCBF increased in more posterior sites when participants were required to make plausibility judgments (they had to determine if a given sentence was plausible) when sentences contained two versus one proposition. These findings provide support for the hypothesis that different aspects of language processing are carried out by different neural mechanisms. However, this evidence does not preclude the possibility that a more general attentional resource mechanism might be necessary to coordinate activity among these different neural pathways, particularly when the total processing load is high or when the task must be carried out under time pressure.

For now, the debate between whether a single or multiple system of working memory resources is needed to support the variety of processes involved in language comprehension remains to be settled. Nevertheless, as we continue to discuss in the chapters to come, neuroimaging techniques offer powerful new tools for exploring these and other theoretical issues pertaining to auditory processing. In particular, we discuss many examples of the way neuroimaging techniques (i.e., functional magnetic resonance imaging [fMRI] and PET scans) are being used to inform and resolve other important theoretical issues pertaining to language comprehension in further chapters. Many of these issues involve the underlying working memory resource requirements of processing different types of words and sentences.

Despite the close relationship between the concepts of working memory resources and mental workload, few investigations and models have specifically incorporated both within the context of the same study. Investigations of working memory and mental workload are generally carried out by scientists in the fields of cognitive psychology and human factors, respectively, with discouragingly few attempts to integrate findings. This gap may be bridged in the near future as both cognitive psychologists and human factors scientists focus increasing attention on the brain mechanisms involved in mental activity. Scientists focusing on an exciting new area within human factors, termed *neuroergonomics* (see Parasuraman, 2003; Parasuraman & Rizzo, 2007) are leading the way in this endeavor. However, for now we turn our focus to key concepts and models essential to the assessment of mental workload.

MENTAL WORKLOAD: KEY CONSTRUCTS

As previously discussed, mental workload is a multidimensional, multifaceted construct (Gopher & Donchin, 1986). Therefore, it is logical to assume that different assessment techniques may be sensitive to different aspects of workload demand. This further suggests that finding one ideal assessment technique is challenging at best and potentially an impossible feat. There are many sources of variation in workload demand, some of which are directly linked to the task of interest and others that are unique to the person or operator performing the task. Complicating this issue is the fact that workload demand may fluctuate rapidly over time. It may be apparent that the demands of a given task change over time and with changes in the operator's experience and skill at the task. Less apparent is that workload may also fluctuate as a function of things such as the operator's strategy and physiological state. In addition, workload may vary as a function of an individual's response to task demands and personal characteristics (Parasuraman & Hancock, 2001). These multifaceted aspects of workload can affect demand singly and in combination, thus providing a continual challenge to workload assessment.

Due to the multidimensional influences, dissociation may occur between the demands of the task and observable human performance. This may be referred to as the underload/overload issue. As a task becomes less demanding, performance may deteriorate. People have a tendency to mentally disengage from a task that is not challenging. Conversely, if a task of relatively low demand becomes more difficult, performance may actually improve as operators focus attention and engage more resources toward task performance. The operators' strategies, skill level, and physical abilities may have dramatic affects on behavioral performance measures. These aspects of assessment are particularly important to keep in mind when attempting to assess the mental workload of auditory processing tasks. For example, in many situations speech processing is highly robust, with low resource demand. Understanding speech may seem so automatic that in some everyday situations listeners may devote little focused attention to the task. (Many college freshmen may mistakenly believe they can adequately pay attention to their professors while also listening to their classmate or perhaps their iPod. The error in this line of reasoning may not become apparent until they perform poorly on the first exam.) Even listeners with compromised abilities, such as people with hearing impairment, may have developed well-practiced compensation strategies that allow them to maintain performance in many everyday listening situations. However, recall our opening example of the child learning arithmetic; performance measures alone may reveal little about the effort involved in the task of processing the auditory material. This is particularly true if there is ample time for performing the given task, a topic we turn to next.

TIME

Time can be both a limiting factor on performance (there is only so much one can do within a given period of time) and a stressor when the demands of the task seem too great for the amount of time available. Therefore, time plays an important role in mental workload whether or not it directly or indirectly affects working memory processes. Several models of mental workload have included time pressure as a major component (Hancock & Caird, 1993; Hancock & Chignell, 1988; Young & Stanton, 2002). For example, Hancock and Chignell (1988) discussed how the perceived time to complete a task interacts with the perceived distance from task completion. As illustrated in Figure 5.1, excessive mental workload can result from either not having

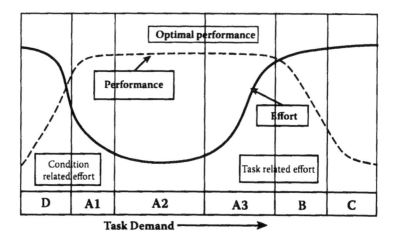

FIGURE 5.1   Mental workload and performance.

enough time to complete the task at hand or from having an extremely high number of tasks to complete within a given time.

The influence of time pressure has been incorporated in a number of subjective mental workload assessment techniques. Another factor affecting mental workload that has been more difficult to incorporate in assessment methods involves the moods, interests, and fatigue level of the person or operator performing the task.

OPERATOR STATE AND STRATEGIES

Operator characteristics (such as a person's mood or fatigue level) also have an impact on workload. To illustrate the influence of these endogenous factors, perhaps everyone can imagine a time when they were performing an everyday task (such as listening to a friend recite the events of the day) and the task seemed more difficult because they were tired, preoccupied, or found the material uninteresting. Despite the low level of resource demand or difficulty inherent in the particular task, one may experience a high level of workload and find it difficult to or be unable to adequately engage sufficient mental resources to attend, comprehend, and respond to the speaker's dialogue. Conversely, when highly motivated and well rested (or when the topic becomes more interesting), the same task may be "easier" to perform.

In work situations, as task load increases, operators may adopt adaptive control strategies to offset performance consequences and to maintain workload within a manageable range (Cnossen, Rothengatter, & Meijman, 2000; Hockey, 1997). Operators may change their performance criteria, offset tasks to other personnel, or engage automation systems to allocate attention to critical task components. Task management strategies include prioritizing and scheduling multiple overlapping tasks and are critical to effective performance in many hazardous occupations (Chou, Madhavan, & Funk, 1996). Unfortunately, empirical evidence indicates that during periods of high mental workload in various critical work environments (i.e., operating rooms and aircraft cockpits), less attention may be focused on the communication task, resulting in shorter utterances, reduced knowledge and background information sharing, and high risk for communication failure (Grommes & Dietrich, 2002). Errors in task management can result in catastrophic consequences, as evidenced by their involvement in a substantial proportion of aviation accidents and incidents (Chou et al., 1996). The relationship between workload and performance may change as a result of prior task demand as well. This relationship has been incorporated into Young and Stanton's (2002) model of mental workload, which is discussed further in this chapter.

COMPENSATORY STRATEGIES

As the mental demands of a task or set of tasks increase, people tend to use compensatory strategies to maintain performance. An early example is the research of Sperandio (1978), who found that air traffic controllers handled the load of an increase in the number of aircraft they had to control by shortening the amount of time they talked to the pilots of the aircraft, while attempting to convey information essential to maintenance of safety. It has also been well documented that older, hearing-impaired listeners will use contextual cues as a compensatory strategy to maintain speech comprehension and communication (Gordon-Salant & Fitzgibbons, 1997; Madden, 1988; Speranza, Daneman, & Schneider, 2000). Similar performance benefits for older adults are observed when context is added to visually degraded sentences (Speranza et al., 2000). However, these compensatory strategies require mental resource expenditure. Thus, despite their potential to preserve communication, the use of compensatory strategies may compromise performance on other concurrent tasks. Alternatively, in resource-demanding situations older adults may lack sufficient spare resource capacity to make use of compensatory strategies. This is more likely when speech is particularly degraded, in the presence of noise, or when presentation rates are high. The additional demands of processing speech may also lead to the use of compensatory strategies for a concurrent task. This may involve engaging automation to assist in monitoring or performing a task that the operator would normally perform to allow communication to take place. In the case of driving an automobile, the driver may choose compensatory strategies in an attempt to offset the additional resource demands of talking on a cellular phone. Unfortunately, these compensatory strategies are frequently not as effective as their users believe.

For example, many drivers will attempt to use compensatory strategies to maintain safe driving while conversing on a cellular phone. There is now considerable evidence that cellular phone conversations increase drivers' mental workload (Haigney et al., 2000; McCarley et al., 2004; Strayer et al., 2003; Strayer & Johnston, 2001). Drivers use a number of compensatory strategies in an attempt to maintain driving performance within safe margins; however, empirical evidence suggests that these strategies may not effectively decrease risk (Haigney et al., 2000). For example, drivers tend to decrease speed when using a mobile phone. However, in simulation studies decreased speed did not result in fewer lane deviations or off-road instances. These findings suggest that drivers may be more confident in their compensation strategies than actual performance would indicate.

Hands-free mobile phones do little to offset the increased demands of communicating while driving (McCarley et al., 2004; Patten, Kircher, Ostlund, & Nilsson, 2004). Research has consistently demonstrated that it is the mental complexity of the conversation (Patten et al., 2004; Strayer & Johnston, 2001) and perhaps the quality of the acoustic signal (Matthews, Legg, & Charlton, 2003) rather than the response characteristics of phone use that represent the greatest hazard to safe driving.

As this driving example illustrates, there are many situations for which understanding the mental workload demands of an auditory task can have important safety consequences. Humans frequently process speech and other auditory information while engaged in other important tasks. We turn now to a discussion of the major methods and techniques that have been developed to assess mental workload.

MENTAL WORKLOAD ASSESSMENT

Numerous investigations have compared methods of assessing mental workload (Baldwin & Coyne, 2005; Baldwin, Freeman, & Coyne, 2004; Brown, 1965; Casali & Wierwille, 1983a, 1984; Verwey & Veltman, 1996; Wierwille & Connor, 1983), and numerous reviews and discussions have been compiled (Baldwin & Coyne, 2005; Casali & Wierwille, 1983a; Eggemeier, Wilson, Kramer, & Damos, 1991; Gopher & Donchin, 1986; Hancock & Desmond, 2001; Knowles, 1963; Moray, 1982; O'Donnell & Eggemeier, 1986; Verwey & Veltman, 1996). At present, no single method appears optimal in all situations; in fact, different techniques may be required simultaneously to assess different aspects of mental workload (Baldwin & Coyne, 2005; Isreal, Chesney, Wickens, & Donchin, 1980; Veltman, 2003). The consensus seems to be that the best assessment methodology depends on the goal of the assessment and the particular operational environment in which the assessment will take place. Three primary categories of workload assessment have been identified, and each of these is discussed.

The three major categories of workload assessment techniques that have been identified in previous literature (O'Donnell & Eggemeier, 1986; Wierwille & Eggemeier, 1993): (a) behavioral measures, which can be further divided into the subcategories of primary measures and secondary task performance measures obtained in a dual-task paradigm; (b) physiological measures (including but not limited to neuroergonomic indices of brain activity); and (c) subjective measures. Considerable debate has ensued concerning the best method of assessment, with benefits and limitations identified for each depending in large part on the intended purpose of measurement, the tasks involved, and the environment in which measurement is taking place. Many of the limitations as well as the benefits associated with each measurement technique are reviewed in the discussion that follows, with an emphasis on issues particularly relevant to assessing the mental workload of auditory processing tasks. Benefits and limitations are discussed in terms of two primary issues involved in assessment: sensitivity and intrusion (O'Donnell & Eggemeier, 1986).

SENSITIVITY AND INTRUSION

Sensitivity refers to the ability of the assessment technique to detect changes in workload requirements imposed by the task under consideration. A technique that fails to detect or reveal workload fluctuations can be said to lack sensitivity. Keep in mind that a particular technique may reflect changes in (be sensitive to) some aspects of workload demand while failing to detect other aspects.

Intrusion refers to the degree to which the assessment technique itself interferes with performance of the task (or set of tasks) under investigation. Assessment techniques are sought that are sensitive to changes in workload fluctuations without interfering with task performance. We begin by discussing behavioral task assessment methodologies.

BEHAVIORAL TASK MEASURES

Behavioral task measures assess workload by examining performance that can be observed. This commonly involves measures such as accuracy, response time (RT), and task completion. Two primary distinctions can be made within the category of behavioral measures. The first is primary task measures, and the second is dual-task or secondary task measures.

Primary Task Measures

Primary task measures involve assessing the task of interest directly. One simply examines performance on the task at hand. This method is certainly the most direct. However, as previously discussed, task demands may change considerably without concomitant changes in observable performance. Conversely, decrements in performance may occur when task demands are too low. Thus, primary task measures, in and of themselves, seldom reflect operator load (Knowles, 1963; O'Donnell & Eggemeier, 1986; Ogden et al., 1979). In other words, primary task measures may assess how well an operator can perform a certain task, but they provide little or no indication of the amount of effort expended by the operator.

A critical limitation of primary task measures is their lack of sensitivity to increased demands until the operator is overloaded (Eggemeier, 1988). In other words, primary task measures may fail to reflect the increased expenditure associated with increasing task difficulty until the operator has reached the edge of his or her operating envelope. The operating envelope can be thought of as the range of resource capacity available for performing a task. A person's operating envelope will vary according to previously mentioned factors, such as skill, ability, and physiological state. When people have reached the edge of their operating envelope, they have exhausted their reserve capacity of resources. Further increases in task demand will thus result in the operator being "overloaded," at which time a sudden drop in performance is detectable. In mental workload assessment, measures that are sensitive to small increases in task demand (prior to resource capacity exhaustion) are necessitated. A method of addressing the lack of sensitivity issue with behavioral measures that has been exceedingly popular in the history of mental workload research is the dual-task paradigm.

Secondary Task Measures

Secondary task measures resulting from implementation of a dual-task paradigm have proven more sensitive to workload fluctuations than primary task measures alone. The theoretical foundation for the dual-task paradigm is the premise that humans have a limited supply of mental resources that can be engaged to meet task demands. Spare mental capacity, then, is the difference between this upper limit and the amount required to perform the task or tasks under investigation. The task of interest (i.e., driving, flying, or comprehending a spoken passage) is designated as the primary task. People are instructed to perform the primary task to the best of their ability. Another task (called the secondary task) that is less important or completely unrelated to the primary task is then used as an index of the spare or reserve resource capacity. The idea is that when a person strives to maintain his or her performance on a primary task, if that task becomes more resource demanding, then fewer resources will be available for performing the secondary task, and thus performance on the secondary task will degrade (Knowles, 1963). If the operator is able to perform well on the secondary task, this is taken to indicate that the primary task is of relatively low resource demand. Conversely, if a person is unable to perform the secondary task and at the same time maintain primary task performance, this is taken to indicate that the primary task is more demanding. The secondary task technique has demonstrated sensitivity to changes in the resource demands of a wide variety of tasks, ranging from driving (Brown & Poulton, 1961; Harms, 1991) to sentence processing (Baldwin & Struckman-Johnson, 2002). Secondary task measures can also be sensitive to individual differences in resource capacity and fluctuating levels of workload.

For example, suppose that Bob performs a given task easily, but it is so difficult for the second person, John, that it takes up nearly all his processing resources, leaving him little or no spare capacity. Now, suppose Bob and John are given another task to perform concurrently and are asked to maintain their performance on the first task, performing the second only to the extent that they may do so without compromising their performance on the first. Bob, who is utilizing fewer resources to perform the first task, has greater spare capacity to devote to the second task and therefore will be able to perform both tasks (providing the second is not too difficult) with relative skill. However, John, who is struggling with the first task, will have few if any spare resources to devote to performing the secondary task. Therefore, John will likely take longer to respond to and make more errors on the secondary task. Stemming from this premise, the dual-task paradigm is a well-established workload assessment technique that involves having the operator perform a second task concurrent with the task of interest. Performance on the secondary task is used as an index of the mental demand imposed by the first or primary task. There are several variations of the basic dual-task paradigm, including the use of loading tasks and adaptive secondary tasks. Discussion of these variations is beyond the scope of our current aim. The interested reader is referred to previous reviews (Knowles, 1963; Ogden et al., 1979; Williges & Wierwille, 1979).

PHYSIOLOGICAL MEASURES

The theoretical foundation of physiological measures of mental workload lies in the assumption that mental effort is accompanied by physiological changes in the individual engaged in the task. For example, researchers make the assumption that exerting greater mental effort will be associated with greater brain activation (Mason, Just, Keller, & Carpenter, 2003). Numerous researchers have advocated the measurement of physiological and neurophysiological changes as indices of mental workload (Just et al., 2003; Kahneman, Tursky, Shapiro, & Crider, 1969; Kramer, 1991; Parasuraman, 2003). Examples of physiological measures of workload include heart rate and heart rate variability, blood pressure, eye blinks, pupil diameter, skin conductance, and many more. Examples of neurophysiological methods that have been used to examine resource expenditure include electroencephalography (EEG) and ERP components, fMRI, and PET scans. While ERP measures can provide a means of examining the time course of different auditory processing stages, brain imaging techniques such as fMRI and PET can provide a powerful tool for examining the region of brain activity and level of brain activation engaged in relation to tasks with different resource requirements. Reviews of some of this work may be found in several resources (Byrne & Parasuraman, 1996; Caplan & Waters, 1999; Gevins et al., 1995; Gopher & Donchin, 1986; Just, Carpenter, Keller, Eddy, et al., 1996;

Kramer, 1991; Wickens, 1990; Wierwille, 1979; Wierwille & Connor, 1983; Wilson & Eggemeier, 1991).

Physiological measurement techniques offer several advantages relative to other forms of assessment. They are objective and, unlike the secondary task techniques discussed previously, physiological measures frequently require no additional overt response or learning on the part of the participant (Isreal, Chesney, et al., 1980). Also, physiological measures may be sensitive to aspects of the task that are not revealed by other forms of assessment. For example, several investigations have found ERP components to be sensitive to increased demands associated with task parameters not demonstrated through performance measures such as RT and accuracy (Baldwin et al., 2004; Kramer et al., 1987). For a helpful discussion on the dissociation between ERP measures and behavioral performance, see the work of Wickens (1990).

Neurophysiological Techniques

Measures of brain function in particular have several advantages over alternative measurement techniques. These advantages include increased sensitivity to both transient and continuous fluctuations in mental demand without the need to introduce an additional task as well as the ability to discern the relative contributions of various brain mechanisms as a result of task dynamics. The ability of neurophysiological measures to provide real-time assessment of mental workload can facilitate both examination of task components and environmental variables leading to compromised performance as well as the potential to develop adaptive automation interfaces to offset workload in high-demand situations.

Neurophysiological techniques vary along several dimensions, including (a) level of invasiveness, (b) spatial resolution capabilities, and (c) temporal resolution capabilities. EEG/ERP methods are relatively low in level of invasiveness and have strong temporal resolution capabilities. However, EEG/ERP methods lack strong spatial resolution capabilities. Conversely, PET scans and fMRI, which have strong spatial resolution capabilities, are highly invasive (lacking portability, which makes them unsuitable for assessing performance in dynamic complex environments). Further, PET scans and fMRI methods lack the temporal resolution capabilities suitable for assessing workload for real-time applications. An additional concern is that scanner equipment required for fMRI is extremely noisy, limiting its use in examining many auditory processing paradigms. Emerging optical brain imaging systems, such as NIRS, offer the combined advantage of low levels of invasiveness, strong temporal resolution, and relatively strong spatial resolution as compared to EEG/ERP techniques. However, further testing is required to determine the suitability of implanting NIRS assessment techniques in complex environments. EEG/ERP techniques have been used extensively for mental workload assessment. We discuss further some applications of the examination of ERP components since this technique has been used extensively not only in mental workload assessment, but also specifically as a tool for investigating the resource requirements and time course of processing various aspects of complex sound.   .

ERP Investigations

The examination of changes in ERP components as a method of understanding human information processing and mental workload, in particular, has had a long history. The P300 ERP component, a positive wave deflection occurring approximately 300 ms after a stimulus event, has been of interest in mental workload assessment dating at least as far back as 1977, when Wickens and colleagues reported its use. The rationale for use of this endogenous component is based on the theory discussed previously that humans have a limited amount of processing resources (Kahneman, 1973; Wickens, 2002).

One of the commonly used paradigms, utilizing an oddball discrimination task, involves presenting two or more stimuli of varying probability. The higher-probability stimulus (e.g., a tone of a given frequency) serves as a distracter, while the infrequent oddball stimulus (e.g., a tone of a different frequency) serves as the target. Within the framework of a limited-capacity processing system, P300 amplitude in an oddball discrimination paradigm is thought to reflect the amount of available attentional resources that can be devoted to a given task or task set. As fewer attentional resources are available (due to the concurrent performance of a task that is increasing in difficulty), amplitude of the P300 response to the target stimuli in an oddball discrimination task can be expected to decrease.

An advantage of assessing workload with an oddball ERP paradigm is that no overt response is required (Baldwin et al., 2004; Isreal, Chesney, et al., 1980). The operator can simply be asked to keep a mental count of the number of targets. Or, in a variation of the oddball paradigm referred to as the irrelevant probe task, the operator can be instructed simply to ignore the irregular stimuli (Kramer, Trejo, & Humphrey, 1995). In an irrelevant probe paradigm, earlier components such as N100 or N200 (negative wave deflections occurring approximately 100 and 200 ms, respectively, after the onset of the stimulus) are typically examined.

ERP components have demonstrated sensitivity to changes in mental workload in a number of operational environments. For example, in an early investigation involving monitoring of a simulated ATC display, the P300 component stemming from an oddball discrimination task successfully differentiated between manipulations of display task demand (Isreal, Wickens, Chesney, & Donchin, 1980). Similarly, Kramer and colleagues found the P300 component to be sensitive to changes in flight task difficulty (Kramer et al., 1987). In a simulated driving environment, Baldwin and colleagues found P300 amplitude to be sensitive to the increased demands of driving in heavy fog versus clear visibility (Baldwin & Coyne, 2005).

In similar fashion, Kramer and colleagues found that N100 and N200 components associated with an irrelevant auditory probe task were sensitive to changes in the demands of a radar monitoring task (Kramer et al., 1995). Additional examples of the auditory probe technique are discussed in subsequent chapters. In particular, the use of ERP techniques for examination of various aspects of speech processing is discussed in Chapter 8.

Despite the stated advantages and demonstrated usefulness of physiological measures, they are not without their disadvantages. They may require extensive equipment costs and training on the part of the experimenter. In addition, the logistics of equipment use pose a challenge to the use of some physiological measures in complex real-world environments (Gevins et al., 1995). Partially in an effort to overcome some of these logistic difficulties, experimenters have often chosen to rely on subjective assessments of mental workload.

SUBJECTIVE TECHNIQUES

Subjective techniques have been utilized extensively as a tool for assessing mental workload (Brookhuis & de Waard, 2001; Lee et al., 2001; Meshkati, Hancock, Rahimi, & Dawes, 1995; Srinivasan & Jovanis, 1997a; Vidulich & Wickens, 1986). They are generally easy to administer and require relatively little preparation or learning by the experimenter or operator. A number of scales have been developed, and their psychometric properties have been compared in numerous investigations (Hendy, Hamilton, & Landry, 1993; Hill et al., 1992; Rubio, Diaz, Martin, & Puente, 2004; Tsang & Velazquez, 1996). One of the most widely used subjective workload scales in human factors research is the NASA Task Load Index (NASA TLX) developed by Hart and Staveland (1988). It has demonstrated sensitivity to changing workload demands in several investigations (Grier et al., 2003; Hill et al., 1992; Reagan & Baldwin, 2006).

Subjective techniques have several advantages, including ease of administration, low cost, and face validity. However, several limitations are also prevalent. For example, as pointed out by Willigies and Wierwille (1979) in an early review, in many cases subjective workload measures are situation specific and may fail to take into account adaptivity, learning, experience, natural ability, and changes in emotional state on the part of the operator. In addition, they pointed out that subjective measures can confuse measurement of mental workload with physical workload.

Vidulich (1988) reviewed several experiments investigating the sensitivity of subjective assessment techniques. In one of his previous studies (Vidulich & Tsang, 1986), a Sternberg memory search task was paired with a secondary task of compensatory tracking. A measure of RT was demonstrated to be sensitive to changes in task difficulty in both single- and dual-task conditions. However, two subjective assessment measures (both were bipolar rating scales) failed to provide a sensitive measure of task difficulty in the dual-task condition. Vidulich (1988) proposed that the presence of the tracking task apparently overwhelmed the subjective distinction between memory search task difficulty levels. Vidulich also noted that the measure of root mean square (RMS) tracking error remained consistent across memory task difficulty conditions.

In a second experiment, involving a tracking task paired with a simple transformation task, Vidulich compared the subjective workload assessment technique (SWAT) and NASA bipolar technique (Vidulich & Tsang, 1986). The two subjective techniques resulted in significantly different scores on one of the conditions. In addition, the findings suggested that both subjective techniques were sensitive to manipulations that influenced perceptual/central processing demands but failed to discriminate between manipulations that influenced response execution demands.

A general lack of sensitivity on the part of subjective methods has been frequently reported in the workload literature. Subjective ratings frequently are sensitive to the increased demands associated with the introduction of an additional task but may fail to indicate when one of the two tasks becomes more demanding (Baldwin & Coyne, 2003, 2005; Baldwin et al., 2004).

SUMMARY

Mental workload is an important construct in human factors research. It also shares many commonalities with concepts used in other disciplines, namely that of working memory resources as used predominantly in the fields of cognitive psychology and cognitive neuroscience. Common themes include the observation that there are limits to the amount of mental effort people have available to perform tasks. Individual differences exist in these capacity limits and will affect performance accordingly. Further, the structure of the tasks being performed (i.e., whether they are primarily visual-spatial or auditory-verbal in nature) will also affect one's ability to perform two tasks at the same time.

Mental workload can be measured in a variety of ways, each having distinct advantages and limitations. The choice of which to use should be driven by the questions to be answered and the constraints of the environment in which workload is to be assessed. Subjective measures are one of the easiest methods to implement but are often insensitive to fluctuating levels of workload found in many tasks. Behavioral measures, particularly within a dual-task paradigm, have led to important theoretical developments and remain a useful assessment technique. Increasingly, however, both human factors practitioners and cognitive psychologists are turning to methods of examining brain function in conjunction with behavioral measures to answer long-standing theoretical questions.

CONCLUDING REMARKS

Regardless of how it is measured, determining the level of mental effort involved in performing various and simultaneous tasks is an important issue and is a major focus of the remaining chapters in this book. Examining the mental workload involved in auditory processing tasks can inform auditory display design, improve communication, and facilitate improved experimental designs in a wide number of disciplines. One area in which consideration of the mental workload of the auditory task is particularly critical is when using auditory tasks in cognitive research. Unfortunately, as discussed in the next chapter, all too frequently this consideration is not given ample attention.

.

6 Auditory Tasks in Cognitive Research

INTRODUCTION

Basic and applied studies in cognition have generally used visual tasks to explore different aspects of human information processing. Nevertheless, although not as numerous, studies using nonverbal auditory tasks have also played an important role in investigations of perceptual and cognitive processing. Such tasks have ranged from simple clicks (to measure the brain stem evoked potential), to tones, musical patterns, environmental sounds, and rhythms. Neuroelectromagnetic responses to nonverbal sounds can be used to examine the integrity of central auditory pathways in infants and other nonverbal populations. Auditory detection tasks have been used to examine what has been termed the *psychological refractory period* (PRP) and as a secondary task index of mental workload. In fact, auditory tasks are frequently used to assess mental workload (Harms, 1986, 1991; Ullsperger, Freude, & Erdmann, 2001; Zeitlin, 1995); cognitive functioning, particularly after brain injury (Allen, Goldstein, & Aldaróndo, 1999; Loring & Larrabee, 2006); and cognitive aging (Schneider & Pichora-Fuller, 2000).

These are only some examples of the many auditory tasks that have been used for more than a century to investigate attention, cognitive processing, and mental workload. As discussed in Chapter 4, the dichotic or selective listening task was a key technique in research on selective attention in the 1950s and 1960s, work that was highly influential in the beginning stages of the "cognitive revolution" (see Broadbent, 1958; C. Cherry, 1953b, 1957; Moray, 1959, 1969; Treisman, 1964a, 1960). Auditory tasks also played a key role in investigations leading to the development of early models of memory (see, e.g., Atkinson & Shiffrin, 1968; Broadbent, 1958; Waugh & Norman, 1965).

More recently, auditory tasks have been used in a variety of cognitive science investigations. Auditory tasks have been used to examine the nature of working memory (Baddeley, 1992; Cocchini, Logie, Della Sala, MacPherson, & Baddeley, 2002; Larsen & Baddeley, 2003); time perception (Brown & Boltz, 2002); age-related changes in cognitive functioning (Baldwin, 2001; Pichora-Fuller & Carson, 2001; Tun, 1998); and the neural mechanisms associated with learning and memory (Jones, Rothbart, & Posner, 2003; Posner & DiGirolamo, 2000). Auditory tasks are also frequently used to examine the mental workload associated with different operational environments, such as driving (Baldwin & Coyne, 2003; Baldwin & Schieber, 1995; Brown, 1965; Harms, 1986, 1991) and aviation (Coyne & Baldwin, 2003; Kramer, Sirevaag, & Braune, 1987; Wickens, Kramer, & Donchin, 1984).

In this chapter, the ways that auditory tasks have been used to investigate cognitive processing are examined, with a particular emphasis on investigations of mental workload. We discuss what these investigations demonstrate about the mental workload of auditory processing, as well as the need to consider the mental workload of auditory processing before reaching definitive conclusions regarding other aspects of cognitive functioning.

Auditory tasks that have been used in cognitive research differ in many ways, as the following examples illustrate. A great deal has been gained from these investigations, although all too frequently, researchers have neglected to consider the mental workload requirements of processing acoustic information. Factors such as individual differences in sensory capabilities challenge interpretation of many of these investigations. For example, as discussed extensively in Chapter 10, older adults with sensory impairments may require more mental effort during early sensory extraction processing stages, leaving a reduced supply of spare attentional effort to be used in the completion of subsequent stages of processing and additional tasks. Neglect of these basic issues, at least in some instances, calls into question the interpretation of the results obtained in these investigations and their implications for cognitive theories.

Another issue with many investigations in which auditory tasks have been used as a means to investigate cognitive processing is that incomplete information is provided on the actual acoustic characteristics of the presented stimuli. In terms of understanding implications for mental workload, particularly from the perspective of sensory-cognitive interaction theory (SCIT; Baldwin, 2002), it is most unfortunate that the presentation level (PL) used to present auditory materials is frequently not reported in the literature. In fact, in most published reports in which auditory issues are not the central focus, PLs are not reported. However, PL has been shown to affect the mental workload of processing speech tasks (Baldwin & Struckman-Johnson, 2002). When applicable, a discussion of the realized or potential impact of acoustic factors on the mental workload of auditory processing and the implications for strengthening understanding of the relevant cognitive mechanisms under investigation are discussed. First, we take a look at the role of auditory and verbal processing in historical developments within psychology.

HISTORICAL BEGINNINGS

The study of auditory processing, and language processing in particular, has informed scientists of the workings of the human brain for centuries. In his book *The Story of Psychology*, Morton Hunt (1993) described the first psychological study as the effort of a seventh-century Egyptian king named Psamtik I (Figure 6.1) to use the innate language abilities of two feral children to discover the identity of the original human race. According to Hunt, the king reasoned that if the children were raised without any exposure to language, then their first words would be those of the first or original race. Unfortunately, the king apparently confused general babbling with actual words and came to the conclusion that the original race was not Egyptian as he had hoped. Nevertheless, as Hunt pointed out, the king's attempt to study the human mind empirically was remarkable for the time.

FIGURE 6.1 Egyptian King Psamtik I.

According to Hunt (1993), verbal abilities were also the focus of Franz Josef Gall's (1758–1828) early interest in the brain and intellectual abilities. Gall purportedly observed, with annoyance, that some of his classmates in both grade school and college appeared to study less and yet achieve better grades. Gall reasoned that their superior performance in relation to their effort might be explained by more developed portions of the front part of their brain. "Evidence" for this came in Gall's observation that these individuals tended to have large, bulging eyes (no doubt being pushed out by the highly developed frontal brain areas). Gall's work in localizing mental functions, which came to be known as phrenology, was controversial largely due to his attempt to document intellectual differences between the races by correlating brain size with intelligence. Despite these spurious associations, Gall made significant contributions to scientific inquiries into the mind-brain relationship. Further, Gall's localization efforts included no less than four separate auditory-verbal areas. These included an area for (a) the memory of words; (b) the sense of language, speech; (c) the sense of sounds, the gift of music; and (d) poetic talent (see http://www.phrenology.com/franzjosephgall.html).

This "localizationist" perspective inspired Bouillaud's (1769–1881) search for the brain areas responsible for speech (see Kaitaro, 2001). Bouillard's work described an understanding that the ability to use and produce speech are two different processes. This theme continued in the work of Paul Broca (1825–1880), who documented the characteristics, symptoms, and subsequent autopsy results from his famous aphasic patient, Leborgne, popularized in subsequent literature as "Tan" since he frequently repeated that utterance (Finger, 1994, pp. 37–38).

In his book *The Origins of Neuroscience: A History of Explorations into Brain Function*, Stanley Finger (1994) pointed out that the tremendous influence that Broca's documentation of the Leborgne case had on the scientific community was likely due to the culmination of several factors. First, not only was Broca highly respected, but

FIGURE 6.2   Carl Wernicke.

also the time was right; it was a sort of tipping point or *zeitgeist,* as Finger called it. Whatever the reason, we see that Broca's documentation of Leborgne's language impairment marked the beginning of a new era in neuroscience.

Another example of the contributions of early auditory research in increasing knowledge of cognitive performance is found in the work of Carl Wernicke (1848–1905) (Figure 6.2). Wernicke is best known for his study of language impairments, or aphasias as they are commonly called. Wernicke documented several case studies of individuals with aphasic symptoms quite opposite those documented by Broca. Broca's account of aphasia involved individuals with an inability to produce speech despite a relatively intact ability to comprehend speech. Conversely, the cases examined by Wernicke included individuals with an inability to comprehend speech. These individuals were able to speak words fluently, although the resulting speech lacked conceptual coherence (Finger, 1994).

Wernicke's observations provided converging evidence for the localization of language function and the existence of separate processes for producing and comprehending speech. Less well known are Wernicke's writings considering the nature of conceptual knowledge. Wernicke believed that language and thought were separate processes (Gage & Hickok, 2005). Therefore, even in his earliest writings he was compelled to discuss the nature of conceptual representations in addition to language representations. As Gage and Hickok discussed, Wernicke's theories regarding cortical mechanisms involved in the representation of conceptual knowledge are strikingly similar to modern accounts by notable cognitive neuroscientists such as Squire (1986) and Damasio (1989). Similar themes appeared in an article by Jonides et al. (2005). The basic similarities between Wernicke's writing over a century ago and more recent publications include the idea that memories are represented in the same neural pathways initially used to encode the stimulus information. For example, processing and storing speech would utilize a broadly distributed set of neural pathways, including the auditory pathways activated by the auditory stimulus. Associations between auditory and visual concepts arise "coincidentally" when two sets of neural pathways are activated at the same time. Future activation of part of one of the pathways is sufficient to activate the entire neural trace in both modalities (see discussion in Gage & Hickok, 2005).

Wernicke's theories of conceptual neural representations seem far advanced for his time. Modern neuroscience techniques currently enable investigators to examine their plausibility. Since we are awaiting further research in these areas, we return to more pedestrian examples of the use of auditory tasks in cognitive research. We begin with an extension of the discussion of dichotic listening tasks from Chapter 4, citing recent applications that have furthered our knowledge of attention. This is followed by a discussion of a variety of auditory tasks that have been used as indices of mental workload. Included in this discussion is a review of common auditorily administered neurophysiological indices and neuropsychological tests designed to assess cortical processing and impairment. The focus here is not to present an exhaustive discussion of the many ways auditory tasks have been used but rather to provide sufficient support for the argument that they have played an integral role in human performance research, past and present. We begin by revisiting the topic of dichotic listening tasks introduced in Chapter 4.

DICHOTIC LISTENING TASKS

Dichotic listening tasks were originally primarily used to examine the nature of selective attention and its role in information processing. More recently, dichotic listening tasks have been used as a measure of cerebral dominance (Rahman, Cockburn, & Govier, 2008); as an index of attentional control (Andersson, Reinvang, Wehling, Hugdahl, & Lundervold, 2008); for examination of new theories of anticipatory and reactive attending (Jones, Johnston, & Puente, 2006; Jones, Moynihan, MacKenzie, & Puente, 2002); and as neuropsychological tests of attention capabilities in individuals suspected to exhibit attentional deficits (Diamond, 2005; Hale, Zaidel, McGough, Phillips, & McCracken, 2006; Shinn, Baran, Moncrieff, & Musiek, 2005).

Considerable empirical evidence indicates that mental effort is required to attend selectively to one auditory message in the presence of competing auditory stimuli (Moray, 1969; Ninio & Kahneman, 1974). The selective listening paradigm was used extensively to investigate attentional functioning during the 1950s and 1960s. The dichotic listening paradigm, in which separate auditory messages are presented to each of the two ears (C. Cherry, 1953a; Moray, 1959; Treisman, 1960, 1964b), was most common. More recently, dichotic listening tasks have been used to examine new theories of anticipatory and reactive attending (Jones et al., 2002, 2006) and as neuropsychological tests of attention capabilities in individuals suspected to exhibit attentional deficits.

In the typical dichotic listening task, listeners are asked to "shadow" or repeat aloud one of the two messages. Following presentation of the messages, listeners would generally be asked to recall the semantic content of the messages, to identify or answer content-related questions (Moray, 1969). Moray pointed out that selective listening tasks are commonly found in many real-world tasks, such as in the air traffic control (ATC) tower, where the ATC operator must selectively attend to one of several simultaneous messages. Similar tasks are also found in classrooms, offices, dispatch headquarters, and medical emergency rooms, to name a few. In the laboratory, another version of the dichotic listening task asks listeners to attend selectively to one of two or more messages and make responses according to some criterion, such as the presence of a word from a particular category (Kahneman, Ben-Ishai, & Lotan, 1973; Ninio & Kahneman, 1974).

A number of generalizations can be made from these investigations (Moray, 1969). The more similar the two messages are, the harder it will be to attend selectively to only one. Messages are more easily separable if they involve distinctly different loudness levels or pitch, if they arrive from distinctly different physical locations (including different ears such as in the dichotic listening paradigm), or even when location is perceptually but not physically different due to experimental manipulations of timing and intensity (Li, Daneman, Qi, & Schneider, 2004). Messages are also more easily separated if the gender of the speakers is different (Egan, Carterette, & Thwing, 1954; Treisman, 1964b).

Moray pointed out that using dichotic presentation to separate two messages (versus monaural presentation) is equivalent to increasing the signal-to-noise ratio (S/N) of the selected message by as much as 30 dB (Egan et al., 1954; Moray, 1969). Egan and colleagues used the articulation score (or the accuracy with which a message could be identified) as the dependent variable. Egan found that when two messages were spoken by the same speaker, were started at the same time, and were presented at the same intensity, the articulation score was 50%. However, using a high-pass filter on either the selected or unselected message improved selective listening, resulting in an articulation score of roughly 70% if the selected message was high-pass filtered and roughly 90% when the rejected message was filtered. These results suggest that greater interference will result when the competing message is similar to the attended message. It is easier to ignore a competing message if it is physically different from the attended message. This finding is important to consider when designing auditory messages in operational environments. As an illustration, in the flight cockpit, where male voices have been traditionally more prevalent, verbal warnings and messages presented in a female voice will be more salient.

ENCODING AND RETRIEVAL PROCESSES

An extensive body of literature now exists on the nature of encoding and retrieval of information in memory. Many of these investigations have utilized aurally presented word tasks (Craik, Govoni, Naveh-Benjamin, & Anderson, 1996). For example, Fergus Craik and colleagues (1996) examined the effects of divided attention during encoding and retrieval of aurally presented words. Their dual-task paradigm indicated that divided attention during the encoding of words significantly disrupted memory for those words, while divided attention during the retrieval of words increased the time needed to respond to a visual reaction time task.

In general, a concurrent task is more disruptive when people are trying to encode speech for later recall, rather than when the concurrent task occurs during retrieval

(Anderson et al., 1998; Naveh-Benjamin, Craik, Guez, & Dori, 1998). Both young and old participants experienced greater dual-task costs during auditory encoding; however, the effects were particularly dramatic for older listeners (Anderson et al., 1998). Interestingly, using the same voice to present words during both encoding and retrieval resulted in significantly better recall of words (Naveh-Benjamin et al., 1998) than using a different voice during retrieval.

The memory enhancement effects of presenting to-be-remembered items in the same voice during encoding and recall exemplifies the multiple memory trace models of speech comprehension. Several models of speech comprehension proposed that parallel memory traces are temporarily stored during speech processing (Ju & Luce, 2006; Luce, Goldinger, Auer, & Vitevitch, 2000; Wingfield & Tun, 1999). The redundant memory traces stemming from acoustic-phonological representation in conjunction with semantic-conceptual representations could be expected to improve memory.

AUDITORY TASK INDICES OF WORKING MEMORY

Working memory is commonly conceptualized as a short-term store for temporarily working with or manipulating information (Baddeley, 1992; Baddeley & Hitch, 1974). The term has essentially replaced the concept of short-term memory in much of the cognitive literature (Lobley, Baddeley, & Gathercole, 2005; Wingfield & Tun, 1999). We discussed the role of working memory in Chapter 4 but return to it in this chapter and focus particularly on some of the many auditory tasks that have been used in investigations of working memory processes. We begin with an extremely influential study that sparked much theoretical discussion and that continues to be the focus of much debate: Daneman and Carpenter's (1980) investigation of listening and reading span.

WORKING MEMORY CAPACITY: COMPLEX SPAN

Daneman and Carpenter published a seminal article in 1980 that addressed the role of individual differences in listening and reading abilities. It was commonly believed at the time that short-term memory must play a role in language comprehension, yet measures of short-term memory capacity (i.e., word span and digit span) appeared to share little relationship with measures of reading and listening comprehension (Daneman & Carpenter, 1980; Daneman & Merikle, 1996). Daneman and Carpenter's paradigm, which they termed the reading span and listening span task, required both processing and storage of verbal information. Rather than simply requiring participants to attempt to recall a string of unrelated words or digits (as in the word span and digit span tasks), the reading or listening span task requires people to process a series of sentences (presented either visually in the case of the reading span or aurally in the case of the listening span) and make a veracity judgment of each sentence while attempting to store the last word from each sentence. Originally, Daneman and Carpenter (1980) had people read a series of unrelated sentences at their own pace and then recall the final word from each sentence.

More recently, the reading span test has been modified somewhat following the example of Engle and colleagues (la Pointe & Engle, 1990; Unsworth & Engle, 2006). In their version, people are required to verify whether a sentence is grammatical or not and then to remember a stimulus item presented immediately after sentence verification. For example, a listener might read a sentence such as: "The senator was glad for the term recess so he could resume his watermelon ties," to which the listener should respond "No" because the sentence does not make grammatical sense. Then, an unrelated stimulus item such as "Yes" is presented for later recall.

Daneman and Carpenter's (1980) reading span and listening span task proved to be much more predictive of language comprehension abilities than digit or word span tasks. Numerous subsequent investigations have confirmed these observations (see a meta-analytical review in Daneman & Merikle, 1996).

LISTENING SPAN

Several investigations have implemented the listening span task in a variety of unique and productive ways. First, as Daneman and Carpenter (1980) originally intended, several investigators utilized the listening span task to examine individual differences in language comprehension in school-age children. For example, listening span tasks are often predictive of reading ability (Swanson & Howell, 2001) and the incidence and severity of learning disabilities (Henry, 2001) in school-age children.

The listening span task may also be predictive of the ability of children of school age to attend to and comprehend academic lessons on a daily basis. Morria and Sarll (2001) examined listening span performance of A-level students in the morning after they had skipped breakfast. Their performance was then compared again after consuming either a glucose drink or a placebo, saccharine-sweetened drink. Twenty minutes after consuming the drink, students who had drunk glucose relative to placebo performed significantly better on the listening span task. Performance in this glucose group improved despite nonsignificant changes in their blood sugar levels. Performance in the placebo group remained unchanged. Results of this investigation underscore the importance of children receiving adequate nutrition and breakfast, in particular (Benton & Jarvis, 2007; Muthayya et al., 2007). The cognitive benefits obtained from eating breakfast can be enhanced in school-age children by also including a midmorning snack, and these benefits are most dramatic for children of low socioeconomic status and those nutritionally at risk (Pollitt, 1995). The impact of morning nutrition on cognitive performance is not limited to children. Young college-age adults showed similar cognitive changes and have demonstrated significantly increased verbal reasoning skills following frequent small meals versus skipping meals altogether or consuming less-frequent larger meals (Hewlett, Smith, & Lucas, 2009). Performance on listening span tasks is also affected by PL, indicating that it is important to evaluate the mental effort involved in carrying out such tasks.

The importance of considering the mental workload of the auditory processing task when assessing listening span can be seen in an investigation of the influence of speech PL (how loudly the speech was presented) on performance (Baldwin & Ash, 2010). A direct intensity relationship was found between presentation intensity and working memory capacity. As intensity increased, assessed listening span capacity also increased. This direct relationship was particularly evident in a group of older (60–80 years), relative to young (18–31 years), listeners.

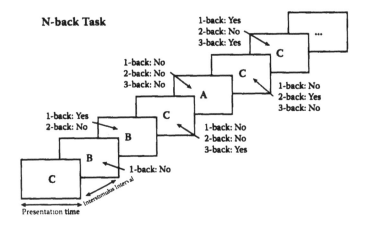

FIGURE 6.3   The *n*-back task.

AUDITORY *N*-BACK TASKS

Another commonly used task that is often presented in the auditory modality is the *n*-back task. The *n*-back task is commonly used to investigate various aspects of working memory. As illustrated in Figure 6.3, participants are asked to monitor an incoming stream of stimuli and make some response (such as indicating the presence or absence of a match to the current stimulus) to items *n* trials previously in the stream. For instance, in a 2-back task, participants are required to make some decision and response in conjunction with the stimulus presented 2 positions back. The difficulty of the task can be manipulated by varying the target position, with the easiest version being the 1-back task (where responses are made to the immediately preceding stimulus) and more difficult versions requiring responses to trials with more intervening stimuli.

Several versions of the *n*-back task have been developed and utilized. Stimuli may be verbal or nonverbal, and people may be asked to monitor either the identity or the location of the stimuli (Owen, McMillan, Laird, & Bullmore, 2005).

Figure 6.4 illustrates a version of the auditory-spatial *n*-back task. Note that by manipulating interaural intensity differences the letters presented seem to come from the left side, the right side, or the middle of the head. Researchers at the University of Helsinki in Finland have been using an auditory-spatial version of the *n*-back task such as the one illustrated in Figure 6.4 to investigate working memory in school-age children (Steenari et al., 2003; Vuontela et al., 2003). Using headphones, they achieved the perception that stimuli were coming from one direction or the other by using interaural intensity differences of approximately 17 dB. For example, to present stimuli that appear to be coming from the left, the left ear sound is presented 17 dB louder than that presented simultaneously to the right ear and vice versa for the simulated right position. Sounds of equal intensity appeared to come from the middle.

This auditory-spatial version of the *n*-back task has been used to examine the effects of sleep quantity and quality on school-age children (Steenari et al., 2003). Steenari and colleagues observed that poorer sleep quality (more activity during sleep as measured by an actigraph) was associated with poorer working memory performance at

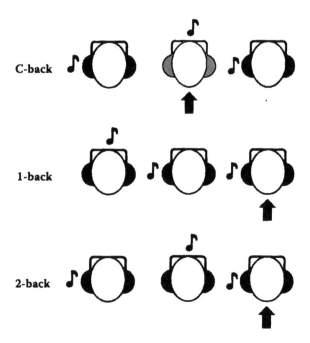

FIGURE 6.4   Auditory-spatial *n*-back task.

all task load levels. Shorter sleep durations were associated with poorer performance on the most difficult version (2-back) of the auditory-spatial task. They concluded that working memory in school age-children was related to both sleep quality and duration.

Using the same version of the auditory-spatial *n*-back task, Vuontela and colleagues (2003) observed that working memory performance improved across the age range of 6 to 13 years. They interpreted this as evidence for the maturation of cognitive mechanisms across this time frame. They also noted that working memory appeared to develop slightly faster in females, at least to age 10 years. Further, they noted that auditory-spatial working memory seems to mature more slowly than its visual counterpart.

Despite the reliance on speeded sensory-perceptual processing in these working memory tasks, all too frequently little if any care is taken to ensure that participants have been equated for sensory acuity. As previously discussed, recent work in our lab indicated that both sensory acuity and stimulus presentation characteristics affected listening span indices of complex span (Baldwin & Ash, in preparation). We return to a discussion of this issue later in this chapter. Now, we turn to another paradigm that has made considerable use of auditory tasks.

PSYCHOLOGICAL REFRACTORY PERIOD

The use of auditory tasks in dual-task studies often involves a phenomenon referred to as the psychological refractory period (PRP). The PRP paradigm requires people to make speeded responses to stimuli from two different tasks in rapid succession (see review in Pashler, 1998a; Welford, 1952). It is generally found that people take longer to respond to the second task when it follows the first task closely in time. Pashler (1998b) gave a helpful analogy to explain the predictions of the PRP paradigm with respect to processing bottlenecks. His analogy involved two people going into a bank, one right after the other. If the two people arrive at almost the same time and it takes the same amount of time for each person to get their banking done as it would have if they had arrived by themselves, then we could say that they accomplished the respective tasks in parallel. If, however, it takes longer for the person who arrives second, then there must be a bottleneck in the process. For example, the bottleneck might be the teller. The second person may need to wait for the first to finish with the teller. In this case, the first person would finish in the same amount of time, but the second one would take longer to accomplish the banking task.

Results of numerous investigations using the PRP paradigm have provided support for a central bottleneck theory of informational processing (Levy & Pashler, 2008; Levy, Pashler, & Boer, 2006; Pashler, 1998b). For example, using a visually presented memory task involving word pairs in conjunction with an auditory tone discrimination task, Pashler observed that memory retrieval was delayed when cue stimuli were presented in close temporal proximity to the auditory task stimuli (Carrier & Pashler, 1995). Results of this auditory-visual task paradigm indicated that retrieving items from memory is subject to a central processing bottleneck. That is, processing auditory stimuli from one task interfered with memory retrieval in a separate task, indicating that both tasks were relying on some shared mechanism and could not therefore be processed in parallel.

In a more recent investigation, Levy, Pashler, and colleagues (Levy & Pashler, 2008; Levy et al., 2006) investigated this relationship in the context of simulated driving while participants performed a concurrent task. It is well known that drivers often engage in a number of extraneous tasks while driving. Researchers sought to see how auditory processing might have an impact on brake response time. In one experiment, participants were asked to make speeded brake responses in a simulated driving task concurrently with a visual or auditory choice response task. Stimulus onset asynchrony (SOA), or the time between presentation of the stimuli from the two tasks, significantly affected brake response times. When the auditory or visual task stimuli were presented in close temporal proximity to the visual stimuli in the brake response task, participants took significantly longer to perform the brake response. The practical implications of this study for driving and performing other extraneous tasks concurrently are important. It appears that performing a visual or auditory task just prior to a brake event will slow response time.

In a subsequent study, Levy and Pashler (2008) used the change-task design in which two tasks are presented, but participants are told to prioritize them such that they are to stop processing and responding to the first task if the second task is presented. For example, as Levy and Pashler pointed out, a person might be performing the low-priority task of listening to the radio while driving, but if a lead car suddenly brakes in front of him or her, attention should be taken away from the listening task and devoted to the task of applying the brakes. People seemed to be able to withhold processing of and thus response to the first task some of the time but not consistently (Logan & Burkell, 1986). Levy and Pashler combined a choice auditory tone task

(Task 1 of low priority) with a brake response time task (high-priority Task 2). First, they observed substantial individual differences in people's ability to withhold making a response to the low-priority auditory tone task during the presence of a lead car braking event (the signal to perform the Task 2 brake responses) despite explicit instructions that the brake response task was to take priority over the tone task. The majority of participants (60%) failed to withhold the tone response over 80% of the time, while two other groups of moderate response rate (40–80% of the time) and low response (under 40%) each represented 20% of the sample. Clearly, people had a difficult time disengaging from the auditory tone-processing task and withholding their response. Levy and Pashler took this as evidence for a central processing bottleneck.

Second, Levy and Pashler (2008) observed that mistakenly performing the auditory task first significantly slowed brake response time. This effect is perhaps not surprising but has important practical implications. It suggests that even when people are clearly aware of the higher priority of a given task, they may be unable to stop themselves from completing a task of lower priority. This failure to inhibit attention toward lower-priority tasks affects their ability to perform a high-priority task. Consider when a driver engaged in a cell phone conversation takes the time to tell the caller that he or she has to end the call rather than just dropping the phone during a critical event. Perhaps the driver may even take the time to explain to the caller why he or she has to hang up. Evidence is clear that performance decrements occur whenever people attempt to perform two different tasks within close temporal succession, although the mechanisms behind this interference continue to be debated (Jentzsch, Leuthold, & Ulrich, 2007; Navon & Miller, 2002; Pashler, Harris, & Nuechterlein, 2008; Schubert, Fischer, & Stelzel, 2008).

Next, we examine the use of auditory tasks in dual-task paradigms for a different purpose: assessing the mental workload associated with performing a task or using a system. In these cases, the auditory task may be called a secondary task or subsidiary task, and performance measures of response time and accuracy are usually assessed. Other investigations aimed at assessing mental workload have used auditory evoked potentials or event-related potentials (ERPs) in response to auditory stimuli. We turn now to a discussion of some of the commonly used auditory tasks in mental workload research.

MENTAL WORKLOAD ASSESSMENT

Mental workload assessment, as described in some detail in Chapter 5, has played an integral role in investigations of human performance. Auditory tasks have frequently been used in these investigations, often because the task or system of interest (i.e., driving or piloting) already placed heavy demands on visual processing channels. In this section, we discuss some of the commonly used auditory secondary tasks and auditory ERP tasks that have been employed in mental workload assessment.

AUDITORY SECONDARY TASKS

Tasks such as logical sentence verification, mental arithmetic, and delayed digit recall, in which the stimuli are presented acoustically, can and have been used as secondary tasks to investigate workload in a wide variety of environments. Auditory secondary tasks have been used to examine new cockpit displays and for examining the affect of environmental factors (such as the difficulty of driving through complex roadways of varying levels of traffic density) on mental workload during driving. For example, a secondary task consisting of auditory mental arithmetic has been found to be sensitive to the mental workload requirements of driving in urban versus rural settings (Harms, 1991). Delayed recall of acoustically presented digits was found to be a sensitive index of changes in mental workload stemming from changes in traffic density and adverse weather (Zeitlin, 1995).

Delayed Digit Recall

The delayed digit recall task has a long history of use, introduced as early as 1959 by Jane Mackworth (1959). It is similar to the n-back task described previously; however, in the delayed digit recall, task digits from 1 to 9 are presented in a random sequence at specific intervals. Participants are then required to say aloud a previously presented digit (Zeitlin, 1995). As with the n-back task, the difficulty of the delayed recall task can be manipulated through specifications of the digit to be recalled. For example, in the easiest condition the participant would simply be required to repeat the last digit presented. A more complex version of the task requires participants to repeat digits presented earlier in the sequence (i.e., two digits before the last digit spoken) or to repeat digits that followed a specified target digit (i.e., the second digit following the last presentation of the number 3 in the sequence).

The delayed digit recall task has been shown to be sensitive to changes in the driving environment in a 4-year investigation of the mental workload of commuters in the New York City area (Zeitlin, 1995). Zeitlin's implementation used a presentation rate of one digit every 2 s for a period of 2 min. Participants were required to say the digit preceding the last digit presented and were required to reach a performance criterion of 98% accuracy in a no-load condition prior to participating in the experimental trials.

Mental Arithmetic

Mental arithmetic tasks have been used extensively as secondary task measures of mental workload. In both laboratory environments (Garvey & Knowles, 1954) and field investigations (Harms, 1991), mental arithmetic tasks have been sensitive to changes in the difficulty of the primary task without disrupting primary task performance. By using auditory inputs for the arithmetic task in combination with spoken responses, there is little chance of direct interference with most visual/motor tasks (Knowles, 1963), making them well suited for a number of operator tasks, such as piloting and driving.

The difficulty of the mental arithmetic task can be varied in many ways. In one version used by Kahneman and colleagues (1969), a prompt, "ready," preceded presentation of three randomized digits. Listeners were instructed to keep each of the digits in memory until a second prompt, "now," was heard. On hearing "now," listeners were asked to add 3 to each of the three preceding digits and verbally report the transformed calculation in order. In one investigation designed specifically to examine the sensitivity and intrusiveness of different workload assessment techniques, the version of the mental arithmetic task modeled after the one previously implemented by Kahneman was not found to be sensitive to changes in flight simulator difficulty manipulated by increasing wind gust disturbance (Wierwille & Connor, 1983). Wierwille and Connor estimated that a sample size of at least 25 participants would be needed for that form of mental arithmetic task to indicate statistically significant sensitivity to the primary flight task difficulty. However, other investigations using a different form of mental arithmetic task have demonstrated sensitivity to primary task difficulty with fewer participants (Baldwin & Schieber, 1995; Harms, 1991).

Baldwin and Schieber (1995) and Harms (1991) have used a version of mental arithmetic in which two-digit numbers are presented and listeners are required to subtract the smaller from the larger of the set (i.e., if "57" is presented, listeners should subtract the 5 from the 7 and verbally report the result, "2"). This subtraction version has been found to be sensitive to changes in primary task driving difficulty in both driving simulation (Baldwin & Schieber, 1995) and on-the-road investigations of driving (Harms, 1991).

Auditory versions of the *n*-back task discussed in this chapter, as well as a wide range of other tasks presented in the auditory modality, have been used in mental workload assessment (see review in Ogden et al., 1979). A specific class of these tasks, those used in conjunction with ERP techniques, is discussed next.

AUDITORY ERP INDICES OF MENTAL WORKLOAD

Auditory ERP paradigms have been developed to examine mental workload in a number of occupational environments. In particular, the P300 component, a positive wave deflection occurring approximately 300 ms after an infrequent or unexpected stimulus event, is frequently found to be sensitive to the attentional processing requirements of a given task or task set. The P300 is said to be an endogenous component reflecting attentional processing. During divided attention tasks, a decrease in the P300 component in response to a secondary task event has been found to be indicative of greater attentional costs associated with a primary task. In other words, as the primary task becomes more mentally demanding, fewer resources are available for processing the secondary task, and this is demonstrated by a P300 of decreased amplitude to the secondary task event (Kramer et al., 1987). Based on this relationship, a technique called the auditory oddball paradigm, which is discussed in the next section, has been used extensively as a physiological index of mental workload.

Auditory Oddball Paradigm

The auditory oddball paradigm capitalizes on the observation that humans are geared to devoting attentional resources to novel stimuli. Novel stimuli can be said to capture attention and therefore result in the utilization of mental resources for processing. In the auditory oddball paradigm, a series of tones is presented. The listener is instructed to ignore all but a distinct target tone, which is presented periodically but considerably less frequently than the standard or distractor tones. For example, the distractor tone may be presented 80% of the time, with the target tone presented in only 20% of the targets. Typically, target tones illicit a P300 component of relatively greater amplitude than is observed for the distractor tones. The observed increased amplitude to the novel stimuli is interpreted as demonstrating that the target tone is being processed. However, as more attentional resources are required by a concurrent task, the relative increase in P300 amplitude to the target tones relative to the distractor tones diminishes. Kramer and colleagues (1987) demonstrated the effectiveness of the auditory oddball paradigm to distinguish between levels of operator workload during simulated flight missions. Specifically, they found that P300 amplitude decreased with increases in flight task difficulty manipulated through wind turbulence and segment (e.g., taking off and landing vs. level flight).

The auditory P300 has also been used in the evaluation of the difficulty of text presented in hypermedia systems (Schultheis & Jameson, 2004). Schultheis and Jameson paired an auditory oddball task with easy and difficult versions of text and measured pupil diameter and the P300 response to the oddball task. P300 amplitude, but not pupil diameter, was significantly reduced for the difficult hypermedia condition. The authors concluded that auditory P300 amplitude and other measures, such as reading speed, may be combined to evaluate the relative ease of use of different hypermedia systems and perhaps even to adaptively vary text difficulty dependent on the cognitive workload experienced by the user.

In another example of the use of the auditory P300, Baldwin and Coyne (2005) found that P300 amplitude was sensitive to the increased difficulty of driving in poor visibility due to fog versus clear visibility, while performance-based and subjective indices were not. They developed analogous auditory and visual versions of the P300 task that could be used to compare displays of different modalities without overtaxing the primary task sensory modality. However, using a cross-modal oddball task did not demonstrate adequate sensitivity in several flight and driving simulation investigations; therefore, caution must be exercised in making recommendations for this procedure. Another auditory version of an ERP task that has been used successfully to assess mental workload is the irrelevant probe task.

Irrelevant Probe Task

In the auditory irrelevant probe task, people are asked to ignore a periodic auditory probe. The probe is thus irrelevant to the task the person is performing. However, ERP responses to these irrelevant probes can sometimes be used effectively as indices of mental workload. For example, the N1 component in response to an auditory irrelevant probe task was sensitive to more difficult sections of a flight task (Kramer et al., 1987). The benefit of the irrelevant probe task compared to many others is that it does not require the operator to perform any other tasks.

These examples clearly point to the utility of the auditory tasks in conjunction with ERP techniques in assessing cognitive workload in a wide variety of domains. The auditory oddball task has also been used in neuropsychological assessment of cognitive status, as discussed next.

AUDITORY NEUROPSYCHOLOGICAL TESTS

Auditory tasks are frequently used in neuropsychological examinations (Sjogren, Christrup, Petersen, & Hojsted, 2005; Tombaugh, 2006; White, Hutchens, & Lubar, 2005). For example, the Paced Auditory Serial Addition Task (PASAT) is a common

neuropsychological test of information processing and working memory, and auditory verbal learning tests (AVLTs), such as the Rey Auditory Verbal Learning Test,
are frequently used to assess cognitive function. Several neuropsychological assessment batteries contain one or more auditory or verbal components. For example, of
the 10 subtests of the Halstead-Reitan Neuropsychological Test Battery (HRNTB),
one (the Rhythm test) relies on auditory processing of pairs of nonverbal sounds. A
second subtest requires recognition of auditorily presented nonsense words.

PACED AUDITORY SERIAL ADDITION TASK

The PASAT was originally developed as a tool for assessing cognitive function and
information processing speed in particular, following traumatic brain injury (for a
review, see Tombaugh, 2006). It involves presenting a series of single-digit numbers
and having the patient or participant sum each two consecutive numbers, reporting
aloud the answers. So, for example, if 5, 7, 3, and 9 were presented, the correct verbal
responses would be 12, 10, and 12. As discussed in a review by Tombaugh (2006),
the PASAT is a commonly used neuropsychological test of attention, but it has been
found to be affected by such factors as age and speech and language abilities. As
discussed in detail in Chapter 10, age-related changes in cognitive performance
void of careful assessment of sensory acuity must be regarded with extreme caution.
Degradations in performance associated with age could be attributed to peripheral
or central hearing mechanisms rather than cognitive deficits. As suggested by SCIT
(Baldwin, 2002) and the effortfulness hypothesis (McCoy et al., 2005; Wingfield,
Tun, & McCoy, 2005), the additional mental effort required to process degraded sensory stimuli may deplete resources from other, later stages of processing. In this way,
declining sensory abilities could exacerbate or be mistaken for cognitive impairments (Baldwin, 2002; Baldwin & Ash, 2010; Baldwin & Struckman-Johnson,
2002; Valentijn et al., 2005).

Clinical evidence for the importance of assessing auditory acuity in AVLTs comes
from work conducted in conjunction with the Maastricht Aging Study (van Boxtel
et al., 2000). Using an auditory verbal learning paradigm and controlling for factors
such as age, educational level, and speed of processing, van Boxtel and colleagues
observed that hearing acuity was a strong predictor of performance on auditorily
administered verbal learning tests.

An additional caveat is that performance on the PASAT often varies considerably
depending on the modality of stimulus presentation. As might be predicted from a
multiple-resource theory perspective (Wickens, 1984), recent evidence indicated that
people may score substantially better when the test is administered in visual rather
than auditory format (Wickens, 1984). This discrepancy in performance obtained
with a visual versus an aural presentation modality suggests that resource competition between mechanisms responsible for encoding the auditory information and
making a verbal response may be at least in part what the PASAT is assessing. This
would suggest the PASAT may be assessing the ability to switch auditory attention
rapidly rather than merely assessing information-processing speed alone.

For example, the PASAT has been used to examine the impact of pharmaceu-
ticals, such as opioids, on cognitive function in patient populations experiencing

chronic pain (Sjogren et al., 2005). The PASAT has also been used as a means of examining working memory processing in clinical populations with attention deficit/hyperactivity disorder (ADHD) (White et al., 2005).

REY AUDITORY VERBAL LEARNING TEST

The Rey Auditory Verbal Learning Test (RAVLT) is one of the most commonly used AVLTs for neuropsychological assessment (Baños, Elliott, & Schmitt, 2005; Poreh, 2005). The RAVLT is also one of the oldest, and its list-learning format has formed the basis for subsequent AVLTs, such as the California Verbal Learning Test, the Wechsler Memory Scale (WMS-III), and the Hopkins Verbal Learning Test. It consists of reading aloud a list of 15 words (List A), asking for free recall, and then presenting the same list four more times with a free recall trial between each. Then, a second list (List B) is presented, followed by free recall. People are then asked to recall List A without it being presented again. In this way, the first recall trial of List A is similar to other immediate memory tests, such as the digit span task. List B is a form of interference. The RAVLT is easy to administer, taking on the order of 10–15 min.

HALSTEAD-REITAN NEUROPSYCHOLOGICAL TEST BATTERY

The HRNTB is commonly used to assess and diagnose neurological impairment due to factors such as brain trauma, mental disorder, and alcoholism (Allen et al., 1999; Horton, 2000; Sweeney, 1999). It consists of several tests and makes use of visual, auditory, and tactile presentation modalities (Reitan & Wolfson, 2000, 2005). The two auditory tests include the Speech-Sounds Reception Test (SSRT) and the Seashore Rhythm Test.

Speech-Sounds Reception Test

The speech-sounds reception test involves auditorily presenting prerecorded nonsense words that rhyme with the "ee" sound. Listeners are required to identify the corresponding letter representations from a set of nonsense words. The test is similar to the Speech Perception in Noise (SPIN) test (Plomp & Mimpen, 1979) and its revised version R-SPIN, except that SPIN tests present the speech sounds in carrier sentences along with varying levels of noise.

Seashore Rhythm Test

The Seashore test was originally part of Carl Seashore's 1939 Measures of Musical Talent test that was administered in an effort to discover musical prodigies. The Seashore Rhythm Test is now included in the HRNTB and consists of discriminating between rhythmic patterns of beats. The modified Seashore Rhythm Test is frequently used as an index of neurological impairment and attentional processing abilities. Taylor-Cooke and Fastenau (2004), for instance, found that the Seashore Rhythm Test distinguished attentional processing deficits in children with epilepsy relative to controls. However, the clinical validity of the Seashore test is sometimes questioned (see review in Sherer, Parsons, Nixon, & Adams, 1991).

There are numerous other auditory tasks that have been used for neuropsychological assessment. But, the list of tasks described in this section should serve to provide an adequate account for the present purposes. We now change to a focus on auditory tasks that have been used in neurophysiological investigations.

NEUROPHYSIOLOGICAL INVESTIGATIONS

As in other arenas, a variety of auditory tasks has been used in neurophysiological investigations of perceptual and cognitive processing. These tasks have ranged from simple auditory clicks that measure brain stem response, to tones, musical patterns, rhythms, words, and sentences.

PREPULSE INHIBITION

Prepulse inhibition (PPI) refers to a reduction in the startle response to a strong sensory stimulus when the stimulus is preceded by a weaker stimulus (Filion & Poje, 2003; Schall & Ward, 1996). Frances Graham (1975) proposed that the PPI might be used as a measure of central processing level. Since that time, PPI has demonstrated sensitivity to a number of clinical pathologies, including schizophrenia and autism.

PPI is a sensitive index of central inhibitory mechanisms or sensorimotor gating (Grillon, Ameli, Charney, Krystal, & Braff, 1992; Perry, Geyer, & Braff, 1999) and central serotonergic functioning (Quednow, Kuhn, Hoenig, Maier, & Wagner, 2004). Generally, patients with schizophrenia demonstrate reduced PPI (Grillon et al., 1992; Kumari, Aasen, & Sharma, 2004; Perry et al., 1999), although Kumari et al. (2004) found that the PPI of female patients with schizophrenia did not differ from normal female controls. Schizophrenia aside, PPI shows consistent sex differences, with females exhibiting lower PPIs than males in both rats and humans (Kumari et al., 2004; Rahman, Kumari, & Wilson, 2003).

Additional neurophysiological investigations relying on auditory stimuli have been used outside clinical settings. One major category includes auditory evoked potentials or ERPs stemming from auditory stimuli.

EVENT-RELATED POTENTIALS

Components of ERPs in response to auditory stimuli have been used extensively as indices of various stages and aspects of human information processing. ERP components, such as the mismatch negativity component (MMN) and N1, have been used to examine the nature of selective attention (see Naatanen & Alho, 2004, for a review). ERP components have also been used to examine a diverse range of issues, including, but not limited to, distractibility, sleep deprivation, alcoholism, dementia, and schizophrenia. I do not attempt to provide a comprehensive list of the various ways that auditory tasks have been used to examine cognitive processing for this task is well beyond our current scope. Rather, my aim is to highlight some of the many ways that auditory tasks have been used and then to further elaborate on the importance of considering the impact of acoustic characteristics on the mental workload requirements of these tasks. For example, P300 amplitude and latency are sensitive to both mental workload and to the amplitude of the stimulus; thus, care must be used when interpreting ERP results in neuropsychological evaluations. In general, P300 latency decreases and amplitude increases as auditory stimulus intensity increases (Polich, Ellerson, & Cohen, 1996), which may unfairly disadvantage older, hearing-impaired listeners. Despite this caution, auditory tasks are frequently used to assess brain function.

Distractibility

Auditory tasks have been used to examine the distractibility of different age groups. For example, ERPs in response to an auditory tone while children were engaged in a visual task have been shown to correlate with behavioral measures of distractibility (Gumenyuk, Korzyukov, Alho, Escera, & Naatanen, 2004). Younger children (8–9 years) were more distracted by irrelevant auditory tones than slightly older children (10–11 years and 12–13 years), as demonstrated by increased response time in the visual task and increased amplitude P300 responses to novel auditory tones.

Diagnostic Uses of MMN

The MMN component is an auditory evoked brain response elicited by any discriminable change in repetitive stimuli. It is assumed to be based on an automatic comparison between the infrequently presented stimulus and an auditory sensory memory trace of the frequent sounds (Alain, Achim, & Woods, 1999; Brattico, Winkler, Naatanen, Paavilainen, & Tervaniemi, 2002; Giard et al., 1995; Koelsch et al., 2001). The MMN typically peaks approximately 100–180 ms after presentation of a stimulus that deviates in any discriminable way from a standard held in auditory sensory memory (Cowan, Winkler, Teder, & Naatanen, 1993). MMN typically increases in amplitude and decreases in latency in relation to the stimulus deviation magnitude (Tiitinen, May, Reinikainen, & Naatanen, 1994). Specifically, as the probability of the deviant stimulus increases or if the repetitive stimulus has greater temporal variation, then MMN is attenuated. Other ERP components that detect novel stimuli (i.e., the N1) increase in amplitude only when a new acoustic element is detected, while MMN occurs with both the addition and the removal of acoustic elements (Cowan et al., 1993).

Auditory Assessments for Nonverbal Individuals

It is believed that the MMN component does not require focused attention on auditory stimuli and does not require an overt verbal or manual response (Naatanen, 1992). However, there is some controversy on this point for it has been shown that under certain conditions MMN amplitude can be modulated by focused attention (Woldorff, Hillyard, Gallen, Hampson, & Bloom, 1998). Nevertheless, there is consensus that MMN generation does not require full attention, and that it can be obtained in conditions when the subject cannot or is unable to pay attention. The MMN is therefore useful in assessing auditory functioning in infants and other nonverbal (i.e., comatose) populations. It has also been used to diagnose or assess impairments related to a number of other factors, discussed subsequently.

Dorsolateral Prefrontal Lesions

For example, the MMN is reduced in individuals with lesions of the dorsolateral prefrontal cortex (DLPFC; see Naatanen & Alho, 2004; Swick, 2005). Alho and colleagues compared MMN responses to standard 1,000 Hz tones and occasional deviants of 1,300 Hz in individuals with DLPFC lesions and their age-matched controls. Individuals with lesions exhibited reduced MMN responses, particularly when the deviant tones were presented to the ear ipsilateral to the lesion (Alho, Woods, Algazi, Knight, & Naatanen, 1994).

Sleep Deprivation

The MMN is sensitive to sleep deprivation. For example, MMN amplitude is present even after prolonged sustained wakefulness. However, Raz and colleagues found that MMN amplitude decreased gradually as participants experienced sustained wakefulness for 24 and 36 hours (Raz, Deouell, & Bentin, 2001).

Auditory tasks have also been used in numerous investigations of age-related changes in cognitive abilities. Because of the well-known relationship between hearing loss and advanced age (Corso, 1963a; Fozard & Gordon-Salant, 2001), considering the mental workload requirements of auditory processing is particularly important in these investigations.

AUDITORY TASKS IN COGNITIVE AGING RESEARCH

Cognitive aging research is a domain in which it is particularly essential that careful consideration is given to the characteristics of the acoustic stimuli to be used and the auditory acuity of the participants involved. It is well documented that aging is accompanied by decreased auditory acuity (Fozard & Gordon-Salant, 2001; Humes & Christopherson, 1991; Kline & Scialfa, 1996, 1997; Rabbitt, 1991; Schieber & Baldwin, 1996). As previously indicated, empirical evidence indicated that the acoustic PL of stimuli has an impact on the mental workload requirements of the processing task (Baldwin & Struckman-Johnson, 2002). Therefore, neglecting the impact on performance of a participant's sensory acuity and the PLs of the auditory stimuli involved can seriously undermine interpretation of any results obtained. Overlooking the potential influence of distracting acoustic environments is also problematic as the performance of older listeners is disrupted considerably more than that of younger adults by adverse listening conditions. Unfortunately, studies using auditory tasks often continue to disregard these basic sensory-acoustic parameters.

Examples of investigations in which researchers neglected to assess or at least report the sensory acuity levels of participants and the PLs used for auditory stimuli are abundant. For example, in a recent study, Federmeier and colleagues (2003) used ERPs to chart the time course of acoustic, lexical access, and word context among young and older participants in a sentence-processing task. Their results indicated that older adults on average took 25 ms longer to process the stimulus in the early sensory stages. They then appeared to "make up" for this lost time in the lexical access stage (400 ms postonset), only ultimately to spend an extra 200 ms to process the message-level contextual factors of a sentence. Their findings provided important cues to differences in the time course of spoken sentence processing between age groups. However, their failure to report either the sensory acuity of their participants or the PLs of their sentences makes it difficult to tease out the full impact of peripheral versus central processing factors. The early processing delay (occurring within the first 200 ms) could be related to age-related changes in the auditory association cortex, as they suggested. Alternatively, this early delay could be due to age-related changes in the outer, middle, or inner ear that result in decreased sensitivity thresholds, thereby resulting in an attenuated signal reaching the auditory association cortex. Therefore, while providing insight into the contribution of sensory and semantic contextual influences on the time course of spoken word processing, their results cannot be as fully appreciated as they might be if age-related differences in pure-tone sensitivity had been taken into account. These age-related sensory-cognitive interactions are discussed in further detail in Chapter 10.

SUMMARY

Auditory tasks are frequently used in cognitive research and, as pointed out in this chapter, many noncognitive factors have an impact on performance of these tasks. In the present chapter, it was only possible to discuss a few of the many auditory tasks that have been used in general cognitive research, research aimed at assessing mental workload, and clinical diagnostic testing. It is hoped that some of the issues raised here help illustrate the importance of understanding the many factors that affect auditory cognition. In particular, prominent issues such as how loudly the stimuli are presented and the hearing abilities of the listeners can have profound influence on performance on these auditory tasks.

CONCLUDING REMARKS

Most of the auditory tasks discussed in the current chapter have involved spoken verbal material. An exception to this was the auditory ERP research involving auditory probes, which are generally tones. In the next chapter, additional attention is devoted to characteristics that influence the mental effort involved in processing nonverbal sounds.

7 Nonverbal Sounds and Workload

INTRODUCTION

Nonverbal sounds are an integral, if underappreciated, part of human experience. When we walk into a room, the echo and reverberations of our footsteps provide important clues to the nature of our environment. The sound of the wind reminds us to wrap up before going outside; sounds can alert us to oncoming vehicles and approaching friends. Nonverbal sounds complement, supplement, and under certain conditions supplant visual information.

Music represents one of the more popular forms of nonverbal sounds. Music perception is a significant cultural accomplishment and may play a fundamental role in many forms of human adaptation. One example is the role of music in facilitating parental commitment through infant bonding and promoting infant mood states associated with optimal growth and development (see Trehub, 2001). Our auditory systems are predisposed to efficient processing of universal musical patterns, suggesting that musical systems (e.g., conventional Western music) have developed to capitalize on efficiencies of the auditory system rather than vice versa. These ideas are explored in more detail in this chapter, which also examines the attention-enhancing role of certain musical structures and the distracting effects of unwanted sound. The chapter focuses on issues such as the mental effort involved in detecting and categorizing sounds, auditory imagery, music perception, noise, and the impact of both music and noise on performance.

In Chapter 3, the topic of auditory stream segregation was introduced, including discussion of some basic properties of perceptual grouping that enable us to distinguish between multiple competing sources of sound. We continue that discussion here by focusing on many of the mechanisms used to organize nonverbal sounds in our environment.

AUDITORY PERCEPTUAL ORGANIZATION

The auditory system uses a number of sophisticated perceptual organizational principles to make sense of the sounds around us, including identifying whether we are hearing one or two different sources and which sounds belong to which source. These principles assist in the process referred to as auditory stream segregation, discussed previously in Chapter 3 (and see Bregman, 1990, for a comprehensive review). These processes make use of regularities in the acoustic information and are briefly reviewed here.

Most sounds outside the laboratory are complex: They are made up of a base or fundamental sound of a particular frequency and a complex blend of harmonics (multiples of the fundamental frequency). For example, a complex sound with a fundamental frequency of 200 Hz would also likely consist of harmonics at 400, 600, 800, and perhaps 1,000 Hz. We make use of these harmonics in segregating acoustic streams. Our auditory systems are capable of engaging in complex spectral analysis and are able to separate multiple concurrent frequencies into their respective sources, partly based on recognizing different harmonic groupings.

Alain, Arnott, and Picton (2001) demonstrated how harmonics are used to segregate auditory streams. They asked listeners to decide whether they were listening to one or two sounds. The sounds presented to listeners consisted of a fundamental frequency of either 200 or 400 Hz and then a series of 12 harmonics that were either all in tune (400, 600, etc.) or with a second harmonic that was mistuned by a small amount (1–16%). The proportion of listeners reporting that they heard two sounds (rather than one) increased as the percentage of mistuning increased. When the second harmonic was mistuned by at least 8%, the overwhelming majority of listeners reported that they were hearing two sounds. This experiment provided convincing evidence that spectral analysis of harmonics can be used to aid auditory stream segregation.

Stimulus onset time, location, and pitch are used in a similar manner. It is highly unlikely that sounds coming from unrelated sources will start and stop at the same time, much less be coming from exactly the same location (Bregman, 1993). Likewise, auditory stream segregation is aided by the fact that when a sound changes its frequency, all of the partials or harmonics of the sound change in the same way. Therefore, if two complex but overlapping series of sounds are being heard and their partials are changing pitch (becoming higher or lower in pitch) at different ratios, the two streams can be easily separated (Bregman, 1993). Pitch continuity or the tendency for these pitch changes to be fluent, gradual changes also aids auditory stream segregation.

AUDITORY OBJECT RECOGNITION

Humans can determine through sound alone much about their environment, including the type of objects it contains and their actions. McAdams (1993) provided a vivid example. He asked us to imagine the sounds associated with a pile of ceramic dinner plates falling off a counter, "tumbling through the air knocking against one another, and finally crashing on to a relatively hard surface upon which all but one of the plates break—the unbroken one is heard turning on the floor and then finally coming to a rest" (p. 146). Forming an auditory image of this event was probably relatively easy for most of us. We are able to determine a surprising amount about an object by sound alone, including what it is: if it is a common object, its size, what it is made of, and its stability, for example.

Auditory object recognition has been studied from two primary theoretical approaches: information processing (Broadbent, 1958) and ecological (Gibson, 1966), with considerably more research investigations framed within the former (McAdams, 1993). Information-processing approaches assume that recognition occurs in a series of stages, beginning with basic acoustic analysis of stimulus properties and then culminating in later cognitive stages of recognition based on stored memory patterns. Ecological acoustics, conversely, assumes that our auditory systems are well adapted for directly perceiving the invariant acoustic information necessary for perception of relevant biologically significant or experientially salient events and objects. From an ecological perspective, it is not necessary to analyze sounds into their constituent parts and match them to memory; rather, the overall structure of the sound is perceived directly. Information-processing approaches assume a memory component.

INFORMATION-PROCESSING APPROACHES

Information processing approaches to auditory object recognition generally assume that recognition begins with analysis of the acoustic features or stimulus properties (McAdams, 1993; Molholm, Ritter, Javitt, & Foxe, 2004; Murray, Camen, Andino, Clarke, & Bovet, 2006), often referred to as data-driven processing. This initial stage includes the translation of information, such as the frequency, intensity, timbre, and duration of an auditory event into a neural code, a process referred to as *transduction*. Subsequent stages rely on auditory grouping, analysis of features, and the matching of auditory patterns with information stored in the auditory lexicon. Depending on the specific theory of prescription, these stages may or may not be interactive—meaning that higher-order stages may shape lower-level analysis and grouping in some theoretical accounts.

The ability to identify specific objects or states of objects is developed with experience. A trained mechanic learns to recognize the sounds of various automobile components: how they differ among different vehicles and whether a component is functioning properly. The layperson may or may not know what component is making the strange clicking or clunking sound when trying to describe it to the mechanic. Yet, this same person is demonstrating that he or she knows the sound is not a normal part of the auditory "soundscape" produced by the vehicle.

Auditory object recognition (for nonverbal sounds) has received considerably less attention than visual object recognition or speech recognition, but recently some important progress has been made. This research has sought to determine the features used to distinguish between different auditory objects, the extent to which auditory objects are processed like visual objects, and whether grouping auditory features into objects helps in their retention (Griffiths & Warren, 2004).

Dyson and Ishfaq (2008) suggested that object-based coding occurs for both auditory and visual stimuli. It has been known for some time that information in visual short-term memory tends to be stored in the form of objects rather than individual features (Duncan, 1984; Luck & Vogel, 1997). Luck and Vogel, for instance, found that we can store about four objects containing at least two features in memory or four individual features. This strongly suggests that visual short-term memory is object based. More recently, Dyson and Ishfaq observed a similar phenomenon with auditory information processing. People were concurrently presented with two different sounds and then asked about features present either within or across the different sounds. People were much faster at responding if the two dimensions in question were from the same object rather than different objects. Such an "object superiority" effect has often been observed in studies of visual perception and attention (Duncan, 1984). Rather than storing sound features independently, this suggests sound features are grouped into auditory objects in much the same way that visual features are stored as objects.

To answer basic questions (e.g., how good people are at recognizing everyday sounds, how long it takes to recognize them, and which features are used), Ballas (1993) examined listeners' perceptual-cognitive ratings of everyday sounds with different acoustical properties. He compared these ratings to their identification responses and the naturally occurring frequency of each sound. Ballas observed that on average it took people roughly 1–7 s to listen to and identify everyday sounds such as a doorbell ringing, a toilet flushing, hammering, and bacon frying. He found that people use a combination of acoustic, perceptual, and cognitive characteristics to identify sounds. Sound identification accuracy was strongly associated with how long it took someone to identify a given sound (identification time). In general, the longer it took people to identify a sound, the less sure they tended to be of the source of the sound and in turn the less likely they were to be accurate. Identification accuracy was significantly associated with a number of bottom-up acoustic features of spectral-temporal properties, including the presence of harmonics, continuous bands, and similar spectral bursts. The presence of harmonics in continuous sound coupled with the number of spectral bursts in noncontinuous sound together was strongly correlated with both identification accuracy and identification time. Acoustic factors such as these could explain about 50% of the variance in identification time. Top-down factors such as familiarity, prior experience, or exposure ratings based on subjective and objective estimates of how frequently the sound is heard and estimates of how many different sources could make the sound were even stronger predictors of identification time. Together, four different aspects of a sound—its spectral, temporal, envelope (e.g., the ratio of burst durations to the total duration), and frequency of occurrence—could account for 75% of the variance in its identification time.

Identification accuracy and time were also associated with the number of possible sources of a given sound, which can also be termed its *source neighborhood* (Ballas, 1993). For example, a clicking noise could be caused by many different things and would tend to take longer for people to attempt to identify, and they would more often be inaccurate.

Aldrich, Hellier, and Edworthy (2009) examined the descriptive and acoustic features that are important to sound identification. Based on past literature, they examined the characteristics of loudness, spectral spread, bandwidth, and pitch, as well as familiarity, in people's judgments of similarity and grouping of complex real-world sounds. They also compared two methodologies: paired comparisons and grouping methods. Considerable overlap was observed between the two methods. People tended to classify sounds based on categories consisting of the source or function of the sound (birds, animals, human nonverbal noises, etc.) somewhat more when using the grouping method than the paired comparison method. Use of acoustic and descriptive features was also observed for both methods. They found that root mean square (RMS) power (a measure of sound intensity) was a particularly important acoustic characteristic in sound classification.

Ballas (1993) found that the power of a sound was correlated not only with loudness but also with perceptual-cognitive characteristics such as hardness, angularity, sharpness, tenseness, and unpleasantness. In fact, strong correlations were found between a number of acoustic aspects and ratings of the perceptual-cognitive characteristics of a sound. For example, the degree of concentration of sound in the octave band centered at 2,500 Hz was strongly related to the relaxed versus tense rating of the sound. The amount of energy in frequencies above 3,150 Hz and between 1,100 and 2,500 Hz was strongly related to ratings on a dull-sharp dimension.

The strong relationship between familiarity with a sound and the speed and accuracy with which it can be named as well as the interaction between bottom-up and top-down factors in sound identification provide support for the information-processing approach to auditory object recognition. Still, others take an ecological approach to auditory object recognition.

ECOLOGICAL ACOUSTICS

The ecological acoustic approach to auditory object recognition takes the different perspective that global symmetrical patterns are the key to recognition. Rather than analyzing individual elements, the complex, presumably invariant inherent symmetrical patterns are recognized directly (Casey, 1998; Gaver, 1993), consistent with a direct perception viewpoint of ecological perception (Gibson, 1966; Warren & Verbrugge, 1984). Two types or categories of listening are usually considered. The first one, and a primary focus for discussion here, is concerned with *everyday listening*—identifying the objects in our natural environment. This can be distinguished from musical listening or what Casey (1998) referred to as *reduced listening*. Note that reduced does not mean easier; it simply means "listening to inherent sound patterns, without regard to their causal identity" (p. 27).

VanDerveer (1979, unpublished data cited in Warren & Verbrugge, 1984) presented everyday sounds (i.e., jingling keys, footsteps) and asked listeners to recognize them. She noted a strong tendency for people to respond with a description of the mechanism making the sound. Acoustic characteristics were described only when the sound could not be identified. This observation can be taken as support for an ecological approach.

There is also some evidence that sounds that would typically require some type of action (i.e., like a ringing phone) activate different cortical regions than sounds that would not typically require action, like piano notes (De Lucia, Camen, Clarke, & Murray, 2009).

From an ecological perspective, more complex sounds are generally easier to identify. Complex sounds have a richer, more informative pattern. For example, imagine that you are trying to identify a noise coming from the next room. The more complex the sound pattern—meaning the more harmonics you are able to detect and the longer the pattern continues—the more likely you are to identify it. Along similar lines, radio stations sometimes play a game of "name that song," playing only the first two or three notes of a song. The more of the song we hear, the more likely it is we will recognize the pitch contour and thus the song being played. The use of acoustical filters that clip or leave out too wide a range of frequencies results in speech that is difficult to understand. Thus, we see that in both information-processing and ecological acoustics explanations, auditory object recognition is facilitated by complexity, a rather counterintuitive phenomenon.

Another important form of nonverbal sound is music. Music processing has much in common with speech processing. It is a culturally rich, possibly biologically adaptive, form of sound used to convey information and emotion. At the same time, music may have dedicated neural circuitry separate from language processing. For example, based largely on patterns of musical abilities in neurologically impaired individuals, Peretz and colleagues (Hyde & Peretz, 2004; Peretz, 2003, 2006) suggested that separate modules may exist for processing lyrics versus language versus melodies. They observed patterns of disability in one of each of these areas within an individual with intact functioning of the other two. So, what specifically is music? The role of music in culture is considered first, and then attempts to answer this question are discussed.

MUSIC

After speech, music is probably the most meaningful and cherished form of sound. Generations define themselves by the music they listen to and create. Couples associate musical pieces with budding romance, and people of all cultures and ages use music in rituals and celebrations and to give form to physical expression through dance. Although the average listener may not put much time or thought into asking how much mental effort is required to listen to a particular musical piece, considerable attention has been focused on music perception (e.g., see Deutsch, 1999; Krumhansl, 1990; Zatorre & Peretz, 2003). One area of investigation that has received considerable popular attention in recent years has been whether music plays a role in developing or increasing intellectual abilities. Phenomena such as the so-called Mozart effect or the notion that listening to particular Mozart compositions can temporarily increase spatial reasoning ability have received considerable attention in the popular media. In this section, we discuss the evidence, or lack thereof, for these claims.

First, we begin with an examination of some of the aspects involved in music perception and musical knowledge. Included in this discussion is a glimpse of the neural pathways and mechanisms involved as well as how we perceptually organize music into patterns for storage and subsequent recognition. This discussion sets the stage for gaining a deeper understanding of how sensory and cognitive processes interact during musical processing to affect performance and cognitive functions.

Music perception and speech perception share many commonalities. We develop a considerable knowledge base of musical information, or a musical lexicon, thought to be much like the mental lexicon used in speech processing. For many of us without or prior to formal musical training, this knowledge base develops implicitly. Much like grammar and syntax, children and adults without formal musical training will implicitly grasp musical rules that would be nearly impossible to express verbally. Both speech and music rely on continuously unfolding temporal sequences and the ability to hold sounds in memory until their patterns can be organized and comprehended. Furthermore, music and speech are both forms of communication and expression.

Music perception differs from other complex auditory processing (i.e., speech perception) in some important ways. For example, while speech perception relies heavily on temporal changes in broadband sounds, music perception depends more on the ability to discriminate slower, more precise changes in frequency (Zatorre, Belin, & Penhume, 2002). As discussed in Chapter 3, Zatorre and his colleagues (2002) observed that temporal resolution is better in the left hemisphere, while the right hemisphere seems more specialized for spectral resolution (essential to music perception). However, pitch and melodic pattern processing make use of bilateral temporal cues (Griffiths et al., 1998; Zatorre, 1998). We return to the topic of musical pitch perception later in the chapter. First, we discuss some additional basic concepts of musical processing, beginning with a short description of what constitutes musical sound.

MUSIC DEFINED

What exactly is music? Like the concept of mental workload, discussed extensively in Chapter 5 and more generally throughout this book, everyone has some idea of what constitutes music versus random sounds (although certainly not everyone would agree on any particular composition). Does simply striking the keys on a piano, for instance, constitute music? What if that striking results from an inanimate object, such as a book or vase, falling on the keys? Most of us would agree that the resulting sound would not constitute music. Could Fido produce music by chasing the cat across the keys? Perhaps these illustrations border on the absurd, but the point is that music possesses something beyond mere sounds and beyond the type or quality of the instrument producing the sound. Structure is one key element of music. Musical scores represent highly organized structures or patterns of sound. Interestingly, there is virtually universal agreement regarding the variations to this pattern that are acceptable, just as there is general agreement on which sounds or notes can be played together to produce a harmonic chord. This general pattern of agreement transcends culture, class, and musical training. In fact, newborns seem to possess the same general preference for musical chords and patterns as adults. Much of this agreement appears to rely on our ability (albeit unconscious) to detect mathematical relationships between sounds, as was noted by the great Greek philosopher, Pythagoras (Figure 7.1).

Pythagoras helped establish the first natural law based on this principle, that a mathematical relationship exists between pitch and the length of a vibrating string (see Ferguson, 2008). Musicians had been using stringed instruments, such as the lyre illustrated in Figure 7.2, for centuries, realizing that sometimes they sounded pleasant and sometimes they did not. But, it was Pythagoras who was able to determine precisely why.

Pythagoras is rumored to have used a box with a single string stretched across to make mathematical calculations and to examine pitch relationships. This device, called a *monochord* (Figure 7.3), was really more of a scientific instrument than

FIGURE 7.1   Pythagoras.

FIGURE 7.2   Lyre.

FIGURE 7.3   Monochord.

a musical instrument, but nonetheless Pythagoras noted that certain string ratios sounded well together, while others did not.

Some 2,000 years later, the relationship between frequencies and pitch became measurable. Most modern tunings now make use of the mathematical properties of frequencies to establish scales. For example, the Western scale has 12 notes, with each successive frequency the frequency of the previous note times a 12th root of 2 (which is approximately 1.059). For all major scales, the first note is the dominant, and the fifth note in the seven-note scale will be the tonic. A ratio of 2:3 exists between the frequencies of the dominant and tonic notes. So, for example, for the key of C major, the dominant starting note is middle C on a piano (with a frequency of 261.6 in the dominant Western tuning system), and the tonic will be G at a frequency of 391.9, with 261.6/391.9 representing roughly a 2:3 ratio.

It is well beyond the scope of this chapter to discuss these mathematical relationships and their relation to musical theory in detail. The reader interested in these topics is referred to excellent works by Deutsch (1999) and Temperley (2001) and Krumhansl (1990). See also an article by Trainor and Trehub (1993) for an accessible introduction to musical theory and discussion of the cycle of fifths in particular. For our present purposes, we are most concerned with the psychoacoustic effects of these mathematical relationships and examine some of the most important factors affecting musical perception and their potential to affect the difficulty (mental workload) of performing this processing task.

MUSICAL STRUCTURE

One of the psychoacoustic effects of the mathematical relationship within musical scores important for understanding musical structure is the concept of scales and octaves.

Octaves

Long before the relationship between vibration rate and frequency was understood, at least as early as the era of Pythagoras around 500 BC, the Greeks understood that musical notes within a given scale or key exhibited mathematical relationships. For example, in Western music, which is the dominant musical form of reference in this chapter, musical notes repeat in patterns of 12 notes within an octave. Notes are given letter names and are generally referenced starting with the middle C (or C4) on a piano. Seven major notes are represented with the letters C, D, E, F, G, A, and B, with five half steps residing between them (see Figure 7.4). The notes that can be played together to form a chord with *consonance* (pleasant harmony) rather than *dissonance* (an unstable sound that implies a need for resolution) have strict mathematical ratios. A ratio of 1:2 represents an octave, such as C4 and C5 (the C note exactly one octave above middle C). Refer to Figure 7.4 for a visual illustration. Other ratios resulting in consonance include 2:3 (called a fifth); 3:4, a fourth; and 4:5, a major third.

Much later, in the 17th century, it was discovered that musical pitch corresponded to the rate of vibration. According to Pierce (1999), Mersenne and Galileo made the discovery independently, with Galileo further suggesting that consonance could be explained by the "agreeable" pulses experienced when sounds were synchronized,

FIGURE 7.4   Piano keyboard with note names and frequencies.

Twinkle, Twinkle Little Star

FIGURE 7.5   Musical notation that maintains contour though played with different notes. (Drawn by Melody Boyleston.)

such that, for instance, when hearing two notes separated by an octave, the tympanic membrane would experience two pulses corresponding to the high note for every one pulse experienced as a result of the low note. Pierce explained that Galileo's position was a rhythmic theory of consonance. However, subsequent research examining the temporal resolution of sound (less than 1 ms when involving localization) did not provide support for the rhythmic theory.

Contour

If the mathematical relationship between notes used to form chords exhibiting consonance was not evidence enough that music is highly structured, perhaps the topic of musical contour will suffice. *Contour* refers to the overall pattern of notes (i.e., pattern of up and down pitches) within a musical piece. Familiar melodies are recognized more by their pattern or contour than by their key or the absolute pitches involved (Edworthy, 1985). Think of a familiar song, such as the children's song, "Twinkle, Twinkle Little Star." Figure 7.5 presents a simple musical notation of the melody from this song. Note that the contour in Figure 7.5 is the same even though the individual notes are completely different. Both melodies are easily recognized as the familiar childhood song, and in fact numerous other variations can be made and recognized as long as the basic contour remains unchanged.

MUSICAL PITCH PERCEPTION

Musical pitch perception differs from acoustical pitch perception. It involves creating the perception of octaves, harmonies, tension buildup and release, and a host of additional properties not found in nonmusical sound. Pitch information, critical to musical processing, is obtained through spectral (frequency) and temporal processing, both early on (at the level of the cochlea) and in cortical areas in and around the primary auditory cortex (A1). Processing of this information can be explained by two different theories of pitch perception, referred to as the place and time code of pitch perception, for spectral and temporal changes, respectively. Griffiths and colleagues (1998) were able to demonstrate the importance of temporal information by examining pitch and melodic pattern perception in stimuli void of spectral information. The unique type of stimuli utilized by Griffiths (p. 426) was "add-same iterated rippled noise (IRNS)," presented through headphones. As Zatorre (1998) described in his discussion of Griffiths' investigation, the stimuli are an outgrowth of a relatively old observation by Huygens, who noted that "the periodic reflections of the noise made by a fountain from the stone steps of a staircase resulted in an audible pitch" (p. 343). This type of perceptible pitch modeled by Griffiths et al. in the lab is generated solely on the basis of analysis of the temporal structure of the sound. It contains no actual spectral cues. Therefore, any pitch perceived must be a result of the analysis of the time code. This observation corresponds nicely with frequency theories of pitch perception.

As a general tutorial, the two dominant theories of pitch perception, once rivals but now recognized as complementary, suggest that our ability to perceive pitch is based on the timing of neural firing and thus is termed the temporal or timing theory. The location along the basilar membrane where maximal stimulation occurs also plays a key role and thus is called the place theory.

We now know that tonotopic organization aids pitch perception in both the cochlea (see review in Pierce, 1999) and the auditory cortex (Liegeois-Chauvel et al., 2003; Qin, Sakai, Chimoto, & Sato, 2005; Rauschecker, Tian, & Hauser, 1995). The relatively early observation that different frequencies result in different patterns of vibration along the basilar membrane led to what was termed the place theory of pitch perception, originally postulated by Helmholtz (1863/1930) and later demonstrated by Bekesy (1960). A major problem for the place theory has always been the issue of the missing fundamental (see discussion in Zatorre, 2005). Recall that despite the presence of multiple harmonics, we perceive the pitch of a sound to be the lowest common denominator of the harmonic frequencies, called the fundamental frequency. In the situation of the missing fundamental, we perceive the fundamental frequency of a set of harmonics even if there is no acoustical power at that frequency (Moore, 1995). Organ makers are said to make use of this auditory illusion to give the perception of extremely low notes that would require organ pipes too large for the accompanying building structure. They simply present the illusion of the lower note by simultaneously playing multiple harmonics of the missing note. (The higher frequencies require shorter pipes.)

An alternative theory that could better account for the phenomenon of the missing fundamental frequency was the timing theory, dating at least to August Seeback in the 1840s (Goldstein, 1973). According to frequency theory, pitch perception occurs in more central processing mechanisms and is a reflection of periodic fluctuations of patterns of neural firings traveling up the auditory nerve. Specifically, patterns in the rate of firing stemming from different harmonics become phase locked (firing at rates of peak compression) for rates up to about 4,000 Hz. This phase-locked pattern or rate of temporal firing is perceived as the fundamental frequency of the sound. It is now widely recognized that pitch perception, essential to both music and speech perception, relies heavily on this temporal coding of frequency that occurs in central processing mechanisms (see review in Moore, 1995).

Now that some of the basic issues regarding what constitutes music and how it is perceived are described, we briefly discuss how we acquire musical knowledge, followed by an examination in greater depth of the impact that music has on performance.

MUSICAL KNOWLEDGE

Implicit musical knowledge consists of knowledge of the relationships between sounds and sound dimensions (such as tonal hierarchies, keys, and patterns of durations) as well as knowledge of musical forms (such as sonatas or gap-fill melodies) that listeners may distinctly recognize even if they are unable to name them (Bigand, 1993; McAdams, 1989). Knowledge of different musical forms is typically learned as part of the acculturation process even in the absence of formal musical training. However, a number of music universals appear to be present in the newborn before any acculturation has taken place.

Music Universals

Newborns and infants have a number of musical abilities, which can be thought of as music universals. For example, infants appear to recognize similarities and differences in melodic contours between two melodies separated by as much as 15 s or with an intervening series of distractor tones (see review in Trehub, 2001). Recognition is established through an ingenious procedure developed by Trehub and her colleagues. Generally, the infant is seated on his or her parent's lap while listening to stimuli. Infants are "trained" to turn toward a Plexiglas screen located at a 45° angle within a specific period of time if they notice a particular auditory change or event. The experimenter then displays a brightly colored toy for a brief period of time to reinforce correct responses. Through this paradigm, Trehub and colleagues have been able to learn a lot about the kinds of auditory stimuli that infants can distinguish. Other investigations of infant musical perceptions have involved preferential looking time (the time after habituation that an infant chooses to look at a particular stimulus object). Pitch contour plays a prominent role in the infant's ability to distinguish between familiar and novel musical patterns. It is also likely to be a key component of "motherese" or "parentese"—the form of infant-directed speech used by parents and other adults to communicate with preverbal children (Kuhl, 2004).

Infant-directed speech consists of melodic patterns, higher fundamental frequencies (F0), and more exaggerated intonations and prosodic features than adult-directed speech. Adults are able to recognize major themes of infant-directed speech (prohibition, approval, comfort, and attention) more accurately than the intent of adult-directed speech even when the linguistic or semantic cues have been removed (Fernald, 1989) or when presented in a language completely unfamiliar to the listeners (Bryant & Barrett, 2007). Since they are not able to use semantic cues, they must rely solely on nonverbal cues such as pitch contours, intensity change, and intonations.

Infants also seem to be able to detect the difference between consonant and dissonant musical patterns, demonstrating a preference for consonant patterns. For example, Zentner and Kagan (1998) observed that infants as young as 4 months of age demonstrated preferential looking and less agitation (fretting and turning away) for consonant melodies versus dissonant melodies.

Newborn infants also appear to have an innate ability to detect rhythmic beats. Sleeping infants notice a missing beat in a rhythmic sequence, as evidenced through an event-related potential (ERP) paradigm (Winkler, Haden, Ladinig, Sziller, & Honing, 2009). The infants, ranging in age from 2 to 3 days old, demonstrated greater mismatch negativity (MMN) for deviant rhythmic sequences relative to standard tonal sequences. The MMN (described in Chapter 3) is an ERP component that is thought to represent the automatic recognition by the brain of a significant difference between successively presented stimuli.

The ability to move in time to music and to detect relative pitch relationships may not be solely a human ability. Evidence suggested that some nonhuman species demonstrated intriguing capabilities in these areas (i.e., see Patel, Iversen, Bregman, & Schulz, 2009). Studies have even found that some nonhuman animals (i.e., cotton-top tamarins) could distinguish between consonant and dissonant tonal stimuli (McDermott & Hauser, 2004). But, unlike humans, the tamarins showed no preference between the two and in fact, given the choice, would prefer quiet to music.

The musical abilities of infants, children, and adults far surpass those of other species. Humans are easily able to recognize a wrong note in a familiar melodic sequence, or even in an unfamiliar sequence if it conforms to their cultural expectations. Even individuals with no formal musical training can often remember hundreds, if not thousands, of melodies and songs and can easily tap in time to nearly any musical rhythm (for a review, see Trehub & Hannon, 2006). Since humans seem so well adapted to musical abilities, it seems reasonable to ask what impact music might have on performance. Does listening to music enhance or degrade performance on other tasks?

MUSIC AND PERFORMANCE

Numerous researchers have examined the influence of music on performance. We begin with a discussion of the effects of background music in general and its impact on both physical and cognitive performance and then turn the focus to empirical investigations of the phenomenon known as the Mozart effect, which has attracted widespread attention in the popular media.

BACKGROUND MUSIC

Numerous investigations have examined the effects of background music on performance (Furnham & Stanley, 2003). Vocal music is generally found to disrupt performance more than instrumental music (Furnham & Stanley, 2003; Salame & Baddeley, 1989), a finding that can be attributed to the irrelevant speech effect discussed previously. However, the performance effects of instrumental music are less clear. Music may increase morale and may work as an incentive to increase productivity for workers. However, much appears to depend on the type of work performed as well as the music played. Minimally, a distinction must be made between mental or cognitive work and physical work. The impact of music on physical performance has received considerable attention in the literature on sport psychology and exercise. Evidence pertaining to the impact of music on factory work and cognitive task performance is even more extensive, dating at least to the early 1930s (see review in Kirkpatrick, 1943). We begin with a discussion of the more recent studies on physical performance and exercise and then discuss cognitive performance.

PHYSICAL PERFORMANCE

The impact of music on physical performance and work productivity has been of interest for some time. Music is often, although not always, found to have an ergogenic effect: It seems to increase the psychophysiological capacity to work while reducing feelings of fatigue and exertion. The effects of music have been of interest to sports psychologists and exercise scientists. Music may directly or indirectly enhance physical performance during exercise (Edworthy & Waring, 2006). Perceived exertion is frequently found to be lower when exercisers are listening to music rather than silence (Karageorghis & Terry, 1997). Likewise, when subjective effort is controlled such that people are performing at a level they perceive to be the same exertion level, they perform significantly more physical work when listening to music than when they are not (Elliott, Carr, & Orme, 2005; Elliott, Carr, & Savage, 2004; Karageorghis & Terry, 1997). In other words, there seems to be a scientific basis for why so many people like to listen to music in the gym. Listening to music may increase the pace of workouts while leaving the exerciser feeling like he or she has exerted less effort.

Synchronous music appears more likely to result in beneficial effects than asynchronous background music. Synchronous music is generally always found to have some beneficial effects, but the effects are generally limited to exercise within the submaximal range (see review in Anshel & Marisi, 1978; Karageorghis & Terry, 1997). Motivational music, which may be either synchronous or asynchronous, appears to result in improved affect, reduced ratings of perceived exertion (RPEs), and improved posttask attitudes (see review in Elliott et al., 2005). As Elliott et al. pointed out, positive affect during exercise and postexercise plays a key role in determining the likelihood that an individual will repeat and maintain a pattern of exercise behavior. However, determining precisely what constitutes "motivational" music has been problematic. Oudeterous music—music that is neither motivational nor nonmotivational (Karageorghis, Terry, & Lane, 1999)—has sometimes been observed to have effects similar to music classified as motivational (Elliott et al., 2005).

Karageorghis et al. (1999) developed a theoretical framework for categorizing the motivational characteristics of music. They proposed a four-factor hierarchical framework: (a) rhythm response, (b) musicality, (c) cultural impact, and (d) association. *Rhythm response*, the most important characteristic in predicting the motivational impact of a particular piece of music, refers primarily to the tempo of the piece in terms of beats per minute (bpm). *Musicality* refers to aspects more traditionally considered musical aspects,

TABLE 7.1
Original BMRI and Instructions

	Not at All Motivating								Extremely Motivating	
1. Familiarity	1	2	3	4	5	6	7	8	9	10
2. Tempo (beat)	1	2	3	4	5	6	7	8	9	10
3. Rhythm	1	2	3	4	5	6	7	8	9	10
4. Lyrics related to physical activity	1	2	3	4	5	6	7	8	9	10
5. Association of music with sport	1	2	3	4	5	6	7	8	9	10
6. Chart success	1	2	3	4	5	6	7	8	9	10
7. Association of music with a film or video	1	2	3	4	5	6	7	8	9	10
8. The artist/s	1	2	3	4	5	6	7	8	9	10
9. Harmony	1	2	3	4	5	6	7	8	9	10
10. Melody	1	2	3	4	5	6	7	8	9	10
11. Stimulative qualities of music	1	2	3	4	5	6	7	8	9	10
12. Danceability	1	2	3	4	5	6	7	8	9	10
13. Date of release	1	2	3	4	5	6	7	8	9	10

including things such as pitch-related elements, harmony and melody. Cultural impact refers to the popularity or pervasiveness of the piece in society; the final, least-important, motivational aspect is the associations the musical piece has for the particular listener.

Karageorghis et al. (1999) turned this four-factor theoretical structure into a 13-item scale they called the Brunel Music Rating Inventory (BMRI). It has been the most widely used method to date of determining the extent to which a piece of music is motivational within the context of exercise (Elliott et al., 2004, 2005). The BMRI has been used by both researchers and exercise professionals (i.e., aerobics instructors) alike. Recent revisions (now called the BMRI-2) have resulted in improved psychometric properties and easier application by nonprofessional exercisers (Karageorghis, Priest, Terry, Chatzisarantis, & Lane, 2006) and colleagues. Table 7.1 presents the original BMRI along with instructions, and Table 7.2 presents the revised BMRI-2 version.

In sum, researchers such as Karageorghis and colleagues have concentrated on developing methods of systematically categorizing musical pieces to facilitate empirical research examining the effects of music on exercise. Additional empirical methods may help to disambiguate the equivocal results of previous investigations. Other efforts have been directed at understanding why music has an impact on exercise.

THEORIES OF THE RELATIONSHIP OF MUSIC TO EXERCISE

Both empirical and anecdotal evidence indicated that fast, loud music seems to enhance exercise performance. Why might this be so? Several different theories have been proposed over the years for the positive influence of music on physical performance. One leading theory is that listening to music simply takes one's mind off the negative aspects of the work (Anshel & Marisi, 1978; Karageorghis & Terry,

TABLE 7.2
Revised BMRI-2

	Strongly Disagree			In Between			Strongly Agree
1. The rhythm of this music would motivate me during exercise	1	2	3	4	5	6	7
2. The style of this music (i.e., rock, dance, jazz, hip-hop, etc.) would motivate me during exercise	1	2	3	4	5	6	7
3. The melody (tune) of this music would motivate me during exercise	1	2	3	4	5	6	7
4. The tempo (speed) of this music would motivate me during exercise	1	2	3	4	5	6	7
5. The sound of the instruments used (i.e., guitar, synthesizer, saxophone, etc.) of this music would motivate me during exercise	1	2	3	4	5	6	7
6. The beat of this music would motivate me during exercise	1	2	3	4	5	6	7

1997). More specifically, the act of listening to music requires part of the person's limited attentional capacity, leaving fewer resources to be devoted to paying attention to signs of physical exertion. As discussed in previous chapters, the idea that humans have a limited amount of attentional resources is long standing and widely held (Broadbent, 1958; Kahneman, 1973; Wickens, 1984, 2002). The attentional resource approach suggests that listening to music has an indirect benefit by literally taking the focus off physical activity, leaving less time for the participant to devote resources to think about and therefore experience feelings of fatigue and exertion.

A second theoretical explanation is that music directly affects psychomotor arousal level (Karageorghis & Terry, 1997). Music may be used to increase the intensity and tempo of the arousal system in preparation for exercise or calm an overanxious person before an important competition. Karageorghis and Terry pointed out that while this appears to be a popular position held by sports psychologists, there has been relatively little empirical attention given to it. The perception that music has an impact on arousal level may result primarily from learned associations between a particular piece of music and a feeling state.

A third explanation for the influence of music on physical state is that people have a natural tendency to respond to the temporal characteristics of the music and thereby synchronize their movement with the temporal beat of a piece. This position is also lacking empirical research, according to Karageorghis and Terry (1997).

It has been suggested that people tend to prefer musical tempi within the range of their heart rate (see discussion in Karageorghis, Jones, & Low, 2006). However, empirical research indicated that the relationship between musical tempo and preference is somewhat more complex. Karageorghis, Jones, and Low (2006) observed that people's preference for tempo depends on their heart rate at any given time. When they are engaged in low or moderate rates of exercise (with corresponding low to medium heart rates), they exhibit a preference for medium- and fast-paced music.

When they are engaged in more vigorous exercise resulting in a fast heart rate, they prefer fast-paced music.

Edworthy and Waring (2006) observed that people exhibited faster treadmill speeds and faster heart rates when listening to fast (200-bpm) versus slow (70-bpm) music. However, for treadmill speed this effect interacted with loudness of the music over time. That is, listening to fast music had little effect on treadmill speed over time if the music was played quietly (60 dB). But, if the fast music was played loudly (80 dB), treadmill speed increased over a 10-min period, although loudness had no impact on heart rate.

In summary, music is generally found to enhance physical performance during exercise either directly or indirectly. Indirectly, music may help keep an exerciser's mind off the physical and often-negative aspects of the workout, thus leading to increased work output at lower subjective levels of perceived exertion. Directly, fast-tempo music may help an exerciser achieve a quicker pace and sustain that pace longer. Next, we focus on the impact of music on cognitive task performance.

MUSIC AND COGNITION

The long history of research into the impact of background music on the perfor-mance of cognitive tasks has met with contradictory and often-controversial results. Music has been found to improve and decrease performance; to distract, annoy, and enhance affective mood; along with a host of other contradictory conclusions. Music can arouse or calm the central nervous system, helping to keep people awake or lull them to sleep. It seems that the impact of music on performance depends on a number of factors, including the type of music (i.e., fast tempo or slow), whether it is played at high or low intensities, the type of task being performed, as well as individual differences in listener characteristics (Day, Lin, Huang, & Chuang, 2009).

In general, music is often found to improve performance of monotonous tasks with low cognitive demands—helping people perform these tasks for longer dura-tions and to feel better about them while doing them (Furnham & Stephenson, 2007). These results are in line with the theory that music facilitates arousal and thus facili-tates short-term increases in attentional resources. Conversely, for more cognitively demanding tasks, like reading comprehension and prose recall, music has more often been found to decrease performance, to increase feelings of distraction, or both (Furnham & Strbac, 2002). Historically, interest in the effects of background music centered around productivity and job satisfaction among factory workers (Kirkpatrick, 1943). This area of research is still important today as the use of personal electronics increases, making music accessible in nearly any environment and as a new genera-tion enters the workforce who are used to music on demand (Day et al., 2009).

MUSIC IN INDUSTRY

Music in industry is, in fact, the title of one of the articles Kirkpatrick published in 1943. There is a strong relationship between music and work that pre-dates elec-tronic forms of music. In preindustrial times, workers sang as they performed their tasks, and the pace of the work informed the pace of the song (Korczynski & Jones,

2006). As Korczynski and Jones pointed out in their review of the history of music in British factories, as work was brought into factories, singing was generally discouraged, and sometimes workers were even fined or otherwise punished for it. Singing was categorized as a leisure-time activity, and work time was sharply distinguished from it. World War II changed this view.

During World War II, efficiency and satisfaction among factory workers was of paramount importance. Fredrick Taylor's Hawthorne studies in the 1920s had demonstrated that subjective factors and perceptions on the part of workers contributed significantly to worker productivity (Kirkpatrick, 1943). A 1943 report in the London journal, *Conditions for Industrial Health and Efficiency,* indicated that worker absenteeism had increased dramatically during the war, nearly tripling among women. Providing music to alleviate fatigue and boredom was seen as one way of combating the absenteeism issue (Unknown, 1943).

Kirkpatrick (1943) and others would attempt to demonstrate empirically that providing music in the workplace could improve morale, relax tensions, and reduce boredom.

Wyatt and Langdon (1938, as cited in Kirkpatrick, 1943) had reported the results of a quasiexperimental design demonstrating productivity increases of up to 6% among factory workers when listening to music for 30- and 45-min periods relative to their normal sound conditions. Other accounts of increased productivity resulting from piped-in background music were even higher.

In this spirit, the British Broadcasting Company (BBC) began airing a program called *Music While You Work.* The program was intentionally aimed at being intellectually accessible to the general public and to provide aesthetically pleasing background music for workers engaged in factory jobs that were all too frequently tedious and monotonous (Korczynski & Jones, 2006). BBC research indicated that by 1945 at least 9,000 factories were broadcasting the program. The music had to be rhythmical and nonvocal; could not be interrupted by announcements; and had to maintain a consistent volume to overcome the noise present in the factory environments (Korczynski & Jones, 2006).

Research available at the time suggested that this type of music did indeed improve productivity and morale among factory workers with largely monotonous, cognitively simple jobs (Kirkpatrick, 1943). The question remained: What impact would music have on more cognitively complex tasks?

Young and Berry (1979) examined the impact of music along with other environmental factors, such as lighting, noise, and landscaping, on performance of realistic office tasks. The office tasks were much more complex than those used in many previous studies. They involved decision making, design, and forms of creative work. Although it was difficult to assess performance outcomes quantitatively, workers expressed a definite preference for music versus no music. The one exception to this is when music was combined with loud background noise. Adding music to the already-noisy background accentuated the undesirable effects of the noise.

Similarly, Oldham, Cummings, Mischel, Schmidtke, and Zhou (1995) found that employees who were allowed to use headsets during a 4-week quasiexperimental investigation improved their performance and had higher organizational satisfaction and better mood states. These findings were particularly strong for workers in relatively simple jobs. However, it is important to note that participants in the Oldham et al. investigation all came from a pool of employees who had expressed an interest in listening to personal music systems at work. In a more controlled laboratory study, Martin, Wogalter, and Forlano (1988) found that background music consisting of an instrumental jazz-rock piece did not improve or disrupt reading comprehension.

A wealth of research has now been conducted to examine the impact of music on cognitive performance. At least two dominant but conflicting theories have been proposed to explain the results of these investigations. Focusing on the influence that music has on attention, Day et al. (2009) aptly referred to these different postulated roles as the distractor versus the arousal inducer. Basing these positions on the classic model of attention presented by Kahneman (1973), music could play the role of distractor, requiring mental resources to process in an obligatory way and thus have a tendency to be a detriment to performance. Conversely, but using the same model, music could induce arousal and therefore temporarily increase the level of available resources, thus having a tendency to improve performance. Both of these views have received support in the literature (Beh & Hirst, 1999; Crawford & Strapp, 1994; Furnham & Strbac, 2002; Jefferies, Smilek, Eich, & Enns, 2008). As mentioned previously, the impact of music appears to depend on a number of factors, including characteristics of the music itself (i.e., tempo and intensity); the listener (i.e., introvert vs. extravert); and the task (i.e., type of working memory resources required and difficulty level). A complete list or discussion of this area of research is well beyond the current scope. Instead, illustrative examples of some of the key findings are presented.

INTENSITY

Music presented at relatively low-intensity levels (i.e., ~55–75 dBA) has been shown to have a positive impact on nonauditory tasks requiring sustained attention or vigilance, such as visual detection tasks (Ayres & Hughes, 1986; Beh & Hirst, 1999; Davies, Lang, & Shackleton, 1973). For example, Beh and Hirst investigated the impact of low- (55-dBA) and high- (85-dBA) intensity background music on a visual reaction time task, a central and peripheral visual vigilance task, and a tracking task designed to simulate some aspects of driving. Participants performed all three tasks at the same time (the high-demand condition) or individually (low-demand condition). Relative to no music, low-intensity music improved performance of the visual reaction time task and responses to both central and peripheral targets in the vigilance task while having no impact on the tracking task. The pattern of results for the higher-intensity music was more complex.

The higher-intensity music in Beh and Hirst's (1999) investigation improved performance in the visual reaction time task, relative to the no-music condition, to the same degree as the low-intensity music. That is, both low- and high-intensity music reduced response time in the visual reaction time task. High-intensity music also reduced response time for central targets in the vigilance task, again similar to the low-intensity music and relative to the no-music condition. Performance differences between high- and low-intensity music, however, were found for the peripheral targets in the vigilance task. In both single- and dual-task conditions (referred to as low and high demand, respectively, by Beh and Hirst), participants demonstrated a vigilance decrement over the 10-min task interval in both the no-music and the high-intensity music condition. Only the low-intensity music condition resulted in stable performance across the task interval.

Using similar task combinations, Ayres and Hughes (1986) found that neither low-intensity (70-dBA) nor high-intensity (107-dBA) music affected performance on a visual search task and a pursuit tracking task. However, the high-intensity music impaired performance on a visual acuity task. From these results, and those of Beh and Hirst (1999) and others, we can tentatively conclude that music presented at low intensities will generally facilitate performance on most tasks. But, higher intensities facilitate performance for some tasks, but for other types of tasks, high-intensity music has either no impact or a detrimental impact on performance. Contradictory results have been found, even when using similar tasks. For example, Turner, Fernandez, and Nelson (1996), using a visual detection task resembling a part-task driving simulation, found that background music played at 70 dBA improved detection performance, while music played at a lower intensity (60 dBA) or higher intensity (80 dBA) decreased detection performance. The equivocal results of these investigations may be due to the use of music with different tempos.

TEMPO

The tempo of a musical piece, often measured in beats per minute (bpm), has an impact on heart rate (Bernardi, Porta, & Sleight, 2006), motor behavior, and cognitive performance and may interact with intensity. North and Hargreaves (1999) compared fast, loud music (140 bpm at 80 dB) to slow, low-intensity music (80 bpm at 60 dB). Performance on a visual motor speed task (simulated racing) was impaired by both a concurrent task load (counting backward by threes) and by listening to fast, loud music. They found that the combination of having to perform the concurrent tasks while listening to the fast, loud music was particularly detrimental to speed maintenance, relative to performance when listening to the slow, low-intensity music. From this, North and Hargreaves concluded that listening to the music and completing the concurrent tasks competed for a limited processing resource.

North and Hargreaves (1999) also noted that musical preference was correlated with performance in the tasks. People liked both pieces of music more when they were not also performing the backward counting task (which also corresponded with when their speed maintenance performance was best). People who listened to the fast, loud music while performing in the demanding concurrent task conditions reported liking it the least, and their performance tended to be the worst. Highest musical preference ratings were obtained in the conditions for which performance was best (the slow, soft music in the single-task condition). These findings corresponded with previous work by North and Hargreaves (1996) and provided support for a link between context and musical preference (Martindale, Moore, & Anderson, 2005; Martindale, Moore, & Borkum, 1990; North & Hargreaves, 1996; Silvia & Brown, 2007). However, because North and Hargreaves manipulated both intensity and tempo at the same time, it is impossible to tease out the relative contribution of each factor.

Brodsky (2001) examined heart rate and performance of participants on a simulated driving task (presented via a commercially available racing game) while they listened to music of three different tempos. Music at a slow tempo (40–70 bpm), medium tempo (85–110 bpm), and fast tempo (120–140 bpm) was played at a consistent intensity level (85 dBA) while participants drove their simulated vehicle through a 90-min course of daylight driving conditions involving both municipal and interstate roadways. In Brodsky's first experiment, no effects were observed as a function of music condition for average heart rate or heart rate fluctuation or for average speed or lane deviations. However, significantly more participants ran through red lights while listening to the fast music (55%) relative to the no-music or slow music conditions (20% and 35%, respectively). In a second experiment in which Brodsky took the speedometer display away from the participant's view, driving speed also increased with increases in music tempo. Faster music was also associated with greater lane deviation, red light violations, and collisions in this second experiment. (Anecdotally, I can relate to this result. My first and only speeding ticket was received in my early 20s just after I had purchased a new CD of music involving a fast tempo. I will not mention its name, but I was playing it quite loudly and apparently not paying much attention to my speedometer when I saw the flashing red lights behind me.)

Relatively few studies have reported examinations of the impact of music tempo on cognitive performance. Those that have contained equivocal results, sometimes within the same investigation (Mayfield & Moss, 1989). For example, Mayfield and Moss reported two investigations in which undergraduate business majors performed mathematical stock market calculations while listening to either no music, slow-tempo music, or fast-tempo music. In their first experiment, they found no effects of music condition on either the accuracy or the speed at which the mathematical calculations were performed. In their second study, using a larger sample, they found that fast-tempo music resulted in a higher level of performance in the calculation task but was also perceived as more distracting than the slow-tempo music.

In a controlled field study, musical tempo had no effect on the time supermarket shoppers took to select and purchase their groceries or on how much money they spent (Herrington & Capella, 1996). However, this study found that the more shoppers liked the music, the longer they tended to shop and the more they tended to spend. Musical preference was not related to either tempo or intensity in their investigation. Milliman (1982) used a more stringent pace criterion involving determining the time it took supermarket shoppers to move from Point A to Point B under the conditions: no music, slow-tempo music, or fast-tempo music. Shoppers exposed to the slow-tempo music exhibited a slower pace than those shoppers exposed to fast-tempo music. A nonsignificant trend ($p < .079$) indicated that the slow-tempo music tended to result in a slower pace than the no-music condition as well. The slow-tempo music also resulted in significantly higher sales figures relative to the fast-tempo music. Interestingly, Milliman observed these results despite the fact that the majority of a random sample of shoppers leaving the store indicated they were not even aware of the music. Of those who did notice the music, no differences in preference were observed between the two types of music.

INDIVIDUAL DIFFERENCES

There also appear to be differences in the effects of music on cognition that are based on individual characteristics. For example, Furnham and Stanley (2003) observed performance impairments when introverts performed a phonological task and a reading comprehension task while listening to instrumental music. Similar results were obtained for extraverts in the phonological task. That is, instrumental music significantly degraded performance on the phonological task for both introverts and extraverts. However, for the reading comprehension task, extraverts performed just as well in the presence of instrumental music as in a silent condition, whereas vocal music significantly disrupted their performance.

In another study, introverts and extraverts performed an immediate or delayed recall task and a reading comprehension task in the presence of either silence or pop music (Furnham & Bradley, 1997). The music had a detrimental effect on immediate recall in both groups, but the introverts experienced significantly more performance impairment relative to extraverts on both the delayed memory and reading comprehension tests.

People can be aware of the impact of music on their performance. When asked about how they used music in their everyday lives, people with different personality types reported using music in different ways (Chamorro-Premuzic & Furnham, 2007). Introverts were more likely to report using music to regulate their mood, while extraverts were more likely to report using it as background. This observation was true for a sample of participants taken from British and American universities (Chamorro-Premuzic & Furnham, 2007) as well as a sample of students from Barcelona (Chamorro-Premuzic, Goma -i-Freixanet, Furnham, & Muro, 2009). In the Barcelona sample, Chamorro-Premuzic et al. also noted that extraverts were more likely to report using music as a background to other activities than were introverts.

It remains to be seen whether people have learned to use music to enhance their mood states, arousal levels, and overall performance or whether their existing states and levels simply reflect both their underlying personality and musical preferences. The current understanding indicates that the impact of music on performance is a complex interplay between the characteristics of the music, such as loudness and tempo, as well as both the activity being performed and characteristics of the listener. What about the impact of different types of music on specific abilities? Is there any validity to the idea that listening to certain types of music, like Mozart compositions, can improve specific forms of reasoning or intelligence?

THE MOZART EFFECT

In 1993, Rauscher, Shaw, and Ky reported the results of a study in the journal *Nature* that sparked considerable discussion among scientists and educators and in the popular media. They had college students listen to 10 min of Mozart's sonata for two pianos in D major (K488) (see Figure 7.6), 10 min of a relaxation tape, or silence. Results of the study by Rauscher and colleagues are shown in Figure 7.7. As illustrated, those who listened to the Mozart piece scored several standard age points higher on a spatial-reasoning component of the Stanford-Binet intelligence test, relative to the other two listening groups (Rauscher et al., 1993). This increase in IQ was roughly equivalent to an increase of 8–9 standard IQ points.

Rausher and colleagues observed that the effect only lasted for the 10–15 min that it took to complete the spatial task. The story was picked up by a *Boston Globe*

FIGURE 7.6 Opening bars of Mozart's K488.

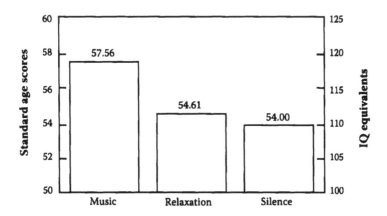

FIGURE 7.7 Mozart effect IQ results. (From Rauscher, F. H., Shaw, G. L., & Ky, K. N. Music and spatial task performance. *Nature, 365*, 611, 1993.)

reporter, who termed it the "Mozart effect," which has since been become a household name and widely known phenomenon, although largely a misunderstood one (Rauscher & Hinton, 2006). It has often been confused with the general statement that "music makes you smarter" (Schellenberg, 2003, p 432).

As Rauscher and others have since argued, (Nantais & Schellenberg, 1999; Rauscher & Hinton, 2006), several points of clarification are warranted. First, the improvement in spatial-temporal abilities reported by Rauscher and her colleagues was considered to be very short lived (on the order of 10–15 min) and resulted from passive listening. This is far different from the long-lasting effects attributed to early music instruction that are sometimes mistakenly referred to as the Mozart effect. In addition, Rauscher and colleagues made no claims that listening to any other type of classical music would improve spatial-temporal skills, and they did not claim that listening to the Mozart piece would improve any other type of ability.

Regardless, publication of Rauscher and colleagues' brief correspondence in 1993 received considerable attention. Shortly thereafter, the governor of Georgia, Zell Miller, in a move that reflected the typical widespread misunderstanding of the findings, funded a program to provide a CD or cassette of classical music (including Mozart's) to all the infants born in the state. Miller is quoted in the January 15, 1998, issue of the *New York Times* as saying, "No one questions that listening to music at a very early age affects the spatial, temporal reasoning that underlies math and engineering and even chess." Despite criticism in the same news article by noted scientist Sandra Trehub, the popular myth that early exposure to classical music could result in long-term boosts in intelligence continued.

Attempts to replicate the findings of Rauscher and colleagues (1993) have met with mixed results (Chabris, 1999; Steele, Bass, & Crook, 1999). Chabris conducted a meta-analysis of 16 studies examining the Mozart effect and concluded that the small amount of enhancement observed in some specific spatial types of tasks was likely due to general arousal, particularly of the right cerebral hemisphere—a brain region critical to the control of both arousal and spatial task performance. Chabris concluded that, regardless, the effect was much smaller (on the order of 2.1 IQ points) than originally reported and was well within the range of average variation for any given individual.

Later researchers suggested that the results might be due, at least in part, to the use by Rauscher and colleagues of a between-subjects design, rather than a within-subjects design. The between-subjects design leaves open the possibility that people were actually experiencing the results of different levels of arousal or changes in mood. Rauscher and colleagues had compared performance after listening to a very lively piece of music to performance after listening to a relaxing piece or complete silence. Differences in general arousal level and mood are quite possible across these different conditions and are a plausible explanation for Rauscher et al.'s (1993) results.

AROUSAL AND PREFERENCE

Some researchers noted that preference for the auditory material (Nantais & Schellenberg, 1999), whether it be Mozart's music or a story, appeared to be the cause of the spatial-temporal benefit. For example, when ratings of enjoyment, arousal, and mood were statistically controlled, the performance benefit of listening to Mozart went away (Thompson, Schellenberg, & Husain, 2001). In other words, listening to Mozart's music only appeared to benefit people who preferred and enjoyed listening to that type of music.

The impact of musical preference has recently been confirmed with a study examining performance on a computer game (Cassity, Henley, & Markley, 2007). They compared performance in a commercially available skateboarding game that involved complex spatial-temporal skills when listening to the regular sound track of the game (*Fight Like a Brave Man* by the Red Hot Chili Peppers) or Mozart's K488. Following performance in the game, they asked participants to rate their preference for different musical genres (two of which used the Red Hot Chili Peppers and Mozart as examples of the genre). They found no evidence for the traditional Mozart effect. In fact, there was a nonsignificant trend in the opposite direction when comparing performance while listening to Mozart versus the Chili Peppers. However, they did observe a significant effect for musical preference, with people tending to perform better when listening to music they liked, a result they noted that was particularly strong for males listening to and preferring heavy metal music. After taking musical preference into account, gender differences in musical impact are seldom observed (Elliott et al., 2005).

Thus, as disappointing as it may be, there is no evidence for an easy shortcut for making our children (male or female) smarter by simply playing them pieces of classical music. However, there is empirical support for the idea that sustained musical experience (in the form of physical and mental practice) can affect neuronal activity patterns.

MUSICAL TRAINING

Extensive musical training may lead to both specialized behavioral and neural processing capabilities. For example, musical conductors appear to have enhanced spatial location capabilities (Munte, Kohlmetz, Nager, & Altenmuller, 2001; Nager, Kohlmetz, Altenmuller, Rodriguez-Fornells, & Munte, 2003), and musical experience results in enhanced memory for musical sequences (Nager et al., 2003; Palmer, 2005). In addition, notable differences in key cortical areas have been observed between musicians and nonmusicians (Ebert, Pantev, Wienbruch, Rockstroth, & Taub, 1995; Pantev et al., 1998). Evidence that musical training can result in changes in cortical functioning (auditory neuroplasticity) comes from an investigation with monkeys (Recanzone, Schreiner, & Merzenich, 1993). Over the course of several weeks, the monkeys were trained to recognize fine-grained changes within a narrow frequency band in tonal patterns. Individual neuronal responses before and after training in the experimental monkeys and control monkeys were compared. The trained monkeys had better behavioral discrimination abilities and significantly greater cortical response for the frequencies involved in the training. Importantly, the extent of their behavioral improvement correlated with both the number of neurons that responded to those frequencies after training and sharpness of their tuning. This provided clear evidence for the potential for musical training to alter cortical functioning, at least in animals. Evidence for auditory neuroplasticity in humans has also been found. Individuals with extensive musical training demonstrated more neuronal activity across a larger cortical area relative to nonmusicians for piano tones but not pure tones (Pantev et al., 1998).

Evidence is mounting to support the position that neuroplasticity resulting from musical training during childhood may significantly affect aspects of cognitive functioning outside the realm of musical processing (Schellenberg, 2005). For example, Patston, Corballis, Hogg, and Tippett (2006) examined visuospatial abilities of musicians and nonmusicians using a line bisection task. (The task requires participants to draw a line through the point they perceive as directly in the middle of a line.) Patston and colleagues found that musicians were significantly more accurate on the task than nonmusicians. Further, while right-handed nonmusicians tended to be more accurate when performing this task with their right hand, Patston's right-handed musicians did not demonstrate this asymmetry. The authors suggested that spatial attention may be more symmetrical in the brains of musicians. However, the results of such studies cannot unequivocally demonstrate that the performance differences are a result of musical training. It could be argued that they result from preexisting differences that might make it more likely that individuals would have the opportunity for extensive musical training—the classic, "which came first, the chicken or the egg" problem.

Evidence that musical training actually changes neural structure and functioning is found in studies involving measurement before and after training among one group with comparison to a control group that receives no musical training. These investigations have been conducted. For example, Pascual-Leone (2001) utilized transcranial magnetic stimulation (TMS) to examine the neural circuitry involved in playing a musical pattern on a keyboard before and after extensive training. TMS involves applying a brief, transient magnetic pulse to a specific cortical area. Depending on the rate at which the pulse is delivered, it can either activate or deactivate the underlying neural circuitry. Deactivating neural signals functionally results in a "temporary lesion." This technique can be used to map precisely which cortical areas are involved (and at what time) in a particular cognitive or motor process.

Pascual-Leone (2001) had adults with no prior musical training practice a sequence of manual keyboard presses with one hand (similar to practicing a scale). One group of participants practiced for 2 h a day for 5 days. Another group continued practicing for an additional length of time. Pascual-Leone compared functional and structural changes in these groups from their initial state to when they had achieved near-perfect levels of performance (i.e., meeting specified sequence and timing criteria). Pascual-Leone interpreted an initial pattern of cortical changes to an unmasking of existing patterns (recruiting existing cortical networks to perform this task, a pattern that was present immediately after practice for the first 3 days). These cortical changes gradually tapered off and were replaced by an emerging pattern that occurred much more slowly (building over the course of several weeks in the group who continued to practice). Pascual-Leone interpreted this slower emerging pattern as the formation of new cortical networks for performing the task. Granted, as exciting as these results are, they primarily relate to the formation of new motor pathways critical to performing complex musical pieces. They do not suggest that simply listening to complex musical pieces can influence the development of cortical pathways that will assist us with other types of complex tasks, as the Mozart effect had suggested. Next, we look at the performance effects of a type of sound that is increasingly present and generally much less pleasant.

NOISE

Examination of the effects of noise on human performance has a long history (Broadbent, 1958, 1971; Gawron, 1982; Kryter, 1960, 1985, 1994; Poulton, 1976, 1977, 1978; Smith & Jones, 1992). The systematic study of the effects of noise on human performance began during World War I (see reviews in Kryter, 1985, and, Matthews et al., 2000) and World War II (Broadbent, 1978; Poulton, 1977). An extensive body of literature pertaining to the effects of noise on human performance now exists. It is beyond the current scope to fully review this work. The interested reader is referred to more comprehensive works that provide broader, more in-depth coverage of the effects of noise (Gawron, 1982; Kryter, 1985, 1994; Smith & Jones, 1992; Tempest, 1985). However, several important aspects of this work, particularly those that have relevance to understanding the effects of noise on mental workload, are discussed.

First, the effects of noise on performance are not straightforward. Noise has been found at times to both improve and interfere with performance. Simple solutions to this puzzling picture (e.g., by examining either the absolute level of noise or the type of noise) have proven elusive (Hygge & Knez, 2001; Smith & Jones, 1992). First, some definitions of noise and key factors influencing its effects are presented. In the simplest sense, noise can be defined as any unwanted sound. However, the term *noise* has also been used to describe random and aperiodic sounds varying in intensity and frequency as well as any sound that interferes with (masks) a desired sound (Smith & Jones, 1992). This means that a sound can be a noise in one situation or environment but not in another. In other words, what constitutes noise, much like what constitutes a weed, is in the eye (or ear) of the beholder.

Two primary factors characterizing noise and its potential effects are intensity and frequency. As discussed in Chapter 2, these factors interact psychophysically. Another key point to be made is that the effects of noise can be discussed in terms of subjective assessments (psychological reactions such as annoyance) and objective assessments (behavioral consequences and physiological responses). Smith and Jones (1992) pointed out that these two assessments may diverge. In other words, a person may find a particular noise highly annoying and yet its presence does not necessarily disrupt performance. Conversely, a person may be enjoying a particular noise (e.g., music), and yet his or her performance may be impaired by the noise. Considerable evidence suggests that subjective and objective assessments often interact as well. In particular, the listener's perceived control over the noise appears to have substantive performance effects. That is, noise is less detrimental to performance if the person perceives that he or she has control over its presence or intensity (Smith & Jones, 1992).

People commonly complain of the detrimental effects of noise on performance in everyday tasks (Smith & Jones, 1992). These complaints are commonly waged against moderately noisy environments (70–90 dBA). However, despite numerous empirical investigations, the precise effects of noise on mental workload remain unclear. Noise has variously improved, impaired, or left unchanged participants' performance on a variety of tasks (Bell, 1978; Davies & Jones, 1975; Gawron, 1982; Hygge & Knez, 2001; Kryter, 1985; Sanders & McCormick, 1993; Smith & Jones, 1992; Smith, Jones, & Broadbent, 1981; Weinstein, 1974, 1977).

In confirmation of the equivocal findings pertaining to the effects of noise, Gawron (1982) reviewed previous research and cited studies with many conflicting results. Two examples cited in the Gawron (1982) report are as follows: Obata, Morita, Hirose, and Matsumoto (1934) found decreased performance speed in the presence of noise, while Davies and Davies (unpublished report cited in Davies, 1968) found increased performance speed in the presence of noise; Hamilton, Hockey, and Rejman (1977) found that noise decreased the number of correct responses, while Park and Payne (1963) found no difference in this measure with a similar noise intensity. Investigators have attempted to explain these contradictory results in numerous ways.

Sanders and McCormick (1993) and others (Smith & Jones, 1992) proposed that these equivocal results can be explained in part by the wide variability of conditions tested in noise experiments. They pointed out that these test conditions have varied regarding intermittent or continuous noise and have used such varying noise sources as tape-recorded machine noise, street sounds, rocket noise, and gibberish. In addition, they pointed out that the tasks used to measure performance have differed dramatically in terms of the relative demands they place on the perceptual, cognitive, memory, and motor capabilities of the participants.

Grant and colleagues (1998) observed that recall of newly studied material was better if the recall situation (noise vs. quiet) matched the study condition. That is, those who studied material in a condition involving background noise recalled more if they were tested in a condition involving background noise relative to quiet. Conversely, those who had studied in quiet performed better if tested in the same quiet conditions, relative to noise. The implications for this form of state-dependent learning are rarely considered in empirical investigations and may be another reason why different studies often report equivocal results.

Kryter (1985) listed 11 different ways that noise has been purported to affect performance. One was the position of Poulton (1976) that noise masked internal speech and other important acoustic cues. The other 10 ranged from noise causing distraction, competing for psychological attention, or causing confusion by conveying irrelevant information to physiological changes and preemption of auditory neural pathways.

Matthews and colleagues (2000) discussed three main ways that noise can affect performance: (a) the disruption of auditory perception (i.e., through either masking or hearing impairment); (b) disruption of postperceptual processing (reduction of attentional resource availability); or (c) indirect stress-related effects such as irritation or annoyance. Each of these avenues is addressed to a limited extent here. However, for the current purposes, emphasis will be placed on further examination of the second causal explanation, that is, that noise may directly affect the level of attentional resources required or available to carry out specific tasks.

Noise is generally thought to have adverse effects on performance. However, in certain circumstances, noise can enhance performance. Enhanced performance in conditions of noise can generally be attributed to physiological changes such as increased arousal levels when operators are fatigued. A brief history of noise research follows.

EARLY NOISE RESEARCH

The Broadbent-Poulton Debate

During World War II, a number of researchers began to investigate the mechanisms behind the detrimental impact of noise on performance, particularly in conditions of sustained attention. Broadbent was a key figure in this early work (Figure 7.8). Broadbent reasoned that noise could directly disrupt performance. According to this position, noise had a distracting effect, which he termed the *internal blink*, that temporarily took attention away from other tasks (Broadbent, 1958; and see discussion in Matthews et al., 2000). This view was later challenged by Poulton and others, who

FIGURE 7.8   Donald Broadbent.

reasoned that noise affected performance by masking acoustic cues and inner speech in particular. Poulton (1976, 1977, 1978) argued that continuous noise and articulatory suppression both prevented listeners from engaging in subvocal rehearsal. Thus, each had similar performance-degrading effects because in both cases they disrupted echoic memory processes.

Poulton followed in the tradition of S. S. Stevens (1972). From a series of experiments conducted during World War II, Stevens had concluded that the negative effects of noise were primarily due to the masking of auditory cues. Poulton (1977) supported the position that the effects of noise were primarily due to masking—the fact that noise covers up important environmental sounds that normally provide informative cues and can further block inner speech (i.e., subvocal rehearsal) that normally would aid performance by supporting verbal working memory. Poulton argued vehemently in published literature against Broadbent's position. Poulton argued that Broadbent had not adequately accounted for the effects of masking in his experiments throughout the 1950s, and his frustration with the lack of acknowledgment of this position was summed well in the last lines of an abstract he published in a review paper in 1977:

> Here, the noise can be said to interfere with or mask inner speech. Yet current explanations of the detrimental effects of continuous intense noise usually follow Broadbent and ignore masking in favor of nonspecific concepts like distraction, the funneling of attention, or overarousal. (p. 977)

Broadbent (1978) responded to Poulton's (1977) review and position statement by pointing out that Poulton had in some cases confused the unit of measurement used (i.e., confusing dB SPL [sound pressure level] with dBA or dBC or vice versa). In others, Broadbent asserted that Poulton had ignored or discounted critical controls that the researchers had put in place to ensure that the effects of masking could be ruled out as the primary cause of noise effects.

Thus, despite some support for Poulton's position (see discussion in Kryter, 1985), the mechanisms behind the impact of noise remained equivocal. Further evidence that contradicts Poulton's argument continued to be found. For instance, Jones (1983) demonstrated that when acoustic cues were eliminated from the task (participants were wearing sound attenuating headphones), the impact of noise was still the same.

The performance effects of noise have also been explained within the context of arousal theory (Broadbent, 1971). If noise increases arousal, then the beneficial performance effects of noise under conditions of sleep deprivation can be explained since the arousal stemming from the noise would serve to help counter low levels of arousal due to sleep deprivation. However, when arousal levels were already too high, then any additional noise would be expected to impair performance. Conversely, noise would also be expected to improve performance in other situations of low arousal (i.e., from extended time on task). However, this hypothesis has not been supported in the literature. As discussed by Matthews et al. (2000) the performance impact (in terms of increased errors) of high noise levels tends to increase with increased time on task.

Arousal theory also leads to predictions of individual differences in the effects of noise. The theory indicates that extraverts, who are generally thought to be underaroused, might benefit from the presence of noise, while introverts, believed to be overaroused, might experience performance decrements from similar levels of noise. Smith and Jones (1992) provided a detailed discussion of this issue.

Broadbent (1979) summarized the results of numerous investigations, concluding that noise affects performance beyond what can be explained by the effects of masking. The precise nature of the effect will depend on a combination of the type of noise and the task being performed—as well as potentially the individual listener (see also Smith, 1989). Broadbent concluded that noise has little or no effect on basic sensory and low-level processes (i.e., visual acuity, contrast sensitivity, eye movements, and simple reaction time). Further, continuous nonverbal noise appears to have little effect on basic working memory tasks such as mental arithmetic. More complex tasks, such as choice serial response tasks, are generally sensitive to the effects of noise, with response times becoming more variable perhaps due to periods of inefficiency or distraction.

The Broadbent-Poulton debate carried on for some time in the early literature until it began to become apparent that neither position adequately explained the equivocal effects of noise on performance. The complex pattern of results across numerous studies indicated that the effects of noise cannot be explained simply by assuming that noise interferes or masks acoustic cues or inner speech, at least not in a passive or mechanistic way (Smith, 1989; Smith & Jones, 1992). In summary, the impact of noise depends not only on the type of noise but also on the task performed. In general, noise often reinforces adoption of a dominant task performance strategy and decreases the efficiency of control processes used to monitor task performance.

Thus, we see that some generalizations pertaining to the effects of noise can be made, although specific effects will depend not only on the task and type of noise but also potentially on characteristics of the individual listener. The next section summarizes and discusses some of the major findings pertaining to the impact that noise may have on mental workload within a resource theory framework.

RESOURCE THEORY FRAMEWORK

Viewing the impact of noise within a resource theory, or mental workload, framework can assist in understanding the effects of noise on performance. For example, Finkelman and Glass (1970) suggested that previous equivocal results pertaining to the effects of noise on performance may be due to the way in which earlier investigators measured performance decrements. Using a mental workload framework, Finkelman and Glass reasoned that when demands imposed by the task and concurrent environmental stress are within the operator's processing capacity, the task can be performed substantially without errors. In fact, some researchers (e.g., Yeh & Wickens, 1988) have suggested that under certain circumstances performance may actually improve as workload increases. For example, when examining the processing demands of relatively easy tasks, an increase in workload may be associated with the operator increasing his or her attention to maintain current levels of performance. This notion would be in line with the classic performance-arousal curve cited in numerous investigations. However, when the limited capacity of an operator's processing resources is exceeded, performance degradation will occur. They termed this the *overload notion*.

Finkelman and Glass (1970) investigated their overload notion by means of a subsidiary task involving delayed digit recall in combination with a primary tracking task involving a vehicular steering simulation. Participants were 23 undergraduate volunteers. Random bursts of 80-dB white noise were presented in either a predictable pattern (9-s burst interpolated with 3-s intervals of silence) or in an unpredictable pattern (bursts of random duration varying in ten 1-s steps between 1 and 9 s). The total duration and ratio of sound/silence was the same for predictable and unpredictable noise conditions.

Finkelman and Glass (1970) found that unpredictable noise had no effect on the primary tracking task but did cause statistically significant performance decrements in the subsidiary recall task. Specifically, the mean number of errors on the subsidiary task was twice as great in the unpredictable noise compared to the predictable noise, suggesting that the unpredictable noise required greater mental effort to ignore relative to the predictable. Time-on-target (TOT) measures of tracking were virtually the same in each noise condition. This suggests that participants were able to concentrate their attentional resources to preserve one task—although possibly at the expense of the second task. Finkelman and Glass concluded that the subsidiary task appeared to be a superior method of detecting performance deficits resulting from environmental noise and interpreted their results within a resource framework. The subsidiary or dual-task technique has also been widely used to examine the impact of noise on task prioritization.

Task Prioritization

The presence of loud noise may cause people to concentrate their attention on tasks of high priority—potentially allowing them to maintain performance on some tasks—while disregarding tasks perceived as less important. Broadbent (1971) explained this by suggesting that noise increases the probability that people will sample information from dominant sources at the expense of nondominant ones.

For example, Hockey and Hamilton (1970) gave participants a short-term word recall task in quiet (55 dB background) and noise (80 dB). Words were presented in one of each of the four corners of the screen. Participants were asked to recall all the words in the correct order; after this, they were unexpectedly asked to recall which corners of the screen the word had been presented. Those performing the task in the presence of noise recalled as many words as those performing the task in quiet. In fact, a marginally significant trend was observed for the noise group to correctly recall more words in the correct order relative to those in the quiet group. However, when asked to provide the location where the word was presented, the quiet group provided the correct location information significantly more often than those in the noise group. Hockey and Hamilton concluded that the noise had induced a high arousal level that resulted in task prioritization. That is, the noise group paid more attention to the most important aspect of the task (word recall in the correct order) but consequently processed less-relevant aspects of the task (the location) to a lesser degree.

In a series of three experiments, Smith (1982) used a paradigm similar to Hockey and Hamilton's (1970) and found supporting results. In his second experiment, Smith extended Hockey and Hamilton's results by manipulating which of the two aspects of the task were given priority—word order or location. An interaction was observed between noise condition and task prioritization such that participants in the noise condition performed better for whichever task was prioritized (order or location), indicating that the previous results could not be explained by something inherent in the task structure but rather relied on prioritization instructions. Smith's third experiment provided further support for this conclusion, indicating that the same pattern of results was observed even when the prioritized task was performed after the less-relevant task.

Together, these findings provide strong support for the position that moderate-to-loud noise results in the focusing of limited resources on the most relevant aspects of a task at the expense of less-relevant aspects—or task prioritization. The next section addresses the impact of lower-intensity noise on cognitive performance.

EVERYDAY NOISE LEVELS

Numerous laboratory studies have demonstrated that acute noise degrades complex task performance (Smith & Jones, 1992). A limited number of studies have demonstrated that even noise present at much lower levels, such as those commonly found in everyday situations, affect performance. The investigation of Weinstein (1974) illustrated this as well as another very important observation. That is, that people may be poor judges of the impact of noise on their performance.

Weinstein (1974) questioned the application of previous studies of the effects of noise on complex activities, noting that past research has most often utilized rather exotic tasks and high noise levels (often above 90 dB). Based on these observations, Weinstein conducted a study that (a) investigated the effects of realistic noise and noise levels (unamplified sounds of a teletype machine) on performance of a familiar task with some real-life significance (proofreading printed material); and (b) examined participants' knowledge of the effect of the noise on their performance.

In line with a resource framework, Weinstein (1974) reasoned that noise would interfere with tasks that were intellectually demanding (such as identifying grammatical errors) while leaving performance on tasks with little cognitive demand (such as identifying spelling errors) unchanged. Participants were 43 undergraduates in an introductory psychology class randomly assigned to either the noise or the quiet condition. Participants were required to proofread texts that were prepared from nonfiction sources and to detect errors introduced at irregular intervals. Errors were of two types (contextual or noncontextual). Contextual errors included errors in grammar, missing words, and incorrect or inappropriate words (e.g., the word *an* instead of the word *and*). The second type of error was noncontextual errors, which included misspelled words and typographical errors. Weinstein reasoned that contextual errors place greater cognitive demand on participants since they could not be detected by examining individual words.

Weinstein (1974) demonstrated that realistic noise levels (66 dB and 70 dB bursts against a background of 55 dB) degraded intellectually demanding tasks. Perhaps even more important is that, when questioned, participants were unaware of the negative impact the noise had on their performance. Overall, participants reported that they felt they had been able to adapt to the noise and thus that it had not affected their performance after its initial onset. Weinstein's behavioral data contradicted this subjective report. These findings underscore the observation yet again that people are often not able to give accurate subjective estimates of the impact of irrelevant sounds on their performance. Now, although the focus of this chapter is on nonverbal sound, it is nevertheless important to include at least some discussion on a type of "noise" that can have particularly detrimental effects on many types of performance. That noise is irrelevant speech.

IRRELEVANT SPEECH AND VISUAL-VERBAL PROCESSING

Examining the affects of irrelevant speech on information processing sheds light on the attentional processes required for storage and transformation of verbal information in memory (Jones & Morris, 1992). Irrelevant speech is the most detrimental form of noise to performance, at least when the task involves processing language in some form or other. Most of the research in this area has examined the impact of irrelevant or unattended speech on performance of tasks requiring the recall of some type of visually presented verbal stimuli, such as digits or words (Jones & Morris, 1992). As an early investigation by Colle and Welsh (1976) demonstrated, irrelevant phonological material, even if it is in a language that is unfamiliar to the listener, disrupts recall of visually presented material. Colle and Welsh also observed that the disruptive effects of irrelevant speech were greatest when the to-be-remembered letters were phonologically distinct (i.e., F, W, L, K, and Q) versus phonologically similar (i.e., E, G, T, P, and C). They reasoned that maintenance of an auditory code would be more helpful for the phonologically distinct letters. Thus, irrelevant speech (or anything else that disrupted the auditory code) might have less of an effect on the phonologically similar letters as people are relying on the auditory code less for these lists.

Salame and Baddeley (1982, 1987) demonstrated that both irrelevant spoken words and nonsense syllables disrupted memory for visually presented digits. As in the experiment by Colle and Welsh (1976), participants were told to ignore the irrelevant sounds. Salame and Baddeley reasoned that the phonological information from the irrelevant material is automatically encoded into the phonological store, where it is represented at a phonemic level rather than a word level. Therefore, phonologically similar material can be nearly as disruptive as actual words. As discussed further in Chapter 9, speech (unlike text and other nonverbal sound) is thought to gain automatic obligatory access to the phonological loop, a prelexical processing component of working memory. This may explain why irrelevant speech is so disruptive to performance of other verbal tasks.

The disruptive effects of irrelevant speech appear to be limited to tasks that require processing of some type of verbal stimuli. Irrelevant verbal material has little to no effect on recall of pitch, as in a tone recall task (Deutsch, 1970).

Until this point, the focus has been on the effects of relatively short periods of noise. What impact does long-term or chronic noise exposure have on performance? This is the focus of the next section.

CHRONIC NOISE EXPOSURE

WORKPLACE NOISE

Noise has also been implicated as an antecedent factor in workplace accidents, both directly—through distraction and fatigue inducement—and indirectly by resulting in both temporary and permanent hearing loss (Picard et al., 2008). Hearing loss, whether temporary or permanent, can prevent perception of important environmental cues to safety (i.e., localizing objects and potential hazards) as well as blocking or reducing perception of warning sounds and speech communications. The need to block unwanted sound can increase cognitive demand in the short term and can exacerbate workplace fatigue in the longer term. Picard et al. conducted an investigation that included over 50,000 males in various occupations between 1983 and 1998 for whom both hearing assessment data and workplace injury records were available. An association was found between accident risk and hearing sensitivity, with increased hearing threshold associated with increased risk of accident exposure even after controlling for age and occupational noise exposure at the time of the hearing test (see discussion in Smith & Jones, 1992). Other investigations have observed associations between workplace noise exposure and increased risk of accidents (Kerr, 1950; Noweir, 1984; Poulton, 1972). However, separating out the higher risk associated with certain types of occupational exposures from noise exposures and other socioeconomic factors can be problematic. For example, in addition to high noise levels, Kerr observed significant correlations between workplace accident rates and several other factors. High accident rates were associated with low intracompany transfer mobility rates, small numbers of female employees and salaried employees, and low promotion probability rates. The difficulty of separating socioeconomic and other confounding variables from the impact of noise reoccurs pertaining to the impact of chronic noise on school-age children, discussed next.

AIRCRAFT NOISE AND CHILDREN

Because short-term noise exposure has been found to disrupt performance on many different cognitive tasks, great concern has been directed at the potential consequences of chronic exposure to aircraft noise among school-age children. Compared to age-matched controls attending schools in quiet neighborhoods, children who live near airports have been shown to have higher blood pressure (Cohen, Evans, Krantz, & Stokols, 1980; Evans, Bullinger, & Hygge, 1998), to have reading comprehension difficulties (Cohen et al., 1980; Green, Pasternack, & Shore, 1982; Stansfeld et al., 2005), to be more likely to fail on cognitive tasks, and to be more likely to give up prematurely (Cohen et al., 1980). Increased noise exposure is associated with a higher probability of reading at least 1 year below grade level (Green et al., 1982).

Negative performance associations are consistently found between chronic noise exposure and cognitive performance. However, it is often difficult to determine the extent to which these negative performance outcomes (i.e., poor reading comprehension abilities and reading below grade level) are a result of the noise exposure, per se, or other socioeconomic factors that are associated with living in high-noise areas (i.e., next to airports, subways, and highly congested roadways). For example, Haines, Stansfeld, Head, and Job (2002) observed significant associations between noise levels and performance on reading and math tests among school-age children. Their sample included approximately 11,000 children from 123 schools. They observed significant associations in chronic noise exposure in a dose-response function (the higher the noise exposure, the greater the performance detriment) to national standardized tests of reading comprehension and mathematics. However, these associations were confounded by socioeconomic factors. Once Haines and colleagues adjusted their model for a measure of social deprivation (the percentage of children in the school eligible for free lunches), the association between noise level and test performance was greatly reduced and no longer statistically significant.

In a more recent cross-national, cross-sectional investigation, Stansfeld and colleagues (2005) controlled many more sociodemographic factors and still observed a significant association between chronic aircraft noise exposure level and reading comprehension. Their sample consisted of nearly 3,000 children from the Netherlands, the United Kingdom, and Spain.

Sociodemographic variables in this investigation included employment status, housing tenure, crowding (the number of people per room in the home), and maternal education level. Both before and after adjusting for each of these variables, significant impairments in reading comprehension scores were obtained as a function of exposure to chronic aircraft noise. The effect sizes of the impact of noise exposures for measures of reading did not differ across countries or socioeconomic status. Specific effects include such observations as a 5 dB increase in noise exposure associated with a reading delay of between 1 and 2 months among 9- and 10-year-old children in the United Kingdom and the Netherlands. Exposure to the aircraft noise was not associated with measures of working memory, prospective memory, or sustained attention.

A unique opportunity to examine the impact of chronic noise exposure on reading comprehension in children was utilized by Hygge, Evans, and Bullinger (2002). They conducted an investigation involving data collection waves before and after the opening of the new Munich International Airport and termination of the old airport. Such a design allowed for a direct experimental test of the effects of airport noise. The sample consisted of children at both airport sites and a control group closely matched for socioeconomic status. They were able to test children in each group individually on an array of cognitive tests, including measures of reading, memory, attention, and speech perception. They obtained measures on each of these tests once before the airport moved and twice afterward. After the airport moved, reading and long-term memory scores declined in the newly exposed group and improved in the formerly exposed group, providing strong causal evidence of the negative but potentially reversible effect of chronic noise on school-age children's cognitive performance.

The impact of noise on performance may not always be detrimental. As discussed in the following section, noise can actually improve performance of certain types of vigilance tasks.

AUDITORY VIGILANCE

The term *vigilance* refers to the ability to sustain attention for an extended period of time (Davies & Tune, 1970; Parasuraman & Davies, 1984; Warm & Alluisi, 1971; Warm, Matthews, & Finomore, 2008). With the increased use of automation in many modern work environments, maintaining vigilant monitoring of system states and controls is an essential part of many jobs. Air traffic control and baggage screening are two commonly cited tasks that require vigilance. It is well documented that maintaining vigilance is hard work (see Grier et al., 2003; See, Howe, Warm, & Dember, 1995; Warm, Parasuraman, & Matthews, 2008). Performance tends to degrade after a relatively short period of time (i.e., after 15–20 min), resulting in increased response time, errors, or both. This performance decline is referred to as the *vigilance decrement*, and it is particularly evident in tasks that have high event rates and that require the operator to maintain an example of the target or signal in memory (Parasuraman, 1979). Performance decrements are generally accompanied by increases in subjective workload ratings as well as physiological changes, such as decreased cerebral blood flow velocity (Hitchcock et al., 2003; Shaw et al., 2009) and increased activation of areas in the right cerebral hemisphere (see review in Parasuraman et al., 1998).

Auditory tasks are less susceptible to the vigilance decrements than visual tasks, but performance declines are still evident (Warm & Alluisi, 1971). For example, Szalma et al. (2004) compared performance and subjective workload ratings on comparable versions of a visual and auditory temporal discrimination task. Participants had to watch or listen for signals of a shorter duration. Despite equating for initial task difficulty using a cross-modal matching technique, performance in the visual version of the task was much worse than for the auditory version. A vigilance decrement was observed for both tasks over the course of four consecutive 10-min vigils, and participants rated the two tasks as equally demanding in terms of workload.

Using the same temporal discrimination task, Shaw et al. (2009) observed similar performance declines for both the auditory and visual versions. Shaw et al. noted that the performance reductions for both modalities were accompanied by similar reductions in cerebral blood flow velocity, particularly in the right hemisphere. They concluded that a supramodal attentional system is responsible for sustained attention, such that neither the visual nor auditory modalities escape the consequences of maintaining vigilance.

Still, the majority of studies that have compared vigilance performance for tasks presented in the auditory versus visual modalities showed at least a slight performance advantage for the auditory task (for a meta-analysis and review, see Koelega, 1992). Further, introverts showed a slight advantage over extraverts in tasks requiring sustained attention, and these performance differences were greater with visual relative to auditory tasks (see Koelega, 1992).

NOISE AND VIGILANCE

A further observation that deserves mention is that mild noise seems to improve vigilance performance, at least in some tasks. This has been attributed to the fact that some amount of stimulation in the environment (i.e., instrumental music) seems to be preferable to sterile environments. Pickett (1978, as cited by Parasuraman, 1986) likened it to being in a "coffee shop atmosphere" rather than a quiet library, where there is some stimulation but little distraction. Background music may aid performance, particularly in jobs that require sustained attention and repetition (Fox, 1971). As discussed previously, noise in the form of music has been shown to improve performance in a visual vigilance task when combined with a simulated driving-like tracking task (Beh & Hirst, 1999). The picture is complex, depending on both the level of noise and individual differences. For example, introverts, hypothesized to have higher base levels of arousal, generally demonstrate performance advantages in vigilance tasks, relative to extraverts, who are hypothesized to have lower base levels of arousal (see Geen, McCown, & Broyles, 1985; Koelega, 1992). Geen et al. found that low levels of noise (tape recordings of a food blender played at 65 dB) improved detection rates among introverts, while the same noise played at a higher intensity level (85 dB) degraded performance. Conversely, extraverts performed worse in the low-intensity noise relative to their performance in the high-intensity noise. In the high-intensity noise, their performance was equal to that of the introverts, but the introverts outperformed the extraverts in the lower-intensity noise levels. In general, mild-to-moderate noise is often found to improve or have no effect on vigilance performance (see reviews in Davies & Tune, 1970; Mirabella & Goldstein, 1967). However, in a more recent review of this area, Koelega (1992) pointed out that these effects are not consistently observed.

SUMMARY

Increased noise exposure is also commonly positively associated with increased annoyance (Clark & Stansfeld, 2007; Stansfeld et al., 2005), which can compound the negative cognitive impact on both children and adults. Given the prevalence of environmental noise in our modern world, its impact on public health, cognition, and quality of life deserves further attention.

In summary, nonverbal sounds—both wanted and unwanted—surround us in the modern world. Regularities in the acoustical signal such as the mathematical relationships between spectral components or harmonics of a sound, as well as changes in stimulus onset time and location, are some of the many cues we use to segment this symphony of sound in the process called auditory scene analysis. Once auditory streams are segregated, sounds appear to be stored based on the objects that created them in a process analogous to visual object-based coding. Our remarkable ability to recognize objects and characteristics of those objects has been explained by the competing theories emphasizing either an information-processing or ecological approach. An extensive body of literature now exists examining the impact of both music and noise on physiological and cognitive performance. Music appears to enhance many types of performance—increasing the effectiveness of fitness workouts while decreasing feelings of fatigue. Music may at times improve productivity and job satisfaction at work and even improve performance on some cognitive tasks, particularly if those tasks are long and repetitive, like vigilance tasks. However, if the music is too loud or the tempo too fast, it is likely to detract from performance, and people will tend to view it as unpleasant. At this point, music can cross the line to become noise—or unwanted sound. Noise has generally been found to have a negative impact on performance on a wide range of cognitive tasks. Workplace noise may increase accident risk, and chronic noise exposure has a detrimental impact on reading comprehension and memory performance among school-age children.

CONCLUDING REMARKS

Many different types of nonverbal sounds form the soundscape around us and have the potential to have an impact on mental workload and performance either positively or negatively. An extensive amount of attention was given to the impact of noise on performance in earlier years. Today, that focus has shifted toward the impact of music on performance. With both, a complex mosaic is found depending on characteristics of the individual, the sound, and the task or tasks to be performed. In the next chapter, attention is directed toward speech processing and the factors that have an impact on the mental effort of this everyday task.

8 Mental Workload and Speech Processing

INTRODUCTION

Each of us engages in communication with others, speaking and listening with seemingly little effort, virtually every day of our lives. Our ability to talk to those around us not only is essential to our survival, but also enriches our lives through social interaction, allows us to pass on knowledge, provides an avenue for entertainment, and permits outlets for creativity and imagination. The ability to communicate ideas, thoughts, and feelings to others through language sets us apart from all other species on the planet. Normally, we are able to process speech—through face-to-face conversations; from the radio, telephones, TV, and other media—all with relative ease and efficiency. But, this apparent effortlessness hides a remarkably complex process that is the focus of this chapter.

Extracting meaningful semantic information from a series of transient acoustical variations is a highly complex and resource-demanding process. Well over a century of research has focused on understanding the mechanisms of speech processing (Bagley, 1900–1901), with considerable accomplishments occurring particularly in the last few decades with the advent of brain-imaging techniques. Today, understanding of our remarkable ability to process speech has been informed by the integration of theories and empirical research from such disciplines as linguistics, psycholinguistics, psychology, cognitive science, computational science, and neuroscience (C. Cherry, 1953a; Friederici, 1999; Norris, 1994; Norris et al., 1995; Plomp & Mimpen, 1979; Treisman, 1964b). Yet, after over a century of research, many of the details of this process remain elusive. As Nygaard and Pisoni (1995) put it:

> If one could identify stretches of the acoustic waveform that correspond to units of perception, then the path from sound to meaning would be clear. However, this correspondence or mapping has proven extremely difficult to find, even after ... years of research on the problem. (p. 63)

This chapter examines current views on the lexical interpretation of acoustic information. The interaction of sensory, acoustical, and cognitive factors is discussed with a focus on their impact on the mental workload of speech processing.

CHAPTER OVERVIEW

To establish a framework for understanding the interaction of sensory-acoustical and perceptual-cognitive factors and their impact on mental workload, we begin by discussing the complexity of speech recognition. Spoken word perception presents unique challenges. Identifying words requires the ability to match a highly variable acoustic pattern, unfolding across time, to representations from our stored repertoire of lexical units. Then, these words (which may be ambiguous—temporarily having several potential meanings) must be held for at least the duration of a sentence until we can comprehend their intended semantic message. How we accomplish this matching of acoustical information to semantic content—referred to as *binding*—has been the subject of numerous investigations in many disciplines, and it forms the cornerstone of most models of speech perception. The challenges of speech processing and the levels of speech processing involved in this binding process are discussed.

Next, an important theoretical debate regarding how speech is processed is presented. We discuss whether speech processing is modular or whether it uses the same general mechanisms involved in nonspeech tasks. These theories have important implications for mental workload and the degree of interference that can be expected when speech processing must be carried out in conjunction with other concurrent tasks.

Major theoretical models of speech perception are then highlighted, with a focus on their implications for how sensory and cognitive factors have an impact on the mental workload of the comprehension task. Finally, a number of issues that have an impact on the mental demand of speech processing are discussed. These issues range from acoustic aspects of the listening environment and speech rate to the types of speech utilized, including real versus synthetic speech. First, we look at the complexity of speech processing—and why it is often more difficult than reading.

SPEECH-PROCESSING CHALLENGES

Unlike the printed word, the speech signal carries a wealth of information beyond straight lexical content. When listening, we are not told explicitly who is speaking and whether an utterance is made angrily or whimsically. We must infer these supralinguistic aspects from the acoustic stimulus itself. Multiple redundancies are processed in parallel to confer meaning on the transient acoustic waveform. Speaker characteristics, emotional tone, situational context, and prosodic cues are a few of the informational features conveyed by spoken language. Each extralinguistical feature has the potential to affect the resource demands of speech processing. These features are integrated in parallel with syntactical knowledge that can be richly complex—as we see from the challenge placed on working memory resources when we process object-relative clauses in a sentence, such as, "The ball that was thrown by the boy rolled into the street."

Consider the many different ways that one can say the word *yes*. It can be said in a matter-of-fact way, indicating simply that one is still listening to the current conversation. Or, it can be used to signify agreement with a wide range of enthusiastic overtures, ranging from the emphatic, "Yes, this is certainly the case, and I agree wholeheartedly," to a mere, "Yes, I see your point." Or, it can even signify agreeable opposition, as in the case of, "Yes, but ... ." So, beyond the obvious speaker-to-speaker variations in pronunciation of specific words, a tremendous amount of information is conveyed by subtle changes in the acoustics of any given word.

The speaker, not the listener, drives the rate of speech processing. With text, the reader has the ability to change the pace as processing difficulty increases—not so with listening. Speakers articulate words and phrases at their own pace, and this pace varies tremendously from person to person (Miller, Grosjean, & Lomanto, 1984). As discussed in more depth in this chapter, the variability of speech rate poses additional challenges to figuring out how to segment the acoustic stream effectively.

Another factor adding to the complexity of speech processing is that, unlike text, listening to speech is often an interactive rather than a passive process. Many everyday situations require us not only to listen to a speaker's meaning but also simultaneously to formulate thoughts and prepare for speech production to facilitate meaningful communication. The task of timing our conversational utterances to begin at appropriate points—called turn-taking—is both resource demanding and critical to successful communication (Sacks, Schegloff, & Jefferson, 1974). Actively engaging in a conversation requires more mental effort than simply passively listening to the radio (Strayer & Johnston, 2001).

Despite the extensive effort that has gone into the development of models of speech perception, considerably less emphasis has been placed on understanding the mental effort involved in speech processing. As Fischler (1998) pointed out, attention and language are "two of the most widely studied aspects of human cognitive skills" (p. 382). Despite this, much less concern has been given to the mental effort required to process speech (Fischler, 1998). The dominant assumption appears to have been that understanding the meaning of speech may be effortful, but that the perceptual process of extracting the speech signal occurs rather automatically. And in fact, for most listeners, in many situations, speech processing appears to require little effort. However, when the listening situation is less than optimal, overall task demands are high, or the message is particularly complex, the effort required to process speech becomes apparent, and the nature of this effort has important implications for understanding human performance in a wide variety of settings.

Aside from the inherent difficulty of processing syntactically complex semantic strings, sensory and additional cognitive factors have been shown to affect the mental workload of speech processing. Degraded listening conditions due to faint signals, noise, reverberation, or competing stimuli (verbal or nonverbal) represent some of the acoustical factors that can increase the mental effort of speech processing. Challenging acoustical situations include listening to a phone conversation when the connection is poor and the speaker is barely audible or trying to follow the conversation of a friend at a noisy, lively party where there are many multiple competing sources of sound and vision. College students asked to assess the meaning of syntactically complex sentences in the presence of noise or a concurrent digit load showed significant performance difficulties (Dick et al., 2001)—even to the extent of resembling the deficits that brain-damaged aphasic patients typically exhibit. Similarly, 18- to 25-year-olds displayed sentence-processing deficits characteristic of much older adults when the sentences they had to process were presented at low presentation levels in conjunction with other demands for their attention—like driving (Baldwin & Struckman-Johnson, 2002).

Cognitive factors may also increase the mental workload of speech processing. For example, speech processing requires greater mental effort in the absence of contextual cues or when the material is difficult to comprehend due to low cohesion or coherence, complex syntactical structure, or the presence of unfamiliar words (Chambers &

Smyth, 1998; Sanders & Gernsbacher, 2004). Several studies have demonstrated that text cohesion—the degree to which a reader must fill in contextual gaps—influences the difficulty of text comprehension (McNamara & Shapiro, 2005). Similar effects are present in discourse or speech comprehension. Ambiguous pronouns, such as in the sentence, "Ivan read the poem to Bob, and then Jill asked him to leave," result in low coherence and leave listeners trying to figure out which "him" has to leave.

Competition for mental resources may also arise from simultaneous visual and psychomotor demands. Even under relatively ideal listening conditions, processing speech imposes attentional processing requirements that can impair performance on other tasks. So, while highly overlearned and relatively automatic tasks, such as maintaining lane position while driving, may not be affected by concurrent auditory processing, more resource-demanding tasks such as speed maintenance and decision making are adversely affected (Brown & Poulton, 1961; see review in Ho & Spence, 2005). Recarte and Nunes (2002) found that, similar to other mentally demanding tasks, simply listening to words for later recall impaired drivers' ability to maintain speed on a highway.

Adding to the complexity of the speech-processing task is the fact that spoken words are not neatly separated into individual components like their printed form. Rather, the acoustic pattern of spoken words is a complex arrangement of sound energy requiring extensive prior knowledge of the language in order to impose phonemic context and ultimately to construct semantic structure and meaning from the continually varying physical stimulus. This challenge is referred to as the *segmentation problem*.

SEGMENTATION

A key issue to resolve in speech perception is how we take the physical stimulus, which is a time-varying continuous signal, and construct a perceptual interpretation of discrete linguistic units (Nygaard & Pisoni, 1995). Unlike the printed word, there are no gaps between spoken words. In fact, although linguistic units such as syllables, phonemes, and words appear to be perceived as discrete units, their corresponding acoustical representations are often continuous and overlapping. Listeners must determine where one word ends and another begins—the segmentation problem (Davis, Marslen-Wilson, & Gaskell, 2002). Our ability to translate the rather continuous physical stream into discrete linguistic units is not completely understood, but segmentation does point to the interaction of sensory-acoustical and perceptual-cognitive factors, that is, distinct acoustical changes and inferences based on expectations stemming from contextual knowledge. Listeners are unable to segment speech in unfamiliar languages. Even trained phoneticians correctly recognize only about 50% of the phonemes spoken in an unfamiliar language (Shockey & Reddy, 1975, as cited by Plomp, 2002).

Cues to Segmentation

Several theoretical explanations have been proposed for how segmentation occurs (Cutler, 1995; Davis et al., 2002; Norris et al., 1995). One early proposal was that segmentation is accomplished through sequential recognition (Cole & Jakimik, 1980; Marslen-Wilson & Welsh, 1978). According to this account, we are able to perceive the start of a new word once we have identified the word preceding it.

However, a strict interpretation of this theoretical perspective would indicate that we must be able to determine the offset of a word before we could tell where the next word begins. Current evidence does not favor this approach. For example, using a gating paradigm, Grosjean (1985) found that roughly 50% of monosyllabic words are not recognized until some time after their offset. That is, the first word is not identified until partway through the next consecutive word.

A second primary theoretical explanation for word segmentation abilities involves lexical competition (Cutler, 1995; McClelland & Elman, 1986). According to this perspective, multiple lexical candidates across word boundaries compete, and the "best fit" is selected to segment the acoustical stream. This perspective fits within the framework of the theoretical tenets of models of speech perception, such as TRACE (McClelland & Elman, 1986) and shortlist (Norris, 1994), discussed later in this chapter.

A third possibility is that word segmentation occurs explicitly in a separate process (Cutler, 1995; Norris et al., 1995). According to this perspective, lexical selection is guided by probability information regarding where a word boundary is likely to be (Norris et al., 1995). Acoustic cues, such as the onset of strong syllables, are used as cues to word segmentation. As pointed out by Norris et al. (1995), most content words (90%) do in fact begin with a strong syllable, and roughly 75% of the time strong syllables are the initial syllables of content words.*

In addition, speakers pronounce phonemes differently depending on the phonemic context. These allophonic cues (phonemic variations) provide listeners with cues to segmentation (Kent & Read, 1992). For example, a phoneme at the beginning of a word will tend to have a longer duration than the same phoneme in later segments of a word (Peterson & Lehiste, 1960). Listeners make use of these cues to determine word boundaries during segmentation (Davis et al., 2002).

These three accounts—the sequential, lexical competition, and separate mechanism theories—may not be mutually exclusive (McQueen, Norris, & Cutler, 1994; Norris et al., 1995). First, most models of speech perception, including the lexical competition models such as TRACE and shortlist, are primarily sequential. These models suggest that once a word has been recognized, alternative lexical candidates are inhibited. Therefore, once an initial word is recognized, subsequent words are easier to identify since they will be segmented from the preceding word, thus reducing the number of lexical alternatives (Norris et al., 1995). Another challenge to spoken word perception is its tremendous variability.

VARIABILITY

The speech signal varies tremendously as a function of both phonetic context and speaker. Different speakers produce substantially different acoustic versions of the same utterance, and any one speaker will produce the same utterance differently depending on the context in which it is spoken. This was illustrated in the example

---

* Acoustic cues also appear to play a role in segmenting speech material at the paragraph level. Waasaf (2007) demonstrated that both speakers and simultaneous interpreters make use of intonation patterns to indicate the start and end of paragraphs, using higher pitches at the start of new paragraphs and lower pitches to signify the end of a paragraph.

of the many ways a person might say the word *yes*. In the words of Plomp (2002), "No two utterances of the same speech fragment are acoustically equal" (p. 94). Our remarkable ability to comprehend speech relies on our ability to map a highly variable signal to an invariant lexical code.

The difficulty inherent in recognizing such a variant signal is illustrated by an investigation conducted by Strange, Verbrugge, Shankweiler, and Edman (1976). They compared the formant frequencies for vowels produced by untrained speakers of different age and gender (5 men, 5 women, and 5 children). Listeners were asked to identify either vowels produced by a single talker or a random mix of vowels spoken by several talkers either in isolation or embedded in a constant consonant-vowel-consonant (CVC) word. Their results indicated that embedding the vowel in the word context greatly improved vowel identification (17% and 9.5% misclassifications in the mixed- and single-talker conditions, respectively, as compared to 42.6% and 31.2% misclassifications in the isolated vowel condition, respectively).

An even more important observation in the study by Strange et al. (1976) was the tremendous variability in the acoustic signal of the vowels when produced by men, women, and children. The overlap in formant frequency for different vowels was so great among the different categories of speakers that configuration for a vowel for one speaker category corresponded with a completely different vowel from another speaker category. These observations illustrate the variability of the speech signal and emphasize the importance of the interaction between sensory and cognitive factors in speech recognition. Although listeners were able to correctly identify vowel sounds better when vowels were produced by the same speaker category, overall recognition was still significantly less than optimal.

Given the degree to which the acoustic pattern of individual words varies from speaker to speaker and even within the same speaker at different times and in different contexts, our ability to comprehend speech is nothing less than remarkable. However, recognizing variant speech signals in everyday contexts requires considerable processing resources. A more variant or degraded speech signal can be expected to require considerably more processing resources or mental workload. The difficulty of learning to understand a new acquaintance with a heavy accent or a young child illustrates how phonemic variance increases the demands of the speech-processing task.

Much of the variability issue we have been discussing has pertained to variation at the phonological level. At a minimum, theories of language find it useful to distinguish between phonological, lexical, and syntactic levels of processing (Jackendoff, 2007). Each of these levels is discussed in turn, beginning with the phonological level. Since phonemic variation is so prevalent, our discussion of processing at the phonological level includes additional examples of variation.

Phonological Level

Depending on such factors as context, emotional tone, and even the personality of the speaker, the acoustical properties of a phoneme or syllable will change quite dramatically (see Kuhl, 2004; Strange et al., 1976). Phonemes are the smallest unit of speech, which if changed would change the meaning of the word. They can be compared to morphemes, which are the smallest unit of speech that has meaning. For example, the /b/ and the /oy/ in boy are both phonemes, although neither has meaning by itself.

Together, they form one morpheme: boy. Adding the single letter /s/ to the word *boy*, making it boys, changes the meaning of the word. In this example, the /s/ in boys is both a phoneme and a special class of morpheme—called a bound or grammatical morpheme. Word endings such as /s/ and /ed/ have meaning but cannot stand alone like free morphemes (i.e., boy and cat). See the work of Carroll (2004) for further details.

Coarticulation

Segments of speech are not typically produced in isolation but rather in phonemic context, meaning in combination with other segments to form syllables, words, and phrases. When produced in combination, speech segments interact or are coproduced in a process called coarticulation (Kent & Read, 1992). Coarticulation specifically refers to situations in which the vocal tract adjusts in such a way that it enables the production of two or more sounds concurrently or in rapid succession. The production of one phoneme can be influenced by either preceding or subsequent phonemes in forward (anticipatory) or backward (retentive) coarticulation. The resulting effect is a change in the acoustical properties of any given phoneme as a function of its phonemic surroundings.

Different acoustic variations for the same phoneme due to coarticulation are called *allophones*. Allophones are cognitively classified as the same phoneme, despite the fact that they differ acoustically. One may be aspirated while another is not, or they may vary substantially in duration. For example, the phoneme /t/ results in quite a different allophone when used in the word *tap* relative to its use in *dot* (Jusczyk, 2003). In other words, depending on the surrounding sounds, the same phoneme is said in different ways. In fact, in certain contexts different phonemes can essentially sound more like each other than individual allophones do. For example, as McLennan, Luce, and Charles-Luce (2005) pointed out, in everyday English conversation, /t/ and /d/ in the middle of a word (such as in water or greedy) often sound similar. Despite this tremendous acoustic/phonemic variability in the speech signal, even infants are capable of classifying speech sounds (see Kuhl, 2004). Infants of only 6 months of age can distinguish vowels of the same category /a/ from /e/ across a wide variety of speakers (Doupe & Kuhl, 1999; Kuhl, 1993).

A number of theories have been presented to illustrate how listeners are able to interpret the many different acoustical versions of the same phoneme. One of the more influential early theoretical perspectives involved the search for invariant acoustic cues (see, e.g., Blumstein & Stevens, 1981; Cole & Scott, 1974). However, the identification of invariant acoustic cues, that is, acoustical cues that remain stable across particular phonemes regardless of context or speaker, has remained elusive (see review in Nygaard & Pisoni, 1995).

A more recent theoretical explanation for the ability to recognize phonemes under conditions of natural speech variability posits the existence of a phonemotopic map in the perisylvian regions in the temporal cortex (Shestakova, Brattico, Soloviev, Klucharev, & Huotilainen, 2004). Making use of magnetoencephalography (MEG), Shestakova and colleagues (2004) examined N1m (the magnetic analog to the N1 event-related potential [ERP] component) responses to three Russian vowels spoken by 450 different speakers. The use of numerous different speakers resulted in variability across phonemes, as would be expected in natural speech. Using a dipole modeling procedure, they identified the average N1m equivalent current dipole

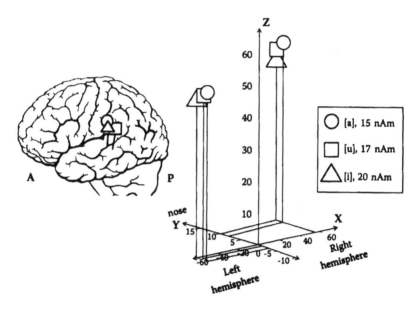

FIGURE 8.1   Results of dipole location for vowel sounds. (Redrawn with permission from Shestakova, A., Brattico, E., Soloviev, A., Klucharev, V., & Huotilainen, M. (2004). Orderly cortical representation of vowel categories presented by multiple exemplars. *Cognitive Brain Research, 21*(3), 342–350.)

(ECD; a source localization technique) peak latency, source strength, and location coordinates for each of the three vowel sounds across individual listeners. As illustrated in a reproduction of their results (see Figure 8.1), greater Euclidean distances were observed for the more spectrally distinct vowels [a] and [i] as well as a statistically significant absolute location difference in the z (inferior-superior) coordinate between these same vowels. The vowel [u], which more closely resembled [i] in spectral characteristics, did not differ with the other vowels in terms of either Euclidean distance or absolute location.

The results of Shestakova and colleagues (2004), along with similar previous findings (Obleser, Elbert, Lahiri, & Eulitz, 2003), provided evidence for the existence of a phonemotopic map in the left and right hemispheric perisylvian regions of the human brain that would aid phonemic recognition.

Phonological processing is also aided by grammatical information. Grammar consists of a set of rules for how phonological units and words can be organized and used to generate meaningful sentences (Carroll, 1994). In all languages, there are certain phonological units that are unacceptable combinations and therefore cannot be used to construct meaningful semantic units. For example, in English the phonological units *p* and *b* are never combined. Therefore, although we can use *p* to form the word *pin* and *b* to form the word *bin*, *pbin* is not a grammatically acceptable combination. These grammatical rules narrow somewhat the number of acceptable sounds in any given language—thus aiding processing at the phonological level. Phonological information must be matched to information in the mental lexicon—our repertoire of stored lexical units.

LEXICAL SELECTION

All theories of speech perception must account for the process of lexical selection (Altmann, 1990; Jackendoff, 1999). At a minimum, this involves inclusion of a *mental lexicon*, a stored repertoire of lexical units and a set of rules by which the lexical units can be combined. Exactly what is contained in the mental lexicon is a point of debate. However, it is acknowledged that any theory of speech must explain how information from the acoustic level is bound with information at the lexical selection level to achieve semantic comprehension (Hagoort, 2003). This process is referred to as *binding*, thus identifying its similarity to combinatorial processes in other modalities. The example of binding in visual processing that is often given is the issue of how color and shape—known to be processed by separate neuronal pathways—are bound together to allow the perception of one coherent object. Speech processing relies on a similar process that is further complicated by the fact that the acoustical input does not occur simultaneously as in vision. Acoustical information to be bound to lexical information occurs over time; therefore, initial information must be held in some type of storage until lexical selection occurs (Davis et al., 2002; Marslen-Wilson, 1987; O'Rourke & Holcomb, 2002). As discussed in Chapter 2, this temporary holding place can be called the phonological store (Baddeley, 1992). The phonological store, a subcomponent of the phonological loop having a duration of 1 to 2 s, is thought to be particularly essential to lexical selection when processing demands are high and to facilitate learning of novel or unfamiliar sound patterns (Baddeley, 1992; Baddeley et al., 1998). Ambiguous, novel, or unfamiliar acoustical information can be held in the phonological store until lexical selection or meaning is assigned. The mental resource requirements of the lexical selection process are affected by both sensory and cognitive factors. For example, when the incoming signal is degraded, the number of potential lexical alternatives will increase, and lexical selection will require greater mental resources (Baldwin, 2007).

SENSORY-LEVEL EFFECTS

The quality of the listening environment plays a critical role in the difficulty of the lexical selection process. Imagine, for example, that a listener is presented with the sentence, "The bats were lying around the dugout after the storm." If the word *bats* is distinctly heard, then for up to 400 ms, both contextually appropriate and inappropriate meanings of the word would be activated, and the most appropriate meaning would be selected after the addition of contextual information. However, suppose the word *bat* was not distinctly heard, but perhaps only the phoneme /_at/ was distinctly heard. Then, words with all possible preceding consonants (i.e., cat, vat, mat, bat) would initially be generated, followed by the activation of all meanings associated with each of the possible words. The automatic activation of such an extensive list of words and subsequent meanings would likely make the lexical selection process more time consuming and mentally demanding. Acoustic factors also have an impact on the duration and strength of the phonological store.

Presentation intensity (how loud or salient a signal is above background noise) has a direct effect on the duration of the echoic trace in phonological storage. Increases in intensity are associated with increased echoic memory duration (Baldwin, 2007). An echoic trace of longer duration may increase the functional capacity of the phonological store, thus facilitating the lexical selection process during a challenging period of high processing demand. Further, the echoic trace is thought to decay rather than extinguish abruptly. Thus, the phonological material available in the initial period of a more persistent echoic trace would be expected to be of greater clarity relative to shorter duration traces. The greater clarity of the phonological material may decrease the mental resource requirements of lexical extraction. Models of spoken word recognition propose that multiple lexical candidates are activated by the initial word segments until alternatives can be ruled out through subsequent acoustical input (Davis et al., 2002). Accordingly, initial segments of greater ambiguity would result in activation of a greater number of lexical alternatives and therefore a more cognitively demanding lexical selection process. Presentation level, a sensory factor, may therefore interact with subsequent perceptual and cognitive stages of auditory processing, potentially compromising performance, particularly when task demands are high.

Cognitive-Level Effects

Cognitive-level effects, including the frequency of the lexical candidate and its distinctiveness, will also influence the resource demand of the lexical selection process. Lexical items that are used more frequently (i.e., the word *back* vs. the word *bade*) are recognized more easily (Luce, Pisoni, & Goldinger, 1990). At the same time, lexical items that have fewer neighbors—similar-sounding words—are recognized more easily. As someone who has played the game Scrabble knows, the lexical neighborhood containing the letter *q* is much smaller than the one containing the letters *s* or *e*. Likewise, after hearing the initial phonemic components *xy* in the word *xylophone*, there are few lexical neighbors to be activated to complete the word. In contrast, hearing *cy* could activate a number of candidates, including cyclone, cyborg, silo, psychology, and so on. Thus, the lexical units themselves will influence the mental demands of the selection process, and the initial acoustic information must be held until the ambiguity can be resolved. Sometimes, this ambiguity still cannot be resolved at the word level, and the potential lexical candidates and their associated meanings must be held for an even longer period of time.

Till and colleagues provided strong evidence for the importance of echoic memory in the lexical-semantic processing stages (Till, Mross, & Kintsch, 1988). They had participants read sentences containing an ambiguous word such as *mint* and then perform a lexical decision task in which they heard words that were (a) semantically related and appropriate to the meaning (money); (b) semantically related but inappropriate to the meaning (candy); or (c) inferentially related (earthquake). The interval between the prime word (mint) and the target word varied from 200 to 1,500 ms. Within the first 400 ms, the meaning for both contextually appropriate and inappropriate words was facilitated. Beyond 400 ms, only meaning for contextually appropriate words was facilitated, suggesting that participants were processing words according to their context while suppressing contextually inappropriate meanings. The ability to discriminate between alternative meanings of ambiguous words based on context relies on the ability to store the auditory information until such decisions can be made.

SENSORY-COGNITIVE INTERACTIONS

The interaction between sensory and perceptual-cognitive processes has particular relevance for examining speech recognition in hearing-impaired and older listeners. Older listeners with elevated hearing thresholds are known to rely on context for speech processing more than their younger counterparts (Marshall, Duke, & Walley, 1996). Marshall and colleagues observed that, in low-context conditions, older listeners required more of the acoustic-phonetic information of a target word to identify words relative to younger listeners. Older participants took longer than their younger counterparts to identify the initial phonemes at the beginning of stimulus words. Once phonemes were recognized and phoneme cohorts were identified, older individuals tended to compensate or make up for lost time by performing the later stages of word recognition (isolating the unique word from among lexical alternatives and accepting this identification) faster than their younger counterparts. The additional cognitive effort that hearing-impaired and older listeners must apply in later stages of word recognition to compensate for initial phoneme identification difficulties could compromise performance when overall task demands are high. During periods of high task demand, a more persistent echoic trace would allow listeners greater flexibility in allocation of limited attentional resources. Additional issues related to the auditory processing abilities of older adults are discussed in Chapter 10.

SYNTACTICAL LEVEL

Syntax refers to the rules governing how words can be used in coherent discourse to convey meaning. In most languages, such as English, sentences must contain a subject (S) and verb (V) and frequently contain an object (O). In these SVO syntactical structures, we see that something or someone does something to something or someone. For example, consider the following simple sentences using the SVO structure:

The boy kicked the ball.
The cat chased the mouse.

Our knowledge of syntactical structure tells us that it was the ball that was kicked and that the boy did the kicking. There is no ambiguity about whether the ball did the kicking. This is true even for the second sentence—in theory, a mouse could chase a cat. However, our knowledge of syntax constrains the interpretation of these sentences.

Syntax involves much more than mere word order. We have no difficulty understanding who is doing what to whom, even if we rearrange the order of the words in the sentence as follows:

The ball was kicked by the boy.
The mouse was chased by the cat.

Syntax plays a more powerful rule as sentences become more complex. For example, consider the following sentences:

The boy who saw the cat kicked the ball.
The cat who was hit by the ball chased the mouse.

Syntax allows us to interpret these sentences correctly. The more complicated the syntax is, the more difficult the utterance will be to comprehend. Consider the following examples from Christianson, Hollingworth, Halliwell, and Ferreira (2001) of what are called "garden path sentences" since they lead the reader or listener down a garden path to a sometimes surprising end:

While the man hunted the deer ran into the woods. (p. 369)
While Anna dressed the baby spit up on the bed. (p. 369)

Several factors determine the difficulty of processing these garden path sentences and affect the likelihood that the reader or listener will misunderstand the sentence or require additional time to reinterpret the sentence. People with smaller working memory capacities take longer to read garden path sentences relative to those with larger capacities (King & Just, 1991; Miyake, Just, & Carpenter, 1994). When encountering ambiguities, people with larger working memory capacities appear to be able to hold multiple meanings in mind simultaneously and therefore require less time to resolve the ambiguities when contextual cues are provided (MacDonald, Just, & Carpenter, 1992; Miyake et al., 1994). However, people in general have more difficulty processing ambiguous or syntactically complex sentences.

Evidence that these complex sentences are more difficult is found in the observation that people take more time to process ambiguous words. Consider the example provided by MacDonald et al. (1992) of the temporarily ambiguous word *warned* in the sentences that follow. When it is initially encountered, it could be either the main verb of the sentence (as in the first sentence), or it may indicate that a relative clause is about to follow (as in the second sentence).

"The experienced soldiers warned about the dangers before the midnight raid."
"The experienced soldiers warned about the dangers conducted the midnight raid."
(p. 61)

Chambers and Smyth (1998) observed that several syntactical and structural properties of discourse affect coherence and assist the listener in interpreting ambiguous words, pronouns in particular. Coherence is increased when an utterance contains a pronoun in the same structural position as the person to whom it refers. For example, Sentence 1 is more coherent than Sentence 2:

1. Mary raced toward Jill and then she unexpectedly came to a stop. (she = Mary)
2. Mary raced toward Jill and then unexpectedly she came to a stop. (she = ?)

The mental resource demands of speech perception will be affected by both sensory and cognitive factors at the phonological, lexical-semantic, and syntactic levels of processing. Effective models of speech perception must account for these influences at all three levels.

APPROACHES TO SPEECH-PROCESSING THEORY

Discussion of several key models of speech processing illustrate how each attempts to deal with the various factors that have an impact on the binding of phonological, lexical, and syntax information. Emphasis is placed on the implications of the theoretical models for mental workload demands. Numerous models have been developed. In general, they share the common assumption that speech processing involves multiple levels of processing (Jackendoff, 1999; Marslen-Wilson, 1987; Massaro, 1982; McClelland & Elman, 1986). The models differ in several key ways. One key point of dissociation is whether they tend to view "speech as special," incorporating a modular view in the tradition of Fodor (1983) and Chomsky (1984) or whether they view speech processing as accomplished by the general cognitive architecture in the tradition of connectionist modelers (McClelland & Elman, 1986; Rumelhart et al., 1986). We see this theme of modularity versus singularity arise in the proposed content of the mental lexicon, particularly whether the rules for lexical combination (syntax) are contained in the same or different architectural structures—dual- versus single-mechanism theories (Joanisse & Seidenberg, 1999, 2005; Ullman, 2001a; Ullman et al., 1997). We also see this controversy over modularity arise in critical debates on the nature of verbal working memory (vWM: Caplan & Waters, 1999; Just & Carpenter, 1992; Just, Carpenter, & Keller, 1996; MacDonald & Christiansen, 2002; Waters & Caplan, 1996).

In addition, key models of spoken word perception can be differentiated by the degree of interactivity presumed to occur between the various levels of processing (Altmann, 1990; Cutler, 1995; Jackendoff, 2007; Marslen-Wilson & Welsh, 1978). All theories must account for the fact that speech comprehension requires listeners to quickly recode a transient acoustic waveform into a more lasting representation by matching it to specific items stored among the many tens of thousands of items in the mental lexicon for further speech processing to occur (Luce et al., 1990). This involves binding information from phonological, lexical-semantic, and syntactical processes (Jackendoff, 2007). The initial stages of speech processing include an acoustic-sensory level at which sounds are registered and a phonological level (at which sounds are recognized as portions of speech). Additional levels, occurring either subsequently or simultaneously—depending on one's view— include lexical selection, word identification, and semantic comprehension, aided by syntax. The registration of sounds at the acoustic-sensory level relies heavily on bottom-up processes. Models of speech processing differ concerning the influence that top-down processes can exert on lower-level processes and the extent to which various processing stages are thought to interact (Friederici, 1999; Jusczyk & Luce, 2002; Nygaard & Pisoni, 1995). For example, Norris's (1994) shortlist model presumes that speech recognition is an entirely bottom-up, data-driven process, while the TRACE model of McClelland and Elman (1986) assumes interactive activation of bottom-up and top-down processes.

We first begin with a discussion of the modularity issue in the form of single-versus dual-mechanism accounts of processing. This ongoing debate has frequently centered on how people process regular and irregular past-tense verbs. It has been referred to at times as the words-and-rules issue.

SINGLE- VERSUS DUAL-MECHANISM THEORIES

The dual- versus single-mechanism issue has generated a productive debate in the literature. Dual-mechanism theories propose that the mental lexicon and the rules for combining its elements are separate processes. More specifically, they postulate separate cognitive and neural mechanisms for learning and knowledge of grammatical rules versus the storage of sound-word associations and semantic meanings (Chomsky, 1995; Friederici, 1999; Pinker, 1994; Pinker & Ullman, 2002; Ullman, 2001b; Ullman et al., 1997).

Much of the debate has involved examining the differences between the way in which regular and irregular past-tense verb forms are learned and retrieved. Regular past-tense forms follow clear rules, like adding -ed to change walk to walked and learn to learned. Irregular forms (which are much less frequent) do not conform to the rules. For example, the irregular past-tense form of go is went and of ring is rang. The dual-mechanism theory proposes that the rule-based forms are processed via the frontal cortex and basal ganglia mechanisms, while the irregulars rely on a mental lexicon involving temporal lobe mechanisms (Ullman, 2001c).

Support for the dual-mechanism theory comes primarily from investigations of children's language acquisition and double dissociations found in different forms of aphasia. For example, neurological evidence provided support for dual-processing mechanisms distinguished by separable functional brain subregions in Broca's area (Newman, Just, Keller, Roth, & Carpenter, 2003; Ullman, 2004; Ullman et al., 1997).

Ullman and colleagues (1997) suggested that language processing relies on both an extensive mental lexicon in which arbitrary sound-meaning pairings of words are stored and a separate rule-based system that they termed the mental grammar. In this model, the mental grammar facilitates the combination of words and phrases according to generative rules (Pinker, 1994; Pinker & Ullman, 2002). As children learn rules (e.g., adding -ed to many verbs results in a past-tense morphology, as in talk and talked), they are able to apply the rule in novel situations, such as to make past-tense forms of novel verbs (e.g., skug can be transformed to skugged). Ullman and colleagues discussed evidence that this rule-based system relies on brain mechanisms (left frontal areas, including Broca's area) separate from the more memory-based repository system, which relies primarily on left temporal cortical areas. In support of the dual-mechanism account, Ullman and colleagues observed neural dissociations between patients with lesions affecting each of the areas of theoretical interest. For example, they reported that patients with frontal lesions often exhibit aggrammatic forms of aphasia characterized by omission or misuse of grammatical morphemes. (Grammatical morphemes include items such as re-, -ed, and -s, morphemes that by themselves have little or no meaning but change the meaning when combined with a lexical morpheme like *work* or *play*). However, their ability to use content words and nouns may be relatively intact. Conversely, individuals with lesions in temporal or parietal areas are more likely to use relatively intact syntactical structure while having profound deficits in accessing semantically appropriate content words. Specifically, patients with posterior aphasias and diseases such as Alzheimer's that affect the general declarative memory system have more difficulties producing irregular past-tense forms relative to regular and novel past-tense verb forms. However, they may actually overuse grammatical rules. Conversely, patients with anterior aphasias and diseases affecting motor areas (such as Parkinson's) typically have more difficulty using rules to produce regular and novel past forms.

Others argued that a dual-mechanism account is not necessary to explain the results of specific neurological impairments (Joanisse & Seidenberg, 1999; Kello, Sibley, & Plaut, 2005; Plaut, 2003). Single-mechanism accounts (which are primarily based on connectionist or neural network models) provide computational evidence that separate mechanisms are not necessary to explain the various patterns of results that are normally taken as support for the dual-mechanism route. Instead, a single computational system suffices (Jackendoff, 2007; Joanisse & Seidenberg, 1999, 2003). Models have been developed that can even account for the specific language deficits observed in different forms of aphasia (e.g., Joanisse & Seidenberg, 1999).

Joanisse and Seidenberg (1999) demonstrated that the past-tense dissociations observed in previous research could be explained within a single connectionist architecture by assuming the network to have both phonological and semantic nodes. They pointed out that there are similarities between many regular and irregular past-tense verbs. For example, like regular past tense, most irregular past-tense verbs still begin with the same letter and often contain the coda (the end portion of a syllable) of the main verb. They provide the following examples: *bake-baked* and *take-took*. Some irregular forms are quite similar to regular forms that are phonemically similar. They provided the examples of *crept* and *slept*, pointing out their similarity to *cropped* and *stepped*. Finally, they pointed out a third pattern that demonstrates an internal form of regularity even though it differs from regular past-tense verbs. Examples of this category are sing-sang, ring-rang, and grow-grew, throw-threw, blow-blew.

These observations led Joanisse and Seidenberg (1999) to suggest that difficulties with phonological processing would result in greater difficulty forming past-tense forms of novel words than of irregular past-tense verbs. Conversely, to construct most irregular past-tense verbs, the entire word must be recognized. Thus, irregular past-tense verb formation would be more affected by semantic difficulties. They pointed out that this corresponds with observations from the two forms of aphasia. Those with Broca's-type aphasia or Parkinson's tend to have greater difficulty with planning articulatory output and have difficulty applying the regular past-tense rule to novel words. Conversely, those with posterior aphasia or Alzheimer's disease tend to have semantic deficits and are more impaired at producing irregular past-tense forms.

Joanisse and Seidenberg (1999) provided support for their position using a connectionist model that contained phonological processing nodes corresponding to acoustic features of speech (which would be used in hearing and speaking) and semantic nodes consisting of each verb to which the phonological nodes must be matched for comprehension or conversion from present to past tense. They mimicked language

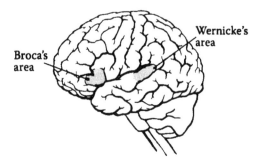

FIGURE 8.2   Broca's and Wernicke's areas. (Drawn by Melody Boyleston.)

acquisition by training the network using a set of present- and past-tense monosyl-labic verbs (containing both the dominant regular form and a smaller proportion of the irregular form). After training, they then damaged the network in one of two ways. First, they damaged only portions of the phonological layers. In the other ver-sion, they damaged only portions of the semantic layer. The resulting performance of the connectionist model mirrored their predictions and the dissociations observed in aphasics. Damage to the phonological layers resulted in more errors when the network attempted to generate past-tense forms of novel words (thus when trying to apply the rule) rather than irregular forms. Conversely, the network demonstrated more errors generating the irregular past-tense forms (relative to novel words) when portions of the semantic layer were damaged.

Joanisse and Seidenberg (2005) found evidence that the phonological character-istics of the past-tense verb may play a larger role in the neural pathways that are activated than whether the verb is regular or irregular. Recall that regular past-tense verbs are thought to rely more on frontal areas, such as Broca's area, while irregular past-tense forms are thought to rely more heavily on posterior temporal areas (Pinker & Ullman, 2002; Ullman et al., 1997) (see Figure 8.2). Using functional magnetic resonance imaging (fMRI), Joanisse and Seidenberg (2005) observed that irregular past-tense forms that shared phonological similarities with regular past-tense forms (e.g., slept, fled, sold) also produced more frontal activation similar to regular forms and relative to irregular forms that were not phonologically similar to regulars (e.g., took, gave) (p. 282). These observations led Joanisse and Seidenberg to the conclusion that dual mechanisms are not necessary to explain the dissociations between regular and irregular past-tense verb formation common to different forms of aphasia.

A debate of similar nature—but perhaps even more germane to the topic of men-tal workload—involves the nature of vWM. On one side of the debate, in the modu-larity or "speech-is-special" camp, are those who propose that language processing makes use of a specialized vWM system that can be distinguished from the more general vWM system used for conscious, controlled processing of verbal information (Caplan & Waters, 1999; Waters & Caplan, 1996). Conversely, others have argued that vWM is a general system that supports all verbal tasks (Just & Carpenter, 1992; Just et al., 1996). We briefly review this debate here, paying particular attention to its implications for positing independent versus general resource pools for verbal tasks. Those who suggest that language has a specialized vWM (Caplan & Waters, 1999;

DeDe et al., 2004; Waters & Caplan, 1996) proposed that this separate system draws on a resource pool largely independent from that of the general working memory system.

SINGLE OR MULTIPLE VERBAL WORKING MEMORY SYSTEMS

A single vWM account was proposed by Just and Carpenter (1992) in reference to a capacity theory of language comprehension. Their capacity theory postulated that individual differences in working memory capacity would affect language comprehension processes in several ways. Notably, working memory capacity would affect the activation and interaction of a greater number of alternatives when presented sentences were ambiguous, particularly regarding syntax. For example, ambiguous sentences were presented, such as, "The waitress was talking about the calamity that occurred before the end of her shift."

Listeners with high working memory capacities demonstrated recognition of the ambiguity, whereas listeners with low working memory capacities were unaffected by the ambiguity. Just and Carpenter (1992) suggested that the individuals with low working memory capacity were able to keep only one possible interpretation in mind and therefore selected that interpretation and were subsequently unaware of alternative interpretations of the sentence.

Just and Carpenter (1992) determined working memory capacity based on performance on Daneman and Carpenter's (1980) reading span task. The reading span task requires participants to read a set of unrelated sentences and subsequently recall the final word from each sentence. Modified versions of the reading span task (including non-language-specific versions such as the operation span) are frequently used to assess working memory capacity (Barrett, Tugade, & Engle, 2004; Engle, Kane, & Tuholski, 1999; Unsworth, Heitz, Schrock, & Engle, 2005). Complex span tasks are thought to involve conscious, controlled processing in addition to temporary storage (see Engle, 2001, for a review) and have been shown to be predictive of a number of cognitive tasks, including reading comprehension (Daneman & Carpenter, 1980); tests of general fluid intelligence (Conway et al., 2002); deductive reasoning (Barrouillet, 1996); and mathematical problem solving (Engle, Tuholski, Laughlin, & Conway, 1999; Passolunghi & Siegel, 2001). See the work of Conway and Kane (2001) for a review. The observation that performance on many language tasks (including reading comprehension and sentence processing) is associated with working memory span task performance provides evidence that they share a common mechanism (Just & Carpenter, 1992).

Conversely, Waters and Caplan (Caplan & Waters, 1999; Waters & Caplan, 1996), in the style of Fodor (1983), postulated that some aspects of language comprehension, specifically those largely unconscious aspects that occur nearly automatically, represent a unique cognitive domain that can reasonably be expected to have developed a unique resource pool at least partially independent from overall general working memory capacity. They called this position the separate-language interpretation resource (SSLR) hypothesis.

Caplan and Waters (1999) contrasted the SSLR theory with a single-resource (SR) theory of researchers, such as that of Just and Carpenter (1992), that would postulate that language processing must draw working memory resources from a single pool. According to SR theory, then, as working memory capacity diminishes, fewer resources are available for language processing. Performing an additional task that requires working memory resources interferes with sentence processing (Just & Carpenter, 1992), thus supporting the conclusion that they compete for resources. In fact, difficulties with syntactical language abilities have been observed when college students are exposed to general stressors such as concurrent digit load, noise, and speeded speech (Bates, Dick, & Wulfeck, 1999).

Additional evidence that speech processing is a form of well-learned auditory pattern perception, rather than a specialized task with its own processing module, comes from both behavioral investigations (see Massaro, 1995, for a review) and neuroimaging research demonstrating that complex environmental sounds and linguistic sounds evoke similar patterns of behavior and largely overlapping areas of brain activation (Dick et al., 2007). In fact, complex musical structures appear to be processed in much the same way as linguistic syntactical constraints in trained musicians (Patel, 2003).

Moving on from questions regarding the possibility of modularity of speech processing, next we examine another distinction between major models of speech perception. Focusing specifically now on models of word recognition, we discuss the relative independence or interactivity proposed between bottom-up and top-down processes.

MODELS OF WORD RECOGNITION

Several influential models of word recognition have been proposed. Each attempts to model the binding of acoustical information with lexical, syntactical information. A fundamental point of departure for the various models is whether they propose that these various stages of speech processing occur serially or in parallel and the degree to which they interact (Altmann, 1990; Cutler, 1995). At one end of the spectrum is the TRACE model (McClelland & Elman, 1986) of speech perception, one of the most widely regarded models of speech perception. TRACE is a computational model that proposes that several levels of processing—distinctive features, phonemes, and words—are simultaneously activated and interactively drive word recognition and speech comprehension. Conversely, computational models such as shortlist (Norris, 1994) propose an entirely bottom-up process. In shortlist, lexical information does not influence phonemic information; the process is strictly serial and feed forward.

Other models propose a limited degree of interaction. The COHORT model (Marslen-Wilson & Welsh, 1978; Marslen-Wilson, 1987) has been revised many times. It contains three primary stages or levels: access, selection, and integration. Originally, the early COHORT model contained a degree of interaction proposing that *access* from the spoken sound to *selection* at the lexical level was strictly bottom up—but that contextual cues from semantic and syntactical stages could influence the *integration* stage, thus assisting in the identification of which specific lexical form was most appropriate for the context. However, later versions of the COHORT model (Marslen-Wilson, 1987) revised this notion of interaction, proposing instead that the entire process proceeded in a bottom-up serial fashion and that stages became integrated but did not interact.

The neighborhood activation model (NAM; Luce & Pisoni, 1998) is similar to the COHORT model in that initial activation is thought to be an entirely bottom-up process. However, unlike the cohort model, NAM proposes that any portion of the word (not only the initial segment) is equally important in driving the activation process (Cutler, 1995). The COHORT model (Marslen-Wilson, 1987) proposed that only initial word phonemes were involved in driving the lexical selection process. Thus, hearing the word *bright* would activate cohorts such as bite, bake, blight, and black, but it would not activate the cohorts light and site. According to the COHORT model, word initial phonemes constrain activation of lexical items much more so than later phonemes; thus, unless the initial phoneme of the word is ambiguous—not heard properly—only cohorts beginning with the same initial phoneme will be activated. Conversely, NAM emphasizes a lexical activation process that involves the entire word segment, a feature it shares with the TRACE model.

The fuzzy logic model of perception (FLMP; Massaro & Cohen, 1991) provides yet another theoretical description of the process of integration between bottom-up and top-down processes. Like TRACE, FLMP assumes that initial sensory-perceptual processing occurs followed by a decision based on a relative goodness of fit or match with lexical candidates. The distinction here is that the FLMP model proposes that these bottom-up and top-down processes occur in parallel but independently (Cutler, 1995). Massaro and Cohen suggested that acoustical and contextual information independently influence perceptual recognition in both speech and reading. Both TRACE and FLMP assume that prelexical and contextual information influence lexical selection; the former simply proposes that the influences interactively occur, while the latter proposes that they exert independent influence.

Evidence from both behavioral studies and computational models overwhelmingly rule out a straight serial-sequential or feed-forward model of speech processing (McQueen, 2005). In the words of McClelland and Elman (1986, p. 4), "It is not only prior context but also subsequent context that influences perception." Subsequent phonemes are used in the segmentation process to determine where one word ends and another begins—meaning that we often do not recognize one word until some time after hearing the next word (Grosjean, 1985). This illustrates how context can have an impact on lexical selection. But, context can even affect our interpretation of acoustical information at the prelexical-phoneme selection level.

Ganong (1980) artificially manipulated the voice onset time (VOT) of acoustic sounds so that they were in the ambiguous zone between two phonemes. Recall that in the ta-da continuum, sounds with a VOT time of less than 30 ms are consistently labeled "d," while sounds with a VOT greater than 40 ms are labeled "t." Ganong constructed words and nonwords such as *dash* and *tash* and *dask* and *task* using ambiguous initial phoneme sounds (d-t) with a VOT of 35 ms. Listeners made use of lexical information to disambiguate the acoustic information, demonstrating a preference for the phoneme that resulted in a word. This demonstrated pattern of interactivity and influence—as proposed in the TRACE and FLMP models of speech processing—indicates that both sensory and cognitive factors will have an impact on the mental workload of speech processing.

PARAMETERS INFLUENCING MENTAL WORKLOAD

A number of different things have an impact on the difficulty of understanding speech. These parameters with an impact on the mental workload of speech processing can be environmental, as in the case of competing talkers or background noise; or a function of the properties of the speech content (i.e., syntactically complex sentences, unfamiliar vocabulary) or the speech signal (i.e., degraded transmissions, compressed signal, synthetic voice). These parameters can be divided into bottom-up and top-down influences. First, an important bottom-up influence—acoustic factors—is discussed.

ACOUSTIC FACTORS

One prediction resulting from current speech-processing models is that the difficulty of the speech-processing task will involve both the quality of the acoustical stimulus and the size of the lexical set from which a selection or decision must be made. At the same time, these two aspects interact: A more degraded acoustic stimulus will increase the number of lexical candidates that are activated and subsequently will increase the demand of speech recognition (see discussion in Baldwin, 2007). Numerous audiological reports have demonstrated that the difficulty of a speech-processing test varies in accordance with both the acoustical properties of the test stimuli and the semantic features of the test stimuli (i.e., test set size and contextual cues) (Carhart, 1965; Miller, Heise, & Lichten, 1951). For example, word recognition is better in the context of sentences, even if the sentences are only syntactically appropriate but are not semantically appropriate (Miller & Isard, 1963).

The acoustic quality of the speech is affected by the presentation level or intensity (loudness) of speech, the acoustic environment in which it is presented (e.g., noise, competing speech), and the sensory capabilities of the listener. Each of these aspects independently and interactively has an impact on the mental workload demands of the speech-processing task.

Presentation Level or Intensity

Presentation level or presentation intensity affects the difficulty of speech processing (Baldwin & Struckman-Johnson, 2002). This is often a neglected aspect of speech perception research. Although researchers have investigated numerous aspects of auditory speech processing, the role of speech intensity beyond mere detectability has received only scant attention. In cognitive research, researchers frequently fail to report the decibel level used for presentation of speech stimuli, making comparisons across investigations problematic.

When presentation levels for speech stimuli are cited, they often seem to be chosen apparently arbitrarily within a range of 50 to 90 dB (Loven & Collins, 1988; Tschopp, Beckenbauer, & Harris, 1991a), perhaps under the assumption that as long as speech is intelligible, it will require equivalent attentional resources for processing across this wide range of (above-threshold) presentation levels. Research by Baldwin and colleagues (Baldwin & Ash, 2010; Baldwin, Lewis, & Morris, 2006; Baldwin & Struckman-Johnson, 2002) contradicted this assumption.

Baldwin and colleagues (Baldwin, Lewis, et al., 2006; Baldwin, May, et al., 2006; Baldwin & Struckman-Johnson, 2002) used a dual-task paradigm to examine the mental workload associated with speech processing in various conditions. As discussed in Chapter 5, dual-task methods provide a more sensitive index of workload than can be obtained with primary task measures—such as speech intelligibility scores—alone. Primary task measures, namely, speech intelligibility quotients, have frequently been the quantitative measure of choice. For example, the effects of numerous physical and environmental parameters on the intelligibility of speech have been explored. Speaks, Karmen, and Benitez (1967) established the presentation levels associated with optimum speech intelligibility in environments with low background noise levels. Speaks et al. (1967) found that percentage correct identification of sentences in a quiet background rose sharply between presentation intensities of 20 and 30 dB. In fact, correct identification rose from 20% to 80% between 20 and 25 dB and reached nearly 100% by 30 dB, remaining constant through presentation intensities of 80 dB before beginning a gradual decline. Examining the impact of vocal force—shouting versus a soft whisper—Pickett (1956) found that speech intelligibility remained relatively constant as speakers increased their vocal force from 55 to 78 dB but diminished above and below this range.

Speech intelligibility has also been examined under less-ideal conditions, such as with degradation due to a chopping circuit that intermittently cuts out portions of the verbal message (Payne et al., 1994) in the presence of noise (Broadbent, 1958; Kryter, 1972, 1985; Smith et al., 1981); for synthetic speech (Robinson & Eberts, 1987; Simpson & Marchionda-Frost, 1984); and under conditions of electronic versus natural amplification (Tschopp, Beckenbauer, & Harris, 1991). Despite the abundance of literature on speech intelligibility under diverse conditions, the potential of signal intensity (from either less-than-ideal listening conditions or degraded sensory capabilities) to alter the mental workload requirements of speech stimuli has received considerably less attention.

Speech Pace

Speech pace can affect the mental workload of speech processing. Pace, as used here, refers to how quickly one verbal item (e.g., a sentence or conversational turn) follows another. It can be distinguished from the term *speech rate*, which refers to the number of syllables per second, a topic taken up in the next section. To distinguish between the two, imagine a situation in which a lively conversation takes place among a group of good friends. The speech pace may be quite rapid, with one person interjecting a statement immediately after or perhaps even during the statement of the previous speaker. Assuming that the speech is not overlapping (when it could arguably be considered competing speech), there is some evidence to suggest that the pace of the speech affects the effort required to process it.

Baldwin and Struckman-Johnson (2002) used a dual-task paradigm for assessing the mental workload of speech processing as a function of presentation intensity and speech pace. Their listeners made sentence verifications for speech played at 45, 55, and 65 dB presented at a rate of either 3 or 5 s between each sentence. Both manipulations—presentation rate and intensity—were shown to affect the mental workload of the speech-processing task. People made more processing errors and took longer to respond when sentences were presented at lower intensities and when the sentences were presented with less time in between (a faster pace). Importantly, though, differences in mental workload were observed only in the dual-task conditions. That is, when the sentence verification task was performed in the presence of a simulated driving task used as a loading task, the extra effort required to process lower-intensity and faster-paced speech was evident. This extra effort was not evident when listeners could devote all of their attentional resources to the speech-processing task. In addition to the pace of speech materials, the rate at which this information is presented is another important factor.

Speech Rate

Speech rate is comprised of a number of different components, including articulation rate and pause rate. Studies have shown that there is tremendous variability both within and between speakers in both the number and duration of pauses and the articulation rate (Miller et al., 1984). Both of these factors have an impact on speech perception (see review in Nygaard & Pisoni, 1995). As previously discussed, we rely on minute changes in the temporal characteristics of speech to make distinctions between phonemic categories, such as to distinguish /pa/ from /ba/. Not surprisingly, then, changes in speech rate affect the mental workload of speech processing.

Speech rate can be measured a number of ways, including the number of syllables or words per minute uttered or the duration of individual syllables and pauses. An early controversy concerned whether speech rate varied primarily because of the number of pauses inserted within utterances or whether the syllable or word utterance rate changed independent of the pauses. Conclusive evidence was provided that both these factors, independently as well as interactively, have effects on speech rate (Miller et al., 1984). Current evidence indicates that speaking rate does vary considerably, within a given individual as a function of factors such as the emotional state of the speaker, the familiarity with the material being spoken, and the workload of the situation. These rate changes have an impact on speech-processing difficulty (see review in Nygaard & Pisoni, 1995).

Researchers commonly distinguish between speech rate and articulation rate. Speech rate includes pauses and is typically measured in words per minute (wpm). Articulation rate excludes pauses (generally of at least 250 ms) and is typically measured in syllables per second (sps) (Goldman-Eisler, 1956; Sturm & Seery, 2007). Average speech rates are around 140–150 wpm with average articulation rates of roughly 4 sps (Venkatagiri, 1999). Slow speech ranges from 136 to 144 wpm with an articulation rate of 3.36–4.19 sps, and fast speech rates are on the order of 172 wpm with an articulation rate range of 5.29–5.9 sps. Contrary to reports in some popular literature, there is no evidence that females have significantly higher speech rates than males (Venkatagiri, 1999).

Changes in speech rate and speech rate variability can have a differential impact on individual aspects of speech processing. For example, as rate increases, vowel durations are shortened more than consonant durations (Gay, 1978), but the relative rate of the difference varies between individual speakers. Phonemes that depend heavily on temporal distinctions, such distinctions made on the basis of VOT, are particularly compromised.

Many sources of evidence indicate that speech rate variability affects the mental workload of speech processing. Variable speech rates decrease word recognition independently and exacerbate processing difficulties when combined with other factors, such as trying to recognize words from multiple speakers (Sommers, Nygaard, & Pisoni, 1994). Listeners who already have problems understanding speech find it even more difficult with increased speech rates. For example, children who speak Black English (BE) have more difficulty comprehending sentences spoken in Standard English (SE) than do children who speak SE. These comprehension difficulties are exacerbated when SE sentences are spoken increasingly faster, whereas such rates do not have a negative impact on SE-speaking children (Nelson & McRoskey, 1978). Similarly, children with language disorders have increased speech-processing difficulties as speech rates increase (Blosser, Weidner, & Dinero, 1976).

Fast speech rates require more mental effort to process and can result in increased processing errors. Importantly, speech production rates tend to increase as workload increases (Brenner, Doherty, & Shipp, 1994), thus potentially further compromising communication. It should be noted that there is some evidence that increasing speech rate artificially through linear time compression may lead to better speech recognition than naturally produced fast speech (Janse, 2004), an issue that is discussed further in reference to designing speech displays in Chapter 11. Both high-workload and high-stress operational environments, unfortunately, can result in rapid speech rates, which may exacerbate communication errors. In one investigation, Morrow and colleagues (Taylor, Yesavage, Morrow, & Dolhert, 1994) observed that during high-workload periods, air traffic controllers may speak to pilots at rates as high as 365 wpm, compared with a normal average closer to 235 wpm. This increased pace was associated with pilots exhibiting nearly 20% more command execution errors. Speech rate is only one of the aspects of a broader category, referred to as prosody or prosodic cues, which both affect and are affected by high-workload situations and have an impact on the mental workload of speech processing.

Prosody

Prosody, or the melodic pattern in speech, greatly influences speech perception (Schirmer, Kotz, & Friederici, 2002). The same word or phrase can have vastly different meanings, depending on prosodic information (or the way it is said). Prosody is comprised of the intonation, loudness, stress patterns, and tempo of speech (Mitchell, Elliott, Barry, Cruttenden, & Woodruff, 2003). Changes in the prosody affect the interpretation of words and sentences. Consider the following example, in which emphasis is placed on the underlined word:

The <u>professor</u> asked the question to the young man in the class.
The professor asked <u>the question</u> to the young man in the class.
The professor asked the question to the <u>young</u> man in the class.
The professor asked the question to the young man <u>in the class</u>.

There are subtle but important differences in the interpretation of each of these sentences depending on their prosody and, in particular, the word that is emphasized. We see that prosodic elements provide information regarding what subjects are important to the current meaning and therefore should be emphasized. In addition, prosodic elements provide important cues to help the listener understand the emotional nature of utterance and detect the presence of intent of humor or sarcasm.

Affective Prosody

One particularly interesting aspect of research pertaining to prosody concerns examinations of how we process the emotional aspects of speech. Specifically, researchers have examined how affective prosody influences comprehension and whether this aspect of the signal is processed independently of the semantic and syntactic content (Berckmoes & Vingerhoets, 2004; Vingerhoets, Berckmoes, & Stroobant, 2003).

Processing emotional aspects of speech results in greater activation of the right cerebral hemisphere, relative to left hemispheric lateralization observed for processing syntactic and semantic aspects (Mitchell et al., 2003; Vingerhoets et al., 2003). This greater right hemispheric activation associated with processing the emotional aspects of speech occurs in both listening (Berckmoes & Vingerhoets, 2004) and reading (Katz, Blasko, & Kazmerski, 2004). Similarly, our brains process humor and sarcasm differently from other nonliteral forms of language. When processing sarcasm, for example, we utilize contextual cues such as knowledge of the speaker's gender and social status in our interpretation of the actual words spoken (Katz et al., 2004).

In particular, Schirmer et al. (2002) found that men and women demonstrate different behavioral and electrophysiological responses to emotional prosody. Women demonstrated a priming effect, responding faster to target words that matched the emotional valence of the prosody of a preceding word. Women also demonstrated smaller N400 responses to match relative to mismatch conditions, which were evident in broadly distributed, nonhemisphere-specific P200 and P300 ERP responses as well. Males did not show these emotional-prosodic priming effects. However, males did demonstrate significantly faster reaction times for positive relative to negative target words. Females did not show this difference. The observation that females often pay more attention to the processing of emotional cues may explain in part the greater bilateral activation among females relative to males during speech-processing tasks discussed previously in Chapter 3.

The processing of emotional content in speech may have an impact on where we direct visual attention. For example, a visual dot presented on the left side of a speaker's face (right hemisphere directed) is discriminated more efficiently when people listen to speech with emotional prosody, whereas the opposite effect (greater discrimination on the right side of the speaker's face) is found when people listen to neutral speech (Thompson, Malloy, & LeBlanc, 2009). It appears that people expect to get more emotional information from processing visual cues on the left side of the face and more speech information from processing cues on the right side of the face (Nicholls, Searle, & Bradshaw, 2004). Importantly, Nicholls and colleagues demonstrated that it was the *expectation* of which side of the face would be most informative rather than where the information actually was. They did this by presenting mirror images of a face and observing decreased McGurk effects to the mirrored image. Expectations play an important role in speech processing. The impact of expectations and context are discussed in greater depth in this chapter.

The focus is turned next to the topic of synthetic speech. One of the primary reasons why even relatively high-quality synthetic speech may be difficult to understand is because of its lack of prosodic cues.

Synthetic Speech

Synthetic speech generation is increasingly common as a method of enhancing automated systems (i.e., computer-aided instruction, information retrieval systems, and guidance through complex menu structures) and as communicative aides for persons with disabilities. Most everyone reading this will have heard synthetic speech in one form or another. It generally relies on text-to-speech (TTS) technologies, which have experienced considerable research and advances in the last several decades. For some individuals, like celebrated astrophysicist Stephen Hawking, TTS speech-generating devices greatly enhance communicative capabilities. Speech-generating devices use TTS technologies to convert words, alphabets, digits, and graphic symbols to audible speech output (Koul, 2003). In other applications, such as for automated menus and navigational aids, synthetic speech greatly reduces the physical storage needs relative to digital recordings of natural speech, thus allowing speech interfaces in a great number of portable devices.

Tremendous progress has been made since the late 1970s in improving the quality of synthetic speech algorithms. Current TTS systems are capable of producing speech sounds that are much more natural and intelligible than their earlier counterparts of the 1970s and early 1990s (Taylor, 2009). However, despite this progress, today's systems are still recognizable as nonhuman, and they still generally require greater mental effort to process than natural speech, particularly in adverse listening conditions (Francis & Nusbaum, 2009; Koul, 2003). So, there is still work to be done.

As Taylor (2009) pointed out, much of the early research focused on making synthetic voices sound more natural without sacrificing intelligibility. The two concepts do not go hand in hand. Some of the very aspects that made early synthetic speech more intelligible (i.e., accurate distinct articulation) also tended to make it sound unnatural and robotic.

Synthetic voices have now become increasingly easier to understand, with intelligibility scores of the better systems nearly rivaling those of natural speech, at least in ideal listening conditions (Koul, 2003), while also sounding more natural. A number of methods have been used to assess the effectiveness or quality of synthetic speech. These range from measuring the intelligibility of individual phonemes and single words to metrics that assess comprehension from processing and perception of more complex forms (i.e., sentences and discourse). Comprehension relates to the ability to form a coherent mental representation and involves higher-level processing (Kintsch & van Dijk, 1978; Koul, 2003). Comprehension measures can be further subdivided into those that are simultaneous (requiring some online recognition or detection during processing) or successive, for which testing occurs immediately after presentation.

In general, intelligibility scores for synthetic speech are lower than for natural speech. In ideal listening conditions, single-word intelligibility scores for natural speech generally range from 97.2% to 99% but are reduced to 81.7% to 96.7% for even the best synthetic speech systems (see review in Koul, 2003). In adverse listening conditions, such as those with poor signal-to-noise ratios, intelligibility scores decrease dramatically for both natural and synthetic speech, but synthetic speech degrades more distinctly (Fucci, Reynolds, Bettagere, & Gonzales, 1995; Koul & Allen, 1993).

Similar results are obtained when perception of synthetic versus natural sentences is compared. Under ideal listening conditions and supportive context, sentence perception is nearly equivalent, although there is considerable variability across synthesizers (see review in Koul, 2003), and synthesized child voices are generally harder to understand than adult voices (Von Berg, Panorska, Uken, & Qeadan, 2009).

When more sensitive indices are used (i.e., sentence verification latency) or when synthetic and natural speech sentences are compared under divided attention conditions, results indicated that the synthetic speech required greater cognitive effort to process. Relative to natural speech, processing synthetic speech is slower and less accurate (Ralston, Pisoni, Lively, Greene, & Mullennix, 1991). Two additional factors that have an impact on the acoustic signal indirectly are discussed next.

Acoustic Environment

The impact of the acoustic environment on mental workload is a topic deserving greater attention. Noise, reverberation, and competing speech have particular deleterious effects on speech intelligibility and can be expected to increase the mental workload of speech processing. However, due to the robust nature of speech processing, not all task paradigms have been sensitive enough to obtain measurable changes in mental workload in support of this position. Urquhart (2003) used a dual-task paradigm to examine the impact of noise (from an army vehicle) on speech processing when stimuli were played at either 83 or 96 dB. Performance on a complex cognitive task increased when speech stimuli were presented at 96 dB relative to 83 dB. However, subjective measures of mental workload obtained from the NASA Task Load Index (NASA TLX; Hart & Staveland, 1988) and the Modified Cooper Harper Rating Scale were not sensitive to these changes in mental workload. Presuming that the decreases in cognitive task performance indicate the task was harder in the presence of noise at the lower (83 dB) condition, the lack of sensitivity of either subjective rating scale is surprising. It is possible that listeners were not aware of the increased workload or that their awareness of the difference was forgotten by the time they completed the rating scales at the end of the experiment.

Another important environmental variable is the number of other simultaneous conversations present. As discussed in Chapter 7 in reference to the impact of noise on workload, irrelevant or competing speech is particularly detrimental to successful speech processing. Open office plans and other work environments where speech must be processed under these conditions increase the mental workload of speech perception. Competing speech can be particularly difficult for those with hearing impairment.

Hearing Impairment

Hearing impairment functionally degrades the quality of the incoming acoustic stimulus. If the acoustical trace is degraded, a larger set of lexical candidates is likely to be activated, making the lexical selection process more difficult. In other words, a hearing-impaired person may have difficulty distinguishing between consonants such as *f, s, th,* and *k,* particularly in conditions of low presentation intensity. When presented with the initial consonant sound in a word such as the *th* in the word *that,* this person could potentially have to select from a larger activated lexical set consisting of *fat, sat,* and *cat* in addition to the lexical versions of *that, than,* and so on. This person would necessarily need to rely more heavily on context to make a lexical selection and might require a longer period of time for the subsequent selection stage. Fletcher (1953) reported that, on average, non-hearing-impaired listeners require 150 to 350 ms to process vowel sounds in normal conversation. Listeners with presbycusis (age-related hearing loss) might take longer to complete the early processing stages by receiving a less-robust preperceptual image with which to begin the processing task. Grosjean (1980; Grosjean & Gee, 1987) found that, in the context of a sentence, lexical decisions regarding the identification of an initial monosyllabic word were often not finalized until presentation of the next noun in the sentence. Grosjean pointed out that this may be a full 500 ms later. A hearing-impaired listener, already experiencing slower phonemic recognition, would then have a larger number of lexical alternatives from which to choose—thus likely causing additional delays in later processing stages and potentially communication breakdown if speech is presented too quickly. This is one way in which hearing impairment may exacerbate or be confused for cognitive impairments among older adults, a topic we take up in Chapter 10. The number of lexical alternatives generated can have an impact on the mental workload of speech processing. The hearing-impaired listener benefits greatly from knowing the general context of the spoken material, much in the same way that an effective title aids reading comprehension. We turn our attention next to these more cognitive-contextual cues.

CONTEXTUAL FACTORS

As with all sensory processing, context is important in speech processing. The same acoustical signal can be perceived as different speech sounds depending on the context in which it is found (Holt & Lotto, 2002). A well-known example is the acoustically similar sentence, "I scream, you scream, we all scream for ice cream." Context is necessary to accurately differentiate between the phonetically similar "I scream" and "ice cream." The importance of context in disambiguating variant speech signals was observed by Shockey and Reddy (1975, as cited by Plomp, 2002) in an investigation of the phonetic recognition abilities of phoneticians listening to a foreign language. The phoneticians could recognize only 56% of the phonetic symbols in the unfamiliar language, and there was great discrepancy in identification of particular phonemes between phoneticians.

Contextual cues reduce the mental workload of speech processing on multiple levels. Lexical or word knowledge assists in the recognition of ambiguous phonemic categories. For example, Ganong (1980) demonstrated that listeners tended to make phonemic categorizations that favored a real word (choosing dash instead of tash or task instead of dask) when presented with ambiguous phonemic information. At the lexical level, words that are more frequent and have fewer lexical neighbors (words that differ by only one phoneme) are recognized more easily than words with high neighborhood densities (Dirks, Takayanagi, & Moshfegh, 2001; Norris, 2006; Slattery, 2009). At the sentence level, word recognition is facilitated by sentence-level context (Revill, Tanenhaus, & Aslin, 2008; Slattery, 2009) and to a lesser degree syntactical information even if it does not make grammatical sense (Miller & Isard, 1963). Contextual effects aid word recognition in all listeners but are differentially beneficial to older listeners (Abada, Baum, & Titone, 2008; Dubno et al., 2008; Laver & Burke, 1993), as discussed further in Chapter 10.

Speaker Familiarity

It is easier to understand the speech of someone with whom we are familiar. A particularly dramatic example of this is how it may be difficult to understand the speech of someone with a heavy accent when we first get to know him or her. However, with greater exposure and familiarity with his or her speech, it becomes easier. Presenting speech material in a familiar or single voice aids word recognition and decreases the mental workload of speech processing. Experiments involving recall of lists of words indicated that recall is better (particularly for items at the beginning of the list) if the list is presented by a single speaker rather than multiple speakers (Martin, Mullennix, Pisoni, & Summers, 1989), and word recognition is greater when presented by a single relative to multiple speakers (Mullennix, Pisoni, & Martin, 1989). The processing of nonlexical information from multiple speakers appears to require attentional resources that must be redirected from the processing of other tasks. Processing speech from familiar voices is less resource demanding than from unfamiliar voices.

In sum, both acoustic and contextual factors can independently and interactively have an impact on the mental workload of speech processing. Adverse listening conditions stemming from low signal intensity, background noise, or hearing impairment increase attentional processing requirements. Fast speech rate and variability in prosodic cues also have a negative impact, while contextual factors serve to reduce ambiguity and decrease processing requirements. In the remainder of this chapter, we examine the impact that processing speech can have on performance of other nonlexical tasks.

APPLICATIONS

Changes in the mental workload requirements of speech processing have implications for how well people can carry out operational tasks at the same time, such as driving, flying, or interacting with computer systems. One area that has received considerable attention in recent years is that of conversing on a mobile phone while driving.

MOBILE PHONES AND DRIVING

It is well documented that driver distraction is a threat to transportation safety (Hancock, Lesch, & Simmons, 2003; Lee et al., 2001; Lee, McGehee, Brown, & Reyes, 2002; Stutts, Reinfurt, Staplin, & Rodgman, 2001). Mobile phone conversations are considerably more distracting than are conversations with passengers

(McEvoy et al., 2005; Strayer et al., 2003). Talking on a mobile phone irrespective of whether it is a handheld or hands-free device slows response time to external events and stimuli by 0.25 s on average (Caird et al., 2008). Conversing on a mobile phone has been found to increase crash risk up to fourfold, and contrary to popular belief, hands-free phone devices offer no significant reduction in crash risk (McEvoy, Stevenson, & Woodward, 2007; Redelmeier & Tibshirani, 1997). Also contrary to popular belief, greater experience using a cell phone while driving does not appear to mitigate the increased crash risk (Cooper & Strayer, 2008).

Degradation in the speech signal is commonplace when conversing on a mobile phone (Kawano, Iwaki, Azuma, Moriwaki, & Hamada, 2005). These degradations can be expected to further increase the processing requirements of the conversation, leaving reduced attentional resources for other tasks such as driving and hazard detection.

When speech processing must be time-shared with a visual task, like driving, processing requirements are reduced when the speech task is presented from the same spatial location where visual attention must be focused (Spence & Read, 2003). This observation provides possible support for a reduction in mental workload and potentially some crash risk reduction when using hands-free mobile devices that utilize the front speakers of the vehicle (i.e., Bluetooth interfaces). However, this conclusion awaits further research.

Speech processing can have detrimental effects on concurrent tasks other than driving. Processing spoken information has been shown to disrupt simple and choice reaction time tasks in both young and old (Tun, Wingfield, & Stine, 1991) and manual tapping tasks (Seth-Smith, Ashton, & McFarland, 1989) and can even disrupt walking in older adults (Lindenberger, Marsiske, & Baltes, 2000).

AIR TRAFFIC CONTROL COMMUNICATIONS

Communication between pilots and air traffic controllers is a special class of speech communications. Failure of this communication process has been attributed as a causal factor in a great number of aviation accidents and incidents (Hawkins, 1987). In an attempt to reduce the number of miscommunication incidents, a number of human factors guidelines have been implemented in this area. The use of restricted vocabulary and implementation of strict adherence to procedural routines are some of the practices that have improved air traffic control communications. This important topic is discussed in more detail in Chapter 11 in a discussion of auditory display design in aviation.

SUMMARY

Extracting meaning from the transient acoustical variations that make up speech is an effortful process. Knowledge of the language, speaker familiarity, semantic context, syntax, and situational expectations are some of the many things used to assist in the interpretation of speech sounds. The effort involved in speech processing can be increased by sensory factors, such as poor signal quality, noise, or hearing impairment, as well as by cognitive factors such as low text cohesion, complex syntax, or the absence of context clues.

Studies of language acquisition and double dissociations found in different forms of aphasia have led some researchers to suggest that language is processed by dual mechanisms, one for the large repertoire of stored lexical units and another for the rules for combining these lexical elements. By this account, arriving at the correct form of a regular past-tense verb (such as dated from date and voted from vote) uses one mechanism with distinct neural circuitry, while arriving at the past tense of an irregular verb (i.e., went from goes, dug from dig) relies on a separate mechanism. Others argued, largely through artificial neural network or connectionist modeling, that separate pathways are not necessary to account for dissociations between regular and irregular past-tense verb formation.

CONCLUDING REMARKS

Sensory and cognitive processes interact to affect the mental workload of speech processing. Up to this point in the book, I have primarily focused on discussing how changes in the auditory environment affect this process, making it more or less effortful. However, more often than not, auditory processing takes place in conjunction with visual processing. This visual processing can be either complementary or contradictory. In the next chapter, I turn to the topic of cross-modal processing, paying particular attention to the interaction of visual and auditory processing.

9 Cross-Modal Influences in Sound and Speech

INTRODUCTION

The human perceptual system has evolved to integrate information from each of the senses (i.e., vision, audition, somatosensory, olfaction, and gustation). Yet, in everyday life, hearing and seeing tend to be treated as separate, independent entities. When we exercise our visual faculty, as when searching for a friend at a crowded party, we think of this task as purely visual. Similarly, having found the friend, when we strain our ears to hear his or her words in the din of the party, we think we are merely exercising our auditory abilities. Most of us tend to think of the senses as independent. After all, many of us know someone who has lost one or the other sense—a friend or relative who is blind or someone who is deaf. But, among those who have both senses, little thought is given to the influence of one sense on the other. As discussed in this chapter, however, vision, audition, and tactile (touch) abilities can have remarkable influences on each other. The senses are interdependent. Hearing a sound helps us direct visual attention to the spatial location of the sound (Perrott, Cisneros, McKinley, & Dangelo, 1996). A visual stimulus can enhance our ability to interpret auditory information (i.e., understanding speech, judging distance). In fact, the auditory cortex can be activated by silent lipreading—when absolutely no sound is present (see Calvert, Bullmore, Brammer, & Campbell, 1997). This chapter examines a number of such cross-modal interactions (primarily between vision and hearing), as well as issues concerning language processing by eye and ear.

The mental workload of speech processing was discussed at length in Chapter 8. Is listening to speech more or less effortful than reading text? Are there certain situations or work environments in which one modality would be processed more efficiently than the other? Does visual processing affect our ability to listen, and vice versa, does listening affect our ability to see? How much benefit does one get from visual cues when listening to speech? These are a few of the questions addressed in this chapter.

Before dealing with these issues, the distinction between *input modality* (i.e., auditory vs. visual) and *information code* (i.e., verbal vs. visual or spatial) must be noted. Input modality is rather straightforward. The form the information is received in—whether auditory, visual, olfactory, or tactile—defines the modality in which stimulus information reaches a person. A closely related term is the *channel* of information. Auditory or acoustic information is received through the auditory channel, visual information through the visual channel, and so on.

Defining the information code is less straightforward. Information taken in through the auditory channel may be processed in a verbal, visual, or spatial code, depending on the type of information. Imagine that a friend verbally describes to you a familiar person or object. Most likely, you will form a visual image of the person or item—a visual code for the information. On the other hand, if the verbal description is of a new apartment (how your friend enters and walks through the living room to get to the kitchen and then down a hallway to get to the office), then a spatial code will likely be formed of the information. Finally, if your friend describes to you what he or she means by the word *independence*, you are most likely to form a verbal code to process the information. As Paivio (1969) and others have discussed, it is relatively difficult to form a visual image of abstract words such as "independence"; therefore, such words are thought to remain in a verbal code during short-term memory retention periods.

In the same way, visual information can be coded verbally—as in the case of text—or verbally and then spatially. An example would be a description that you provide of the layout of your apartment in a written letter. Visual information in the form of pictures or patterns may be processed using a visual code. There is currently ongoing debate (see Bernstein, Auer, & Moore, 2004) regarding whether audiovisual (AV) speech information is combined into a common format or code (an amodal linguistic code) or whether AV information retains modality-specific representations that may be linked or associated at higher processing levels. However, regardless of the specific level of the mechanism responsible, cross-modal speech interactions have important implications for human performance in many applied settings. For the purposes of this chapter, a neutral stance is taken regarding the verbal code of processing, leaving a precise definition of the nature of this code to await the outcome of further research. There are also other forms of auditory codes in addition to the verbal one described (i.e., rhythmic, melodic, etc.), but for the present purposes, the focus here is on the dominant codes in the working memory literature: verbal, visual, and spatial (Baddeley, 2003; Deyzac, Logie, & Denis, 2006; Logie, 1986; Shah & Miyake, 1996; Smith & Jonides, 1997). We focus on these codes as they have received considerably more attention in the relevant literature and would seem to have the greatest potential to be involved in cross-modal interaction effects.

Four main topic areas are discussed in this chapter. First, evidence for the distinction between visual and verbal codes is presented, followed by a discussion of important issues pertaining to processing language by eye or by ear (text vs. speech). The third major section deals with cross-modal links, including linguistic, spatial, and multisensory facilitation of response. The fourth and final section discusses dual-task investigations involving tasks presented in auditory and other modalities, focusing on the impact of task modality on either facilitating or disrupting task performance. As discussed in the next section, the existence of both visual and verbal codes in human memory is now well documented, and each code has unique processing characteristics.

VERBAL AND VISUAL CODES

Many researchers have examined how people process information presented through visual and auditory channels (Paivio, 1969; Paivio & Csapo, 1971). Early evidence for a distinction in processing mechanisms between the two presentation formats can be found in the dual-coding theory of Paivio (1969), as well as by Brooks (1968). Further evidence for multiple coding systems was later presented by Santa (1977) and Wickens (1980, 1984) and from the results of numerous studies of dual-task performance, particularly those involving cross-modality interference as discussed in the following material.

In the late 1960s and early 1970s, there was controversy over how information (i.e., pictures, words) was stored in long-term memory. Some thought that information of all types was stored in the form of a set of propositions (Pylyshyn, 1973). Evidence for a visual code as an alternative explanation came from research with mental rotation (Shepard & Metzler, 1971) and from Paivio's demonstrations that material is remembered better if people are able to encode it both visually and verbally (e.g., Paivio, 1969; Paivio & Csapo, 1971). Santa (1977) provided further evidence for a distinction between verbal and visual-spatial processing in an experiment involving the presentation of geometric shapes and visually presented words. Participants viewed an array, which in one condition consisted of three geometric shapes, two objects above and one below (Figure 9.1(a)). In a second condition, three words were presented visually, two words above and one below. See an example of the type of stimuli used by Santa in Figure 9.1. Following presentation of the arrays, participants saw a series of test arrays presented either in the same configuration (two above and one below) (Figure 9.1(c)) or linearly (all in the same row) (Figure 9.1(d)) and were required to determine if the test arrays consisted of the same elements as the original display regardless of order. Examination of response times for making a determination revealed a clear interaction between the geometric and word conditions and the format of the test array. Participants were faster in making a determination in the geometric condition if the test array preserved the same spatial configuration. However, in the word condition participants were faster when the test arrays presented words in a linear fashion rather than preserving the spatial configuration. These results suggest that the geometric shapes were processed using a visual code, while the words—although presented visually—were processed using a verbal code.

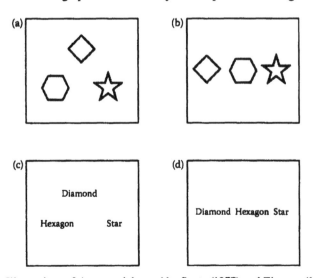

FIGURE 9.1 Illustrations of the materials used by Santa (1977) and Zimmer (1998).

Zimmer (1998, Experiment 2) replicated Santa's (1977) results. In a variation of Santa's paradigm, Zimmer included trials in which the shapes and words were occasionally intermixed on the same trials (Experiment 1). Participants were slower to respond to incongruent trials in these "mixed" trials just as they had been in the geometric shape trials, suggesting that isolated words may also be coded visually rather than only verbally. Note that so far we have been discussing visual presentation of material that may be coded either visually or verbally. Zimmer showed that in at least some cases text could be processed visually. What happens when the material to be processed is inherently verbal (i.e., words), but it is presented visually as text versus aurally as speech? This issue is addressed next, and it has considerable importance to a number of applied domains. For example, air traffic controllers normally communicate with pilots via the speech channel. Miscommunications can have disastrous consequences and have been cited as a primary factor in many aviation accidents (Hawkins, 1987). As our national air space becomes increasingly congested, technological solutions are sought to improve both efficiency and safety. In the effort to develop advanced next-generation (NexGen) technologies to accomplish this goal (Joint Planning and Development Office, 2010), examination of text versus speech displays plays a critical role. Scientists are asking whether we could reduce communication transmission errors by presenting air traffic control (ATC) information in text format using a system such as controller-pilot data link (CPDL). How might presenting ATC information via text rather than speech change the pilot and air traffic controllers' workload in an environment that already places heavy demands on visual processing resources?

LANGUAGE BY EAR AND BY EYE

Processing and retention of verbal information is essential to many everyday tasks. Considerable research has been directed toward illuminating the processes and structures associated with verbal information processing (see Healy & McNamara, 1996 for an extensive review). A key theoretical assumption is that working memory consists of multiple processing streams with unique neurophysiological mechanisms for each sensory modality (Baddeley, 1992; Just et al., 2003; Posner et al., 1988; Smith & Jonides, 1997), but see the work of Rosenblum (2004) for an alternative opinion. Sensory systems are thought to have access to unique processing structures or mechanisms with the potential to function relatively independently, although subjected to a total overall resource limitation due to an attentional control mechanism, such as the central executive proposed by Baddeley and Hitch (1974). However, a key theoretical question is the extent to which verbal processing is regulated by presentation modality constraints or by a more general phonological code, irrespective of presentation format. In other words, is text processed differently from speech?

EVIDENCE FOR SEPARATE CODES

Considerable evidence indicates that language is processed differently when it is heard than when it is read. The conclusion is that auditory and visual language processing draw from different pools of resources. Evidence in support of this assumption comes from several sources, including (a) neuroimaging findings showing that reading words activates distinctly different brain regions than listening to words (Posner et al., 1988); (b) serial recall paradigms demonstrating differential primacy and recency effects for word lists that are seen as opposed to those that are heard (Crowder, 1972); (c) studies of patients with neurological damage resulting in quite specific linguistic impairments (Gathercole, 1994); and (d) differential patterns of performance for text versus speech tasks in dual-task paradigms. Each of these lines of evidence is briefly explored in this chapter.

Physiological Evidence

Posner and colleagues (1988) provided strong evidence that visual and auditory word identification occurs in distinctly separate neural brain areas. Although recognition of visually presented words has generally been thought to occur through a series of stages involving visual detection, phonological followed by articulatory encoding, and then semantic identification, Posner et al. (1988) indicated that, for highly skilled readers, the phonological stage may be bypassed, and processing progresses straight from visual detection to semantic identification of words. They based this conclusion on the finding that visual word recognition does not lead to increased activation of the temporoparietal cortex areas commonly associated with phonological processing. Posner and colleagues examined cerebral blood flow via positron emission tomography (PET) while participants passively viewed nouns. The passive viewing of nouns led to increased activation of five areas, all located within the occipital lobe. Adding complexity to the reading task, such as requiring complex naming or semantic analysis, did not activate any additional posterior sites. Posner and colleagues concluded that visually specific coding takes place entirely within the occipital lobes. Conversely, auditorily presented words did not activate these occipital areas but rather activated left temporoparietal areas.

Specifically, the phonological store is thought to reside in the perisylvian region of the left hemisphere (Figure 9.2) (Baddeley et al., 1998; Paulesu et al., 1993), and the articulatory rehearsal component is thought to reside in Broca's area (Chiu et al., 2005; Paulesu et al., 1993). These areas of the phonological loop appear not to be required in highly skilled readers processing familiar words. Baddeley and colleagues (1998) suggested that reliance on visual representations of words may help account for the normal language abilities and eventual acquisition of normal vocabularies in adults with severe phonological memory impairments. Additional evidence that text versus speech material is processed differently comes from serial recall studies.

Primacy and Recency in Serial Recall

Primacy effects, or superior recall for items at the beginning of a list, and recency effects, superior recall for items at the end of a list, have been the subject of numerous investigations (Glanzer & Cunitz, 1966; Murdock, 1962; Routh, 1971, 1976; Sharps, Price, & Bence, 1996). Recency effects are stronger for auditory as opposed to visual information (Corballis, 1966; Laughery & Pinkus, 1966; Madigan, 1971; Routh, 1971, 1976). Routh (1976) demonstrated that auditory as opposed to visual presentation resulted in a stronger prerecency effect as well as a stronger recency effect.

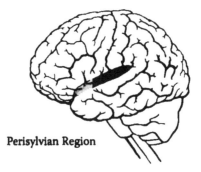

Perisylvian Region

FIGURE 9.2 Perisylvian region of the left hemisphere where the hypothetical phonological store is thought to reside. (Drawn by Melody Boyleston.)

Further, Routh (1976) demonstrated that the addition of a concurrent task during presentation had greater negative effects on recall of visually presented digits than on auditorily presented digits. For visually presented items, recall errors increased linearly as the difficulty level of the concurrent task increased. This observation supports the hypothesis that speech has automatic access to the phonological loop, while text (at least for most people) requires an additional processing step of converting the orthographic information into a verbal code. This automatic access can result in an advantage for speech in serial recall tasks.

Acoustic Advantage

People appear to know intuitively that there is an acoustic advantage when trying to hold information in an active state (i.e., in short-term or working memory). They will generally try to use an auditory code for the information if possible. This usually takes the form of subvocal rehearsal. For example, if I am trying to remember a phone number long enough to dial it, I—like most other people—will "say" the number silently in my head. This silent speech is referred to as *subvocal rehearsal*. Conrad (1964) documented the preferred status of the auditory code in his observations that visually presented letters were more likely to be confused with their rhyming counterparts during recall rather than the letters that they looked like. He termed this effect "acoustic confusion." Thus, for example, a visually presented *T* was more likely to be confused with the acoustically similar but visually dissimilar *D* than it would be with a letter such as *F*, which is distinct phonologically. In fact, the tendency to use an auditory code is so strong it can sometimes cause problems. This preference may become a maladaptive strategy in advanced age as people continue to rely on a verbal code even when it would be more beneficial not to (Hartley et al., 2001).

In most everyday situations, hearing verbal material appears to be beneficial to short-term recall. Crowder (1970) provided empirical evidence for this acoustic advantage in a recall paradigm in which three presentation conditions were used. In all three conditions, digits were presented visually to participants. However, in the first condition, participants were instructed simply to read each digit silently to him- or herself. In a second condition, participants were instructed to say each digit aloud, and in a third condition, an experimenter spoke each digit in synchrony with the visual presentation. There was a clear recall advantage for the two conditions involving vocalization, regardless of whether it was generated by the participants or the experimenter. The recall advantage, not surprisingly, was limited to a significant recency effect, while recall for items in prerecency serial orders was not affected.

Neuropsychological Evidence

Neuropsychological studies of such disorders as aphasia and dyslexia also provide evidence for the existence of separate mechanisms for processing text versus speech (Gathercole, 1994). A case study of patient P.V. provided early evidence for a dissociation between reading and speech processing (Vallar & Baddeley, 1984). P.V. had suffered a stroke that damaged her left hemisphere during early adulthood. Although she could converse normally, she demonstrated specific significant language difficulties. When performing a neuropsychological test administered through spoken commands, P.V. was profoundly impaired, particularly for more complex commands, (i.e., "Touch the yellow square inside the white box."). P.V.'s performance improved dramatically when she was allowed to read the commands. Tests of P.V.'s memory span also showed significant dissociations for spoken versus read digits, with her memory span for digits that were read much higher. This remarkable case not only provided evidence for distinctions between processing speech and text but also helped Baddeley and colleagues refine their model of working memory.

Further evidence for the distinction between text and speech processes comes from the extensive literature on dyslexia. At least two primary forms of dyslexia have been observed (Erskine & Seymour, 2005; Hulme & Snowling, 1988). One involves difficulty with orthography, while the other primarily involves difficulties with phonology. Individuals with greater orthographic difficulties, often called a dyseidetic form or surface form of dyslexia, demonstrate greater impairment with word identification, particularly irregular words, such as *sew*, that are not pronounced according to grapheme to phoneme rules, such as the words *new* or *few* (Erskine & Seymour, 2005). Conversely, individuals with phonological dyslexia demonstrate greater difficulty with decoding pronounceable nonwords (such as *tew*), long unfamiliar words, and other phonological tasks (Erskine & Seymour, 2005; Hulme & Snowling, 1988). Similarly, individuals with phonological aphasia have significant verbal processing deficits and yet may have spared object identification performance (Rohrer et al., 2009).

Further support for the position that reading versus listening may draw, at least in part, from separate working memory mechanisms with independent pools of processing resources comes from dual-task language investigations.

Time-Sharing Text versus Speech

In many situations it will be easier to time-share speech versus text-based tasks. As previously discussed, speech is believed to gain automatic access to the phonological store, regardless of the mental workload demands imposed by a concurrent task (Gathercole, 1994; Routh, 1971; Salame & Baddeley, 1982). This is thought to be one of the main reasons why irrelevant speech is so disruptive, as discussed in Chapter 7. Text does not appear to have the advantage of automatic transfer. Verbal information presented visually (text) requires an extra processing step to convert the orthographic form (generally by means of the subvocal rehearsal mechanism) into a phonological form. Therefore, according to Gathercole (1994), a concurrent memory load can be expected to disrupt text processing to a greater degree than would be expected for speech processing. In situations with a high mental workload, presenting verbal information as speech rather than text can be expected to have processing advantages, particularly in environments with heavy visual processing loads, such as driving (Dingus et al., 1998; Dingus, Hulse, Mollenhauer, & Fleischman, 1997) and aviation (Risser, Scerbo, Baldwin, & McNamara, 2004, 2006).

A slight benefit was observed for simulated ATC procedural commands provided by speech versus text in a series of investigations examining differing types of working memory interference (Risser et al., 2003, 2006; Scerbo, Risser, Baldwin, & McNamara, 2003). Risser and colleagues provided participants with multiple procedural commands while they concurrently performed interference tasks designed to disrupt central executive, visuospatial, or verbal working memory. Commands were either spoken or read. Both central executive and verbal working memory interference tasks were more disruptive than visual interference regardless of modality. This was predicted on the basis that commands would evoke verbal processing regardless of their presentation modality. However, the pattern of disruption was greater for commands that were read rather than heard. Baldwin and colleagues interpreted these results as further evidence that reading requires an additional processing step and therefore is more susceptible to performance impairments in conditions of divided attention.

To this point in this chapter, the focus has primarily been on distinctions—between verbal and visual codes and between processing differences between speech and text. In the next section, the discussion changes to ways in which auditory and visual information may enhance performance through multisensory integration or cross-modal links.

CROSS-MODAL LINKS

This section discusses a number of examples of how auditory and visual information complement and potentially enhance information processing and reduce mental workload. Three important areas of multisensory integration are discussed. The first pertains to AV speech, the next to cross-modal spatial attention, and the third to instances when providing information in two modalities (i.e., auditory and visual) enhances performance even when the information in one modality can be viewed as redundant.

As previously discussed regarding the McGurk effect in Chapter 8, visual and auditory information are often combined during speech processing. Visual cues can reduce the mental workload of speech processing, particularly in noisy or other environments where the speech signal is acoustically degraded. Considerable information can be obtained visually not only by observing nonverbal body language, but also by watching the movements of the speaker's lips and face.

AUDIOVISUAL SPEECH

In naturalistic settings, visual and auditory information are often used together in speech perception (Massaro, 1987). Speech reading, or lipreading as it was more commonly referred to in the past, makes use of the visible features of a face, particularly aspects of the lips in relation to the cheeks, chin, and nose (Calvert et al., 1997). Initially of interest primarily in the study of hearing-impaired populations, speech reading has gained greater attention for its relevance to understanding speech in noisy environments and for its ecological relevance for speech perception in everyday contexts. Speech reading improves auditory discrimination, even under normal listening conditions (Arnold & Hill, 2001; Calvert et al., 1999; Calvert & Campbell, 2003).

Being able to watch a speaker can greatly reduce the mental workload associated with listening and improve speech recognition in adverse listening environments (Ross, Saint-Amour, Leavitt, Javitt, & Foxe, 2007). Behavioral investigations have demonstrated that in noisy listening environments the ability to watch the speaker is comparable to raising the signal-to-noise ratio (SNR) by approximately 11 dB (see review in Bernstein et al., 2004). Speech reading, or viewing a speaker's articulatory movements, can improve word recognition performance across a wide range of SNRs but appears to have maximal benefit at intermediate levels of SNR of around −12 dB for young normal-hearing listeners (Ross et al., 2007). This is in contrast to earlier estimates that proposed an "inverse effectiveness" for multisensory integration—essentially proposing that visual speech cues would have increasing benefit the more that SNRs are degraded.

Early investigations of speech reading (Erber, 1969; Sumby & Irwin, 1954) used a limited set of words and typically provided listeners with a checklist with which to attempt their word identifications. Using such a restricted set of words in combination with the checklist likely led to overestimations of the benefits of visible speech cues at levels at which auditory information was of limited benefit (see discussion in Ross et al., 2007). But, the fact that people use speech-reading cues, particularly in adverse listening conditions, is well documented (Campbell, 1986; McGurk & MacDonald, 1976). Research suggested that people tend to pay particular attention to the right side of a speaker's mouth (Campbell, 1986; Nicholls et al., 2004). This is indicated by a reduced McGurk effect (binding of visual and auditory speech information leading to the illusory misperception of a sound) when people view mirror images of a talking face (Nicholls et al., 2004) and the finding that people are more accurate at identifying phonemes spoken by the right side of a chimeric image (Campbell, 1986). Although the right side of the mouth has also been shown to exhibit more expressive verbal information, that mirrored images reduce the McGurk effect suggests that people are not merely paying attention to visual cues with the most information. Rather, people have apparently learned to expect greater information from attention to the right side, an example of a top-down influence discussed in Chapter 8.

Interestingly, even speech reading in the absence of auditory stimuli activates the auditory cortex (Calvert et al., 1997; Calvert & Campbell, 2003) and cortical language areas such as Broca's and Wernicke's areas and Heschl's gyrus (Hauser, Chomsky, & Fitch, 2002), at least in normal-hearing individuals. (For a discussion of differences in neural circuitry during speech reading between normal-hearing and deaf individuals, see MacSweeney et al., 2002.) Calvert et al. (1997) observed that portions of the superior temporal gyri normally activated by speech sounds were activated by silent lip movements in normal-hearing participants, but only when the lip movements were associated with actual speech sounds. Mouth and lip movements not associated with making speech sounds did not activate these areas. Pseudospeech, plausible lip movements that did not form actual words, activated the same auditory cortical areas. This indicates that cross-modal speech associations are likely occurring at an early prelexical stage during phonetic classification. Early cross-modal integration of auditory and visual information is also observed in a nonlinguistic phenomenon.

CROSS-MODAL SPATIAL ATTENTION

In the natural world, sounds often serve the purpose of directing our visual attention. A sudden sound heard coming from the left naturally directs our gaze in that direction. Laboratory studies have shown that a visual cue at a particular location in space facilitates response to an auditory stimulus at the same spatial location (Spence, Nicholls, Gillespie, & Driver, 1998). Furthermore, tactile discriminations between stimuli presented to the left and right hand can be made faster when preceded by a visual or auditory cue on the same side (Spence et al., 1998), and tactile cues facilitate response to auditory and visual targets presented on the ipsilateral (same) side as the cue. These multisensory cueing enhancements—which provide evidence for cross-modal links in spatial attention—have now been found for every combination of auditory, visual, and tactile stimuli (see chapters in Spence & Driver, 2004). These cross-modal links have important implications for human performance and display design.

Cross-modal links suggest that a common spatially referenced attentional system may be shared among sensory modalities. This contrasts with modality-specific information-processing models (Wickens, 1980, 1984; and see Bernstein et al., 2004, for a modality-specific account of language processing) that postulate separate attentional resources for individual sensory systems (Ho & Spence, 2005). A shared system view, on the other hand, predicts either enhancement or disruption during multimodal processing depending on the coherency of the information coming from the two sensory channels. Conversely, modality-specific accounts would propose relative independence between the two processing channels, at least at early sensory processing levels. Note, however, that, as Bernstein and colleagues pointed out, the McGurk effect—in which a person seen mouthing a particular phoneme (say /ga/) dubbed with a different phoneme (/ba/) is heard as saying a third, different sound (/da/)—demonstrates that AV speech interaction can be explained by both common format and modality-specific accounts of speech binding. We return to a discussion of the McGurk effect later in the chapter, concentrating at present on spatial processing.

Cross-modal links in spatial attention have some important implications for applied settings, such as in the task of driving. Providing an alert or cue in one modality can speed spatial orientation to that location, thus facilitating detection of a stimulus in a different modality. Thus, mounting evidence suggests that auditory and haptic or tactile alerts may provide effective collision warnings.

Cross-modal spatial cueing (providing an auditory or haptic-tactile cue) in the direction of an impending collision event greatly speeds spatial orienting, event detection, and crash avoidance. For example, cross-modal links have been shown to facilitate detection and discrimination of hazards in collision avoidance situations during simulated driving. Ho and Spence (2005) conducted a series of investigations examining the relative impact of spatially predictive auditory cues on visual hazard detection in a dual-task paradigm involving high visual demand. Drivers were engaged in a simulated car-following task along with a rapid serial visual presentation (RSVP) task. From time to time, they were presented with traffic "hazards." The hazards consisted of either an apparent rapid closure rate between their vehicle and the one in front (potential front-to-rear-end collision situation) or a fast-approaching vehicle in their rearview mirror (potential tailgating-type collision situation). Drivers received an auditory warning (either neutral or nonverbal but spatially predictive or verbal and spatially predictive) prior to critical detection events. Specifically, in their series of five investigations, they examined auditory cues that were (a) neutral nonspatially predictive cues (i.e., a car horn sound coming from a speaker under the driver's seat); (b) nonverbal spatially predictive cues (car horn coming from spatially relevant area); and (c) verbal spatially predictive cues (the word *front* or *back* coming from either a neutral location or a spatially predictive location). Spatially predictive cues enhanced hazard detection in all instances. The redundant semantically and spatially predictive cue was particularly effective. That is, Ho and Spence's participants were faster and at least as accurate at detecting hazards appropriately (determining if there really was a hazard event and whether it was coming from the front or the back) when the auditory warning location matched the location of the hazard. Note that participants always had to look forward to assess the situation. Simulated events behind the driver could only be observed by looking at the rearview mirror (located in front of the driver). Thus, it is particularly remarkable that in the case of the car approaching swiftly from behind, the congruent auditory cue of *back* facilitated correct detection of the event requiring a forward glance. The realistic simulation of the driving setting probably facilitated this effect. Driving is a highly practiced task for most adults, and looking in the rearview mirror to examine the rear scene is likely a relatively automatic response. The main point for the present purposes is that cross-modal links have been shown to have important implications for many applied settings, including the task of driving and collision avoidance. The combined spatially and semantically predictive cue was particularly helpful in directing attention, a finding that is relevant to the design of hazard warning systems and target detection paradigms for aviation and military applications.

Providing spatial audio cues have also been shown to improve the visual search in tasks other than driving (Perrott et al., 1996; Tannen, Nelson, Bolia, Warm, & Dember, 2004). Tannen et al. had participants locate visual targets of differing detection difficulty while engaged in a simulated flight task. They proved spatial audio cues greatly enhanced detection of visual targets (relative to a no-cueing condition), particularly for the targets most difficult to detect, although spatial audio cues in this case were not significantly better than visual or combined visual and auditory cues. In addition to cross-modal spatial cueing, spatial-linguistic interactions have been observed.

Spatial-Linguistic Interactions

Driver and Spence (1994) conducted a novel set of experiments examining the impact of cross-modal spatial attention on speech processing. They used a dichotic listening task similar to the ones used by Cherry (1953) and Moray (1959) and described in previous chapters. The novel part was that, in addition to shadowing one of the two messages, participants had to perform an unrelated visual detection task or fixate on a visual screen in one of two locations (Experiment 2). The relevant visual array and the relevant speech were positioned in front of the participant to the right or left in congruent (same-side) or incongruent (opposite-side) locations. They observed that it was harder to attend to the relevant speech stream if it was coming from an incongruent location. In other words, people found it easier to shadow speech coming from the same location as the relevant visual display, even if the visual display was completely unrelated to the speech stimuli. In the same series, they also found that participants received more benefit from congruent visual speech when it was presented from the same location (vs. incongruent) as the speech stream they were shadowing (Experiment 1). Thus, regardless of whether the visual task was related (as in the visible speech) or unrelated (as in the visual detection task), these findings indicate that it is easier to process speech if it comes from the same spatial location to which visual attention is directed.

Spence and Read (2003) examined the implications of these results within the context of driving. They varied the location of a speech stimulus while participants were either performing a shadowing task by itself or in combination with a simulated driving task. Shadowing performance was always better when the speech seemed to come from directly in front of the listener, and this performance enhancement was particularly evident in the more demanding situation of performing the shadowing task in combination with the driving task. In the case of the dual-task driving condition (driving while shadowing), their results indicated that it was easier to process the speech when it was coming from where their visual attention was directed. This finding suggests that centrally located speakers (or simulating a central location for a pilot wearing a headset) might reduce the workload of processing auditory navigational messages, traffic advisories, and other similar information. Since we generally tend to look at the people we are talking to if possible, the improved shadowing performance in Spence and Read's shadowing-only condition is not that surprising. Does this suggest that auditory in-vehicle information should always come from where the operator is likely to be looking? This question and additional issues pertaining to the design of auditory displays and spatial audio for collision avoidance warnings are explored in Chapter 11, "Auditory Display Design." At this point, a third example of cross-modal links and their potential impact in applied settings is discussed. As discussed next, in some cases redundant information can enhance performance.

REDUNDANT TARGET EFFECT

In early studies of attention and information processing, researchers often compared peoples' reaction times to visual or auditory stimuli. It was soon discovered that when people were provided information in both modalities for the same target, performance was enhanced. This has come to be referred to as the redundant target effect (RTE). The well-documented RTE is the observation that providing redundant visual and auditory information often improves accuracy and reaction time in target detection paradigms over that which can be achieved by either auditory or visual information alone (Colquhoun, 1975; Miller, 1991; Sinnett, Soto-Faraco, & Spence, 2008).

Two early explanations for the RTE were that either (a) energy summation occurred such that the two modalities resulted in a more intense level of energy than either alone or (b) preparation enhancement such that one modality served as an alerting cue for the other (see Bernstein, 1970; Nickerson, 1973). Miller (1982) observed that the facilitation required identification and classification of stimuli in each modality rather than merely summation of sensory energy by using concurrent distracting stimuli. Trials in which concurrent distractor stimuli were present would have theoretically resulted in an equivalent level of sensory energy, yet they did not facilitate performance. Race models were proposed as an alternative. According to the race model, response in a given trial was a result of information from only one modality—either visual or auditory—and on any given trial either one could be responsible for the response. It was simply a matter of which was perceived first. In this way, over the course of several trials average response time for cross-modal trials would be expected to be faster than for either modality alone (see discussion in Miller, 1991). Subsequent research indicated that the number and nature of fast responses in cross-modal trials could not be explained based on the race model. Rather, multisensory integration must be occurring. Both modalities must be influencing responses at the same time, a phenomenon referred to as *intersensory facilitation* (Nickerson, 1973; Sinnett et al., 2008).

RTEs also occur in more applied settings and are more pronounced in dual-task conditions than single-task situations. Levy and Pashler (2008) examined the redundant modality information in a simulated driving context. Both a redundant tactile cue (Experiment 1) and a redundant auditory cue (Experiment 2) improved brake response times (the visual cue was the onset of the brake lights of a lead car) in both single- and dual-task conditions. Importantly, providing a redundant tactile or auditory cue also significantly increased the likelihood that people would prioritize the braking task in the dual-task trials. Participants had been instructed that the braking task was most important, and that they should not even make a response to the other concurrent task (distinguishing between one and two tones) if it occurred at the same time as a braking event. Yet, despite these instructions, a significant number of people failed to prioritize the driving task sufficiently to withhold making a response to the tone task at least some of the time. In these cases, responses to the tone task were made at the expense of brake response time. That is, not surprisingly, when participants first responded to the tone task before braking, brake response times were significantly delayed. The implications of these results are clear. Even when a task such as driving should clearly be prioritized, people may still complete a second, less-important task first (like responding to a caller on a cell phone) at the expense of performance in the driving task. Providing a redundant cue for the lead car braking increased the probability that people prioritized the driving/braking task—a result that has important implications for designing effective collision-warning systems as discussed in Chapter 11. Other driving simulation investigations have also noted superior hazard avoidance behavior for combined auditory and visual warning over either modality alone (Kramer, Cassavaugh, Horrey, Becic, & Mayhugh, 2007).

Sinnett et al. (2008) demonstrated that intersensory competition rather than facilitation can occur in some circumstances. They devised a novel set of experiments in which the stimuli were the same but the instructions given to participants emphasized either simple speeded detection or speeded modality discrimination (Experiment 1). Intersensory facilitation was observed when participants were asked to make a simple speeded response (0.5% misses in bimodal trials relative to 3.0% and 3.6% for unimodal visual and auditory trials, respectively). When asked to make a speeded modality discrimination, response times for bimodal trials were significantly longer than either visual or auditory unimodal trials.

In the final section of this chapter, additional examples of times when auditory and visual information either enhance or disrupt performance are provided. Next, the focus is on when the tasks presented in different modalities must be time-shared. So, unique information is coming in from each modality; therefore, it must be processed separately rather than integrated.

TIME-SHARING AUDITORY AND VISUAL TASKS

There are many situations for which people need or want to perform two or more tasks at the same time. For example, we listen to music or the radio and have conversations while walking or driving, read notes and other written material during teleconferences, and listen to music while studying, exercising, and completing daily activities. Our ability to time-share two or more tasks has received considerable attention in the scientific literature (see Pashler, 1994; Schumacher et al., 2001; Wickens, 2002).

Wickens (1980, 1984, 2002) examined the influence of task presentation modality on time-sharing efficiency in a number of investigations. Wickens pointed out that time-sharing efficiency for concurrent tasks differed by more than simply the difficulty of each task. Difficulty for individual tasks can be assessed by a participant's ability to perform the task by itself under single-task conditions. Two tasks requiring similar processing resources can be time-shared less proficiently than a different set of tasks that, although more difficult, rely on different processing resources. Much like Baddeley's model, Wickens proposed that a single-channel model of processing cannot adequately explain the existing evidence on patterns of dual-task interference. Wickens's multiple-resource theory (MRT) proposed in 1980 was intended primarily to account for information processing during multiple-task performance and to allow prediction of performance breakdown in high-workload situations (Wickens, 2002). According to MRT, time-sharing efficiency, as observed through performance of each task, is moderated by the information-processing channel (auditory or visual), the type of code (spatial or verbal), and processing stages (encoding and comprehension or response selection). In later years, Wickens (2002) added a fourth dimension distinguishing between focal and ambient visual processing. Each of these dichotomies has important implications for the topics discussed in this chapter. An emphasis here is placed on the information-processing channel, otherwise known as the presentation modality.

Two tasks presented in different modalities (visual-auditory) tend to be time-shared more efficiently than two tasks presented in the same modality (auditory-auditory or visual-visual) (Wickens, 1984). For example, imagine the scenario of a pilot maintaining control of an aircraft while navigating, communicating, and avoiding hazards. What is the best method of presenting different types of information to the pilot? Numerous cockpit displays have been developed to assist the pilot with these tasks, and MRT has played an important role in their development.

To illustrate the visual-verbal dichotomy, imagine that you are driving and receiving directions to a new destination. Will it be easier for you to read directions while driving or hear them? Empirical evidence indicated that it is easier to time-share the visual driving task with vocal guidance rather than visual map directions (Streeter et al., 1985). In another example, Wickens and Colcombe (2007) showed that an auditory cockpit display of traffic information (CDTI) alerting system was processed more efficiently than a visual CDTI when traffic monitoring was time-shared with a visual tracking task simulating flight control.

Sonification (presenting complex information in the form of continuous auditory graphs) has been shown to improve visual monitoring of anesthetized patients' status in a simulated operation room compared to monitoring the same information displayed visually (Watson & Sanderson, 2004). Sonification, an important relatively new form of auditory display, is discussed further in Chapter 11. Many more examples can be found of improved time-sharing efficiency following the application of cross-modal task pairing, in line with the predictions of MRT (Horrey & Wickens, 2004; Levy et al., 2006; Sodnik et al., 2008).

An important exception to the benefit of time-sharing cross-modal task pairings has been observed in certain circumstances and is worthy of discussion (Latorella, 1998; Wickens & Liu, 1988). Wickens and Liu (1988) pointed out the specifics of these circumstances. One example is when a nonurgent discrete auditory task is time-shared with a continuous visual task. For example, Helleberg and Wickens (2003) found that discrete auditory messages simulating ATC communications disrupted ongoing performance of a visual flight task more than text-based communications. The onset of the auditory stimulus can sometimes preempt or disrupt processing of the continuous visual task at inopportune times. In these situations, providing a visual alert that can be monitored at times deemed appropriate by the operator may be time-shared more efficiently.

In the CDTI example described (Wickens & Colcombe, 2007), evidence for a negative auditory preemption effect was observed when the auditory alert provided excessive false alarms (Experiment 2). In these instances, the pilot's attention was drawn away from the flight control task unnecessarily by false alarms with the auditory alert condition, and tracking performance suffered more than when the display with a high false alarm rate was presented in the visual modality. Despite these important exceptions, performing cross-modal task pairings (i.e., an auditory task concurrently with a visual task) generally results in lower mental workload levels and improved performance relative to intramodal task pairings (i.e., two auditory tasks or two visual tasks). However, that said, make no mistake that this in any way suggests that the two tasks will not interfere with each other as long as they are presented in different modalities. Numerous examples of auditory tasks disrupting visual tasks abound.

PERFECT TIME-SHARING: A MYTH

The previous discussion focused on only one aspect of MRT, presentation modality. As long as tasks require effortful processing (regardless of their presentation modality), there is still potential for performance on one or both tasks to decline. It is important to reiterate this point. The previous discussion should not be taken to imply that a verbal task (such as talking on a cell phone) would not be expected to interfere with a visual task (like driving a car). MRT and related theories simply postulate that the pattern and degree of dual-task interference will vary with different task pairings; thus, talking on a cell phone can be expected to be *less* disruptive than reading text, but both activities can be disruptive. In fact, there are numerous examples in the literature of an auditory task disrupting performance on a visual task.

Performing an auditory task has been noted to cause inattentional blindness during simulated driving tasks (McCarley et al., 2004; Pizzighello & Bressan, 2008; Strayer & Drews, 2007; Strayer et al., 2003). Inattentional blindness is a phenomenon in which people are looking at an object or scene but fail to notice some critical aspect because their attention is directed elsewhere. In one investigation, people were talking on a hands-free cell phone while engaged in simulated driving (Strayer et al., 2003). Compared to when they simply drove, people in the cell phone condition recalled significantly fewer roadside signs despite the fact that eye-tracking data indicated that they had looked at the signs.

There is abundant evidence that conversing on a cellular or mobile phone impairs driving performance despite the fact that one is primarily an auditory-vocal task and the other primarily a visual-manual task (Caird et al., 2008; McCarley et al., 2004; Rakauskas, Gugerty, & Ward, 2004; Reed & Green, 1999). As discussed previously in Chapter 8, using a cellular phone while driving dramatically increases the chance of motor vehicle collisions (Lam, 2002; McEvoy et al., 2005, 2007).

Jolicoeur (1999) proposed that the dual-task interference effects of an auditory task on visual information stems from a more central interference mechanism that disrupts visual encoding, thereby preventing short-term consolidation of visual information into working memory. Jolicoeur observed that an auditory tone discrimination task disrupted recall of visually presented letters only when the presentation of the auditory tone followed presentation of the letters within 250 ms or less (when participants would be expected to be attempting to encode the letters). When the tones were presented 600 ms after letter presentation, recall performance was not disrupted. When the tone discrimination task was made more difficult (distinguishing between four tones versus two tones), the effects of presentation delay were even more pronounced. This central mechanism is equivalent to the processing stage component in Wickens' MRT model and coincides with the psychological refractory period (Pashler, 1998b). Jolicoeur and others (Brisson & Jolicoeur, 2007; Brisson, Leblanc, & Jolicoeur, 2009; Levy et al., 2006; Pashler, 1994; Tombu & Jolicoeur, 2004; Welford, 1952) argued for the existence of a centralized bottleneck to explain the interference effects observed with concurrent auditory and visual tasks (but see Meyer & Kieras, 1997; Schumacher et al., 2001, for alternative explanations).

Posner et al. (1988) also suggested that the visual-spatial attentional system is connected to a more general attentional system involved in language processing. Evidence for this came from the observation that participants were less able to benefit from a visual cue in a spatial location task when they were simultaneously required to pay close attention to spoken words. Regardless of where the processing interference occurs, speech processing can interfere with visual tasks and visual-spatial attention.

SUMMARY

Information can be processed in different ways (i.e., using a visual, verbal, or spatial code) regardless of the modality in which it is presented. Language is processed, at least initially, using a verbal code regardless of whether it is presented visually in the form of text or aurally in the form of speech. Forming a verbal code from text requires an extra processing step; therefore, reading, relative to listening, tends to be more disrupted by concurrent working memory demands.

Cross-modal links facilitate speech processing, spatial location, and speeded response in normal circumstances. Being able to see a speaker talking improves speech intelligibility, particularly for hard-of-hearing individuals or when SNRs are poor. Similarly, providing auditory cues speeds visual target detection and localization.

When performing two tasks at once, performance is generally better if one task is presented through the visual channel while the other is presented through the auditory channel. There are some exceptions to this rule. But, as predicted by MRT, generally it is more difficult to watch two things at the same time or hear two things at the same time than it is to divide attention across our sensory systems. Time-sharing efficiency is affected by more than the modality in which the information is presented. The code (i.e., verbal vs. spatial) of processing plays a major role. It is more difficult to time-share two tasks that both require verbal or spatial processing codes than it is to time-share tasks when one requires a verbal code and the other a spatial code. Regardless of the input modality and processing code, time-sharing efficiency is governed by a centralized limit. Theoretical differences arise over the precise nature of this difference (is it a structural bottleneck, a central attentional resource limitation?), but nearly all information-processing accounts agree that there is an upper limit to the number of things people can process at any given time.

CONCLUDING REMARKS

Information coming in through the visual channel generally facilitates and supports auditory processing. Hearing a sound helps guide our visual attention toward the spatial location of the source of the sound, thus aiding detection and identification of objects in our environment. Providing complementary or supplementary information through the auditory channel can serve to offset some of the visual processing load in visually demanding tasks, such as driving, flying, and monitoring visual displays. The topic of using the auditory channel in informational displays is discussed more fully in Chapter 11. The facilitatory effects of complementary visual and auditory information are particularly beneficial to understanding speech in degraded listening conditions or when the listener is experiencing hearing loss such as that which accompanies the normal aging process. In the next chapter, the impact of aging and age-related hearing loss on auditory cognition is discussed.

10 Auditory Processing and the Older Adult

INTRODUCTION

Older adults often say they have difficulty understanding speech even though they may say they can hear it. Does this difficulty come from changes in hearing abilities or cognitive skills? Thus far, the chapters in this book have discussed various aspects of auditory cognition as they apply to the performance of healthy young adults with normal hearing. In this chapter, the same auditory processes are examined, but now the emphasis is placed on one group in particular—older adults.

There are many reasons for such a focus on the older adult. A general but important rationale is that the population as a whole is aging, in both the United States and worldwide, so there are many more people over the age of 60 than ever before, with the numbers projected to rise rapidly over the next several decades. Auditory processing plays an essential role in many social and economic areas. Mandatory retirement policies are being revised in many countries, and retirement ages are also increasing, resulting in many more adults over the age of 60 remaining in the workforce. Older adults are prone to auditory-processing difficulties, so understanding the nature of such difficulties and how to mitigate them are important issues. The topic takes on added importance given that, irrespective of retirement policies, increasing numbers of older adults are choosing for economic reasons to remain in the workforce rather than retire. It is understood that for these older workers to be productive and be financially able to support themselves late in life, they will need to pay attention not only to their physical health, but also to their cognitive functioning, including their auditory cognitive abilities.

AGE-RELATED CHANGES IN AUDITORY PROCESSING

Older adults often report difficulties understanding speech and other forms of complex auditory information (Fozard & Gordon-Salant, 2001; Schneider, Daneman, Murphy, & See, 2000; Schneider, Daneman, & Pichora-Fuller, 2002; WGSUA, 1988). Some may report that their hearing is fine. They can hear conversations, they just do not understand them. Since speech communication is such an essential part of daily functioning in many everyday tasks and is often critical to success in both social and occupational situations, age-related hearing difficulties are of global concern. Communication in noisy or reverberant environments and processing synthetic, compressed, or rapid speech—common in many modern interfaces—poses a particular challenge for the older adult. As a result of decreasing speech-processing abilities, older adults may experience anxiety, frustration, and the tendency to withdraw or be excluded from social interactions at home and work. Compounding the problem is that others may use a simplified form of speech with older adults, referred to as *elderspeak* (Caporael, 1981), which is comparable to baby talk—often grossly oversimplified relative to the cognitive capabilities of the older listener.

Given these factors, it is not surprising that hearing impairment among older adults is associated with increased rates of social isolation, depression, and cognitive dysfunction (see discussion in Frisina & Frisina, 1997). Whether hearing impairment plays a causal role or is an exacerbating factor in cognitive dysfunction is a matter of debate. But, it is clear that communication difficulties cause distress to many older adults.

Older adults are also more likely to miss important sounds in their everyday and working environments, such as warning sounds. For example, Slawinski and MacNeil (2002) observed that, compared to the young, older listeners required higher intensities (an average of 10 dB higher) to detect warnings (car horns and police sirens), both with and without the presence of background noise. Older adults with a history of occupational or recreational noise exposure are particularly at risk of missing these critical signals. Taking into consideration the capabilities of older adults is recognized as an important aspect of auditory in-vehicle display design (Baldwin, 2002). Fortunately, the design changes that decrease the mental workload of auditory processing for older adults generally also improve performance among their younger counterparts.

It is important to understand the interaction of sensory/peripheral factors with higher-order cognitive factors in contributing to the speech-processing difficulties of older adults. Such an understanding can assist in facilitating improvements in the productivity and overall quality of life for this fast-growing segment of the population (Arlinger, Lunner, Lyxell, & Pichora-Fuller, 2009; Baldwin, 2002; Baldwin & Struckman-Johnson, 2002; Schneider et al., 2002). This chapter discusses key theoretical issues, topics of debate, and empirical investigations aimed at understanding this complex issue.

PERCEPTUAL/COGNITIVE CONTRIBUTIONS

Age-related changes occur in both hearing abilities (Corso, 1963a, 1963b; Fozard, 1990; Schieber & Baldwin, 1996; Schneider & Pichora-Fuller, 2000) and cognitive processes such as attention, working memory capacity, verbal recall, and speed of processing (Anderson et al., 1998; Blumenthal & Madden, 1988; Craik, 1977; Kemper, 2006; Lindenberger & Baltes, 1994; Park & Payer, 2006; Waters & Caplan, 2001). A growing body of evidence has documented the existence of strong associations between sensory and cognitive processing abilities (Baldwin & Ash, 2010; Baldwin & Struckman-Johnson, 2002; Lindenberger & Baltes, 1994, 1997; McCoy et al., 2005; Pichora-Fuller, Schneider, & Daneman, 1995; Rabbitt, 1968; Schneider et al., 2002; Scialfa, 2002; Tun, McCoy, & Wingfield, 2009), further complicating our understanding of the mechanisms responsible for age-related changes in speech-processing and general cognitive abilities (see reviews in Burke & Shafto, 2008; Schneider, Pichora-Fuller, & Daneman, 2010).

Given the age-related changes to both sensory and cognitive processes, it is remarkable that older adults maintain their speech understanding capabilities as long as they do (Wingfield & Grossman, 2006). However, mounting evidence indicates that maintenance of speech understanding stems from increased reliance on top-down processing mechanisms (Baldwin & Ash, 2010; Tun et al., 2009). In turn, the increased reliance on top-down processing requires more mental resources than bottom-up sensory processes, and therefore it can be expected that these compensatory processes will come at the cost of other capabilities (i.e., investing resources into remembering complex discourse and other higher-level processes). Factors such as multiple competing sources of sound, background noise, and reverberation exacerbate the hearing difficulties of older adults (Gordon-Salant & Fitzgibbons, 1995b; Hargus & Gordon-Salant, 1995; Tun, 1998; Tun, O'Kane, & Wingfield, 2002). Fortunately, a number of things can be done to improve the auditory-processing performance of older adults. This chapter examines the performance gains that may be achieved by providing structural support to older listeners, such as contextual cues, improved signal-to-noise (S/N) ratios, and direct visual contact with the speaker.

Throughout this discussion, particular attention is paid to understanding the relative contribution of peripheral versus central age-related changes and their relative impact on auditory information processing (Baldwin & Struckman-Johnson, 2002; Schneider et al., 2002). First, an overview of some of the age-related changes that occur to the auditory system is presented, followed by a discussion of how these changes have an impact on hearing acuity. Next, major theories of age-related changes in higher-order cognitive processes are presented, followed by a discussion of how these processes interact to affect the mental effort older adults must exert in auditory-processing tasks.

AGE-RELATED CHANGES IN HEARING

A wide range of changes in sensory functioning accompanies advanced age (see Fozard, 1990; Fozard & Gordon-Salant, 2001; Kline & Scialfa, 1997; Schieber & Baldwin, 1996). Most of these age-related sensory changes occur gradually, potentially leaving the older adult unaware of the extent of his or her impairment. For example, Rabbitt (1991a) found that older participants tended to self-report high levels of visual abilities, and that these self-reports showed little correlation to actual measures of visual acuity. Gradual hearing loss is also likely to go unnoticed or be underestimated, at least initially.

In addition to normal age-related changes, the incidence of pathological conditions leading to rapid deterioration of sensory and cognitive processes also increases with age. Fundamental age-related changes in audition are briefly reviewed here. More in-depth discussion of age-related changes in both visual and auditory functioning can be found in a number of reviews (see Fozard, 1990; Fozard & Gordon-Salant, 2001; Gates, Cooper, Kannel, & Miller, 1990; Kline & Scialfa, 1997; Schieber & Baldwin, 1996).

PRESBYCUSIS

Age-related decline in hearing acuity, called *presbyacousia*, or more commonly *presbycusis* (Corso, 1963b), is a well-documented aspect of the normal aging process

(Corso, 1963b; Etholm & Belal, 1974). Presbycusis likely results from age-related degeneration of anatomical structures supporting hearing. However, there are considerable individual differences in such changes, and some, such as degeneration of the middle ear joints, appear to have little impact on sound transmission (Etholm & Belal, 1974). Common disorders among older listeners include eczema of the auditory canal, atrophic scars, thickening and immobility of the tympanic membrane, the presence of excessive cerumen (earwax), and loss of transparency and retraction (medial displacement) of the tympanic membrane (Corso, 1963b). Note that retraction also occurs whenever the pressure changes abruptly between the ambient pressure and the middle ear side of the tympanic membrane, such as during the sudden decrease in altitude experienced during air travel. The normal tympanic membrane is transparent. Loss of transparency of the tympanic membrane is common among children with otitis media (a bacterial infection of the middle ear); however, in younger people this condition generally clears up within a few months. Among older adults, the condition is persistent.

At least four types of presbycusis have been identified (see review by Schieber & Baldwin, 1996). A major contributing factor in presbycusis is thought to be loss of hair cells in the inner ear (see discussions in Corso, 1981; Olsho, Harkins, & Lenhardt, 1985). A possible genetic link for presbycusis has been suggested.* The application of molecular biological, genomic, and proteomic techniques to improve understanding of the auditory system is a relatively new but emerging area of research that holds great promise for understanding inherited causes of deafness (Jamesdaniel, Salvi, & Coling, 2009).

In addition to these peripheral age-related changes in hearing abilities, considerable evidence indicates that advanced age is also associated with changes to more central auditory mechanisms.

Central Presbycusis

Age-related degenerative changes to central auditory pathways are classified under the rubric central presbycusis. The listening problems experienced by older adults are often greater than would be expected based on peripheral hearing assessments and are greater than in younger adults with similarly elevated pure-tone thresholds (Frisina & Frisina, 1997). Plomp and Mimpen (1979) noted that speech reception thresholds (SRTs) in noise are often worse for older listeners than would be predicted on the basis of SRTs obtained in quiet conditions. These findings indicate that reductions in central auditory-processing abilities, common among older listeners, also play a significant role in age-related changes in speech processing in addition to peripheral problems (Bellis, 2002; Frisina & Frisina, 1997).

---

* Specifically, a particular molecule, cadherin 23 (CDH23), has been implicated. CDH23 is thought to play a role in the mechanotransduction activity of inner ear hair cells (Siemens et al., 2004). Mutations in the CDH23 gene are associated with deafness and age-related hearing loss. As Siemens and colleagues (2004) pointed out, CDH23 may play a significant role in regulating the mechanotransduction channel of the tip link. This tip link, only 150–200 nm long, is the area thought to regulate the mechanically gated ion channels in hair cells. A link has been observed between aquaporins (aquaporin 4 [Aqp4] expression in particular, a protein that regulates water and ion flux across cell membranes in the cochlea and inferior colliculus) and auditory threshold changes associated with both age and mild-versus-severe presbycusis in mice (Christensen, D'Souza, Zhu, & Frisina, 2009).

Frisina and Frisina (1997) found that older adults with good cognitive function could make use of contextual cues as well as or better than their younger counterparts to overcome the adverse affects of peripheral impairments in a speech-processing task. In support of this, Lunner (2003) observed a strong association between speech recognition in noise and working memory, indicating that individuals with higher working memory span scores were better at compensating for the negative impact of noise.

However, even after controlling for audibility and cognitive functioning, many investigations indicated that older listeners may still experience greater difficulty processing speech in noise, relative to in quiet (Frisina & Frisina, 1997; Plomp & Mimpen, 1979). This suggests that central processing issues, such as temporal resolution, may play a fundamental role in age-related communication difficulties.

Despite the attention that has been placed on age-related changes in auditory processing, the relative contribution of peripheral versus central mechanisms in accounting for these changes is still unclear. This issue is discussed in more detail in subsequent sections. First, however, we turn attention toward common age-related hearing difficulties. The hearing difficulties of older adults can be attributed to changes in peripheral, central pathways and normal cognitive changes that accompany advanced age. These causal factors tend to overlap, making their independent contribution difficult to ascertain. For ease of exposition, these factors are discussed individually.

PURE-TONE THRESHOLD ELEVATION

Hearing loss is one of the most prevalent disabling chronic conditions in adults over age 60. Of this age group, 41% report hearing difficulties (Gates et al., 1990), and the prevalence of difficulties increases with advanced age. Sensorineural hearing loss results in increased pure-tone thresholds, difficulties in speech perception, and greater susceptibility to the adverse effects of environmental factors such as noise and reverberation (Jenstad & Souza, 2007). Age-related decreases in hearing acuity are prevalent first at high frequencies—important for speech perception—and then progressively spread to lower frequencies (Brant & Fozard, 1990; Corso, 1963b). Early investigations (Corso, 1963b) revealed that men typically begin experiencing hearing loss at an earlier age than women (32 vs. 37 years) and have greater high-frequency loss. At the time, this finding was attributed to the likelihood that males have higher exposure to occupational noise than females. More recent research indicated that this differential pattern of hearing loss continues to be observed, with males exhibiting greater and more rapid loss than females at most ages and frequencies; the greater hearing loss of males occurs even for those working in low-noise occupations and in the absence of evidence of noise-induced hearing loss (Pearson et al., 1995). Noise exposure does not appear to have as significant an impact on rate of hearing loss as previously thought (Lee, Matthews, Dubno, & Mills, 2005).

Once presbycusis strikes women, it may progress at a faster rate, at least at some frequencies. Corso (1963b) found that between 51 and 57 years of age, women's lower-frequency thresholds tended to be higher than men's. More recent longitudinal studies (Lee et al., 2005) have found that females' threshold increase at 1 kHz tends to be slower than males', but that their rate of threshold increase from 6 to 12 kHz is significantly faster than males. Irrespective of absolute pure-tone thresholds, on average males' word recognition abilities are lower and decline more rapidly than females' in old age (Gates et al., 1990). The greater word recognition difficulties that males experience are thought to be related to marked high-frequency hearing loss (Amos & Humes, 2007; Gates et al., 1990).

As illustrated in Figure 10.1, hearing loss increases with age. On average, people in their 60s have pure-tone detection thresholds that are about 13 dB higher than those of people in their 20s (Corso, 1963a). Note that this decrease in sensory acuity, though substantial, does not generally meet the criteria for a diagnosis of hearing impairment. Mild hearing loss is defined as a hearing level (HL) of 25–40 dB averaged across frequencies.* Therefore, in addition to the high incidence of clinical hearing loss among older adults, the incidence of "normal" or subclinical threshold elevation is virtually ubiquitous. The important performance effects of these subclinical age-related changes have come under empirical scrutiny (Baldwin, 2001, 2007; Baldwin & Struckman-Johnson, 2002) and are discussed in a subsequent section.

On average, pure-tone thresholds across the 0.25 to 12 kHz range increase by an average of 1 dB per year from age 60 on (Lee et al., 2005). Brant and Fozard (1990) pointed out that, among males, threshold elevation in the primary speech frequencies (0.5–2.0 kHz) is much lower, at about 0.3–0.4 dB per year through age 60 years, but then elevation rates in those frequencies accelerate to reach an increase of between 1.2 and 1.4 dB per year between the ages of 80 and 95 years. It is commonly known that the frequency range between 0.5 and 2.0 kHz is essential for speech perception. However, higher frequencies also play a vital role in speech intelligibility, particularly for distinguishing between consonants such as *f, s, th,* and *k* (Botwinick, 1984; Villaume, Brown, & Darling, 1994). The incidence and extent of this type of hearing impairment is particularly high for some groups of older adults who have experienced high degrees of occupational noise exposure, such as older pilots (Beringer, Harris, & Joseph, 1998). Decreased ability to hear high frequencies can also impair older persons' ability to hear critical auditory cues, such as aircraft alarms.

Along with declining pure-tone sensory acuity, a number of changes occur in more central auditory-processing centers that can compound the effects of elevated thresholds and further compromise auditory processing in older adults.

CENTRAL AUDITORY-PROCESSING CHANGES

An increasing number of older adults experience auditory-processing disorders (APDs) stemming from damage or degeneration of central auditory pathways (Golding, Carter, Mitchell, & Hood, 2004). An APD or central auditory-processing disorders (CAPDs) may be present despite evidence of normal hearing sensitivity thresholds. An ADP can occur at any age, but its prevalence increases with advanced age (Golding et al., 2004). Incidence rate estimates generated from large-scale population studies for older adults varied considerably, from as low as 22.6% (Cooper

---

* For a detailed description of audiometric testing procedures and classifications, refer to publications and standards of the ASHA and Lloyd and Kaplan (Lloyd & Kaplan, 1978).

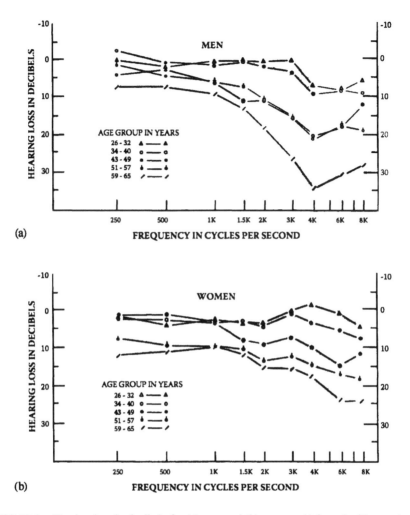

FIGURE 10.1 Hearing loss in decibels for (a) men and (b) women. (Adapted with permission from Corso, J. F. (1963b). Age and sex differences in pure-tone thresholds. *Archives of Otolaryngology, 77*, 385–405.)

& Gates, 1991) to over 76% (Golding et al., 2004). An APD is characterized by a discrepancy between audiometric test results and speech-processing abilities and may be present with or without concomitant hearing loss. An APD manifests itself differently in men and women, often striking men in their late 20s and women not until immediately after menopause (Bellis, 2002).

Gender and Auditory-Processing Disorder

In addition to the later onset of APD in women, there are generally gender differences in the way an APD manifests itself. In men, an APD is associated with greater difficulty hearing, particularly processing speech, in noise as well as greater difficulty understanding precisely *what* is being said (Bellis, 2002). An APD in women, on the other hand, tends to manifest itself in difficulties understanding how something is being said. For example, postmenopausal women with the sudden onset of an APD may find themselves having difficulty understanding the emotional content of messages; they may miss "the hidden meaning of conversations, and have difficulty appreciating the subtle aspects of humor and sarcasm" (Bellis, 2002, p. 151). Understanding the emotional content of spoken conversation relies on the right hemisphere to a greater extent than the left, and the APDs of postmenopausal women tend to be right hemisphere dysfunctions (Bellis & Wilber, 2001). The cause of these menopause-related changes requires further investigation. One explanation that is receiving increasing attention but remains controversial assigns a causal mechanism to the dramatic changes in hormone levels that accompany menopause.

There is some evidence that both hearing (Swanson & Dengerink, 1988) and cognitive (Maki & Resnick, 2001) abilities in young women fluctuate with the menstrual cycle, generally being less keen during the menstrual phase when estrogen levels are at their lowest. Among women in their 50s (who are generally in menopause), estrogen may play a neuroprotective role for both hearing and cognitive abilities (Duka, Tasker, & McGowan, 2000). Hormone replacement therapy (HRT) has shown promise for offsetting a number of performance decrements on verbal- and auditory-processing measures, such as auditory oddball discrimination in an event-related potential (ERP) paradigm (Anderer et al., 2003, 2004), and surgical menopause without HRT is associated with decrements in these same kinds of auditory measures 3 and 6 months postoperation (Farrag, Khedr, Abdel-Aleem, & Rageh, 2002). At the same time, HRT consisting of both progestin in combination with estrogen has been linked to decrements in a number of auditory abilities, including increased (worse) pure-tone thresholds and decreased speech perception in noise. As Greenwood and Parasuraman (2012) pointed out, examining the impact of HRT on cognitive function is challenging since women who choose estrogen treatment are also generally better educated, younger, and healthier than those who do not. Further muddying the waters is the fact that investigations often use different types of hormones. For example, one large-scale clinical trial of high impact utilized an equine-derived (from horses rather than human) estrogen treatment. Little evidence was found in this investigation for positive benefits from HRT, but some negative outcomes related to increased risk of various medical conditions were observed. In addition, equivocal results regarding the impact of estrogen on cognitive abilities in randomized trials may depend in part on the age of participants or time since last menstrual period (Sherwin, 2007, 2009). Regardless, the observations that central auditory-processing changes tend to affect men at much earlier ages than women and women tend to experience sudden changes around the time of menopause point to the potential role that hormones may play in central auditory abilities.

In addition to changes in the peripheral and central auditory pathways, consistent evidence indicated that older adults demonstrate decrements in higher-order cognitive-processing abilities, such as encoding and retrieving items from memory, attention, and executive process control. These changes have the potential to have an impact on auditory processing.

AGE-RELATED COGNITIVE CHANGES

Age-related changes in cognitive processing have also been implicated in the auditory-processing difficulties of older adults. Older adults consistently demonstrate compromised performance on a wide range of cognitive and intellectual tasks. It is beyond the scope of the current chapter to discuss this topic in detail. The interested reader is referred to one of the many excellent reviews on this issue (see reviews in Cerella, 1985; Craik, 1977; Perfect & Maylor, 2000; Salthouse, 1999). Here, three of the leading classes of models of cognitive aging are briefly discussed. These theories have particular relevance to understanding the speech-processing difficulties of older adults. The theories are the generalized slowing models and models postulating reduced availability of specific higher-order resources, such as working memory capacity and inhibitory processes (see also a discussion of how these model classes may relate to language processing in Federmeier et al., 2003).

GENERALIZED SLOWING

An important theory of age-related differences in cognitive performance suggests that much can be explained by global changes in the rate of processing between young and old individuals. For example, Cerella (1985) examined a number of investigations citing age-related decrements in cognitive performance and found a linear function between speed of processing and performance. This and other similar investigations (Cerella, 1991) have led to models that propose that age-related performance decrements can be attributed to "generalized slowing of the central nervous system uniformly affecting all information processes" (p. 215). Cerella's (1985) early calculations indicated that the slope or overall slowing was smaller for sensorimotor processes than for higher-order cognitive processes. Subsequent research conducted by Salthouse (1994) indicated that perceptual speed, rather than age per se, accounts for a substantial amount of variance in cognitive performance comparisons between young and old. As an example of how generalized slowing might affect speech processing, Salthouse suggested that if older adults take longer to process information at each stage, there is a greater chance that the initial material will be degraded or lost by the time later segments are encountered. Think of trying to understand a conversation at a noisy ball game. The extra time spent trying to decipher an unusual or particularly difficult phrase might result in earlier speech segments being forgotten.

Other theories of cognitive aging point to changes in more specific processes either alone or in combination, rather than generalized slowing. Park and colleagues (2002) reported that general continuous decline starting in the 20s occurs for all processing-intensive tasks. They showed that there is a steady decline in performance of tasks that rely on processing speed, working memory, attentional control, or inhibitory processes. Either alone or in combination with changes in other processes, age-related changes in working memory capacity are frequently implicated.

WORKING MEMORY REDUCTIONS

Over the last several decades, working memory has been associated with many processes and abilities, including but not limited to language comprehension, decision making, problem solving, and even general intelligence (Baddeley & Hitch, 1974; Daneman & Carpenter, 1980; Daneman & Merikle, 1996). As Reuter-Lorenz and Sylvester (2005) pointed out, it is not surprising, given the number of abilities that are thought to be associated with working memory, that it has been implicated as a general causal mechanism of age-related cognitive decline. Older adults are frequently found to have impaired working memory capabilities (Park et al., 2002: Salthouse, 2003). Speech processing places heavy demands on working memory storage and processing mechanisms; therefore, any decrements in this resource would be expected to have a negative impact on speech processing. Research carried out by Baldwin and Ash (2010) indicated that the extra effort required to process a degraded speech signal further challenges both the working memory capacity and assessment of that capacity in older adults. This topic is discussed further in a subsequent section.

INHIBITORY DEFICITS

Age-related deficits in the ability to inhibit irrelevant task information are another prominent explanation for the cognitive processing difficulties of older adults (Hasher, Quig, & May, 1997; Hasher & Zacks, 1979; Kane, Hasher, Stoltzfus, & Zacks, 1994). Consistent with this perspective, older adults are more susceptible to the negative effects of proactive interference (Bowles & Salthouse, 2003). This form of interference occurs when previously learned information disrupts learning and memory of subsequent information. For example, when performing several trials of a working memory span task, older individuals were much more likely to include items learned in previous trials, thus reducing their performance on each successive trial (Bowles & Salthouse, 2003). Older adults also had more difficulty than younger adults clearing speech information that was no longer relevant out of working memory, and the irrelevant content could impair performance on subsequent tasks (Hasher et al., 1997).

There is evidence supporting each of these models of age-related changes in cognitive processing, and they may each play some role in the speech-processing difficulties of older adults. Age-related changes in peripheral hearing, central auditory pathways, and cognitive processing may all individually and interactively have an impact on the speech processing of older adults. How such age-related changes lead to challenges to speech processing is the focus of the next section.

AGE-RELATED CHANGES IN SPEECH PROCESSING

As noted, older adults are much more likely than their younger counterparts to experience difficulties in spoken word recognition (Gordon-Salant, 1986; Marshall et al., 1996), particularly in the presence of adverse listening conditions, such as noise (Bergman, 1971; Gordon-Salant & Fitzgibbons, 1995b; Li et al., 2004; Pichora-Fuller et al., 1995; Tun, 1998). Speech-processing difficulties among older listeners are frequently evident despite the absence of clinically significant hearing loss (Schneider et al., 2002), which suggests the existence of APDs or that subclinical hearing loss plays a larger role in auditory difficulties than previously thought (Baldwin & Struckman-Johnson, 2002).

The extent of the impact of speech-processing difficulties on everyday functioning is not well understood. To illustrate the potential importance of this issue, consider the observations of Schneider and colleagues (2000). They had younger and older adults listen to prose passages and then answer a series of questions regarding each passage. The questions pertained to either concrete details presented in the passage or integrative questions that required listeners to make inferences regarding the overall "gist" of the passage. First, note how this task resembles many communication situations in everyday life. We are frequently presented with a string of sentences (i.e., a spoken paragraph, passage, or story) and then expected not only to comprehend and remember the specifics of what was said but also to understand the general gist of the communication (i.e., inferring the tone, criticality, potential sarcasm, or humor).

Both young and older listeners in Schneider et al.'s (2000) investigation heard the passage at the same presentation level, and all listeners had clinically normal hearing abilities (which they defined as < 30 dB HL for frequencies up to 3 kHz). Not surprisingly, regardless of age, participants correctly answered more of the detail questions than the integrative questions. However, compared to the young, older listeners also answered fewer questions correctly overall and made relatively more errors on the integrative questions than the detail questions. This finding could be interpreted as reflecting an age-related cognitive impairment in speech processing per se. However, the researchers went on in a second investigation to compare performance in the same task when the presentation level of the sentences was adjusted as a function of the individual listeners' pure-tone threshold. Using a fixed sensation level method in which stimuli were presented at a fixed level (often 50 dB) above threshold, the age differences in performance virtually disappeared. While both younger and older listeners still had more difficulty with the integrative questions than the detail questions, the performance function was essentially equivalent for both groups. These findings provide strong evidence for the impact of subclinical hearing loss on speech comprehension, a topic we return to in this chapter.

Elevated hearing thresholds account for a substantial portion of the speech-processing difficulties of older adults (Humes & Christopherson, 1991; Humes et al., 1994). However, many older adults experience speech-processing difficulties beyond what would be predicted by increased pure-tone thresholds alone (Frisina & Frisina, 1997; Humes & Christopherson, 1991). In an effort to examine the relative contribution of peripheral and central mechanisms in the speech-processing difficulties of older adults, Frisina and Frisina (1997) conducted research on young and elderly listeners with normal hearing and older listeners with varying levels of hearing impairment. In quiet conditions, the speech perception abilities of the young did not differ from the normal-hearing older listeners on measures of spondee word recognition and recognition of target words in sentences of high and low contextual support. (Spondee words are two-syllable words that have equal stress on both syllables.) Performance on these tasks was reduced for the hearing-impaired groups relative to the young and old listeners with normal hearing abilities. In the presence of noise, even older adults with normal hearing demonstrated decrements in speech-processing performance relative to the young listeners (Frisina & Frisina, 1997; Pichora-Fuller et al., 1995). These decrements in understanding speech are particularly evident when context is not available to aid recognition. This observation suggests that pure-tone threshold elevation alone may not account for all speech-processing difficulties experienced by older adults.

It was mentioned previously that age-related high-frequency hearing loss can make it difficult for older listeners to distinguish between certain consonants. There is also evidence that the rapid changes in spectral cues that are characteristic of the speech signal pose considerable challenges for older listeners.

TEMPORAL PROCESSING DEFICITS

Effective speech processing relies on the listener's ability to discern rapid changes in spectral cues. As discussed in the chapter on language processing, speech variability stemming from speaker and situational characteristics and phonemic variance due to coarticulation place tremendous importance on the temporal processing capabilities of listeners. For example, the ability to detect voice-onset-time (VOT) differences on the order of 10 ms (within a 0 to 40 ms range) enables a listener to distinguish between the syllables *da* and *ta* (Blumstein, Myers, & Rissman, 2005).

Neurophysiological evidence indicated that older listeners have difficulty processing rapid temporal changes in speech cues (Tremblay, Piskosz, & Souza, 2002). Young adults are able to make use of temporal cues, such as a 20 to 40 ms difference in VOT between the phonemes /p/ and /b/, which enables them to distinguish between words such as pill versus bill and pet versus bet. Older adults have greater difficulty resolving these subtle temporal differences. Illustrating this phenomenon, Tremblay and colleagues observed lower d'* scores for older listeners relative to younger listeners trying to discern between phoneme pairs that differed by VOT times of 10 ms. This difficulty was accompanied by delays in early sensory-attentional components of the ERP components elicited by the phonemes. For both young and older listeners, latency of the N1 component increased with increases in VOT. However, for older listeners, this increased N1 latency was particularly pronounced, relative to younger listeners, at VOTs of 30 ms or greater. Compared to younger listeners, older adults exhibited a delayed P2 component at all VOTs. Recall that both verbal and nonverbal auditory stimuli elicit a negative deflection in the ERP peaking approximately 100 ms after stimulus onset (the N1 component) followed by a positive deflection peaking around 200 ms poststimulus (the P2 component). Both N1 and P2 are modulated by the physical characteristics of a sound and by attention (see discussion in Federmeier et al., 2003) but are thought to reflect distinct processes (Martin & Boothroyd, 2000; Tremblay et al., 2002).

Tremblay and colleagues (2002) proposed several potential explanations for why these early sensory components may be delayed in older adults. Similar to proposals by Schneider and Pichora-Fuller (2001), they suggested that aging may be associated with loss of neural synchrony. Older auditory neurons may exhibit a decreased ability to synchronously time-lock on the initial consonant burst and then subsequently exhibit a larger response to the onset of voicing in phonemes with relatively long

---

* Here, d' is a measure of the sensitivity of an observer to distinguish between one condition and another or to determine a signal against a background of noise. Lower d' scores indicate lower sensitivity.

VOTs. Alternatively, delays may be caused from prolonged physiological recovery from forward masking or age-related slowing in refractory processes.

Consistent with the findings of Tremblay and colleagues (2002), Federmeier and colleagues (2003) observed delays in the N1 and P2 ERP components in older listeners in response to sentence initial words in a sentence-processing task. Interestingly, although these early components were delayed on average by 25 ms, the later N400 component reflecting semantic processing (Kutas & Hillyard, 1980) was not delayed in older listeners relative to the young. Both young and old listeners had higher-amplitude N400 responses to anomalous rather than congruent sentence final words, as was typically found (Curran, Tucker, Kutas, & Posner, 1993; Kutas & Hillyard, 1984) at equivalent latencies. The N400 component in response to semantically incongruous sentence final words was attenuated but not delayed in older adults (Faustmann, Murdoch, Finnigan, & Copland, 2007; Federmeier, Mclennan, De Ochoa, & Kutas, 2002; Federmeier et al., 2003).

The observation that early acoustic stages of processing are delayed indicates that older adults experience difficulty with early signal extraction processes. But, since later semantic components are not delayed, these investigations suggested that rather than experiencing generalized slowing (which would delay each stage additively), older adults may actually be making up for time lost during early stages by performing higher-order stages more quickly. They may be compensating during intermediate stages of lexical selection to complete semantic stages within the same time frame as younger listeners. Such a strategy of "making up for lost time" would allow older listeners to compensate for many speech-processing difficulties; however, the compensatory process would be expected to require mental effort and be capacity limited (Baldwin, 2002; McCoy et al., 2005; Pichora-Fuller et al., 1995). Compensatory processes would be expected to occur at the expense of other processing tasks (such as rehearsal and storage).

This issue is discussed in more detail as it is fundamental to understanding the speech-processing difficulties of older adults. But first, a more detailed discussion is provided of the impact of adverse listening conditions found in many recreational and occupational settings on speech comprehension in older adults.

ADVERSE LISTENING CONDITIONS

Adverse listening conditions such as noise, reverberation, and speeded or compressed speech are problematic for older adults. The presence of more than one adverse listening characteristic (i.e., noise and reverberation) is particularly detrimental to the speech comprehension of older listeners (Jenstad & Souza, 2007).

Noise

Noise has particularly deleterious effects on the performance of older adults. Because everyday speech processing frequently takes place in noisy, distracting environments, comprehension suffers. Word recognition scores obtained in the presence of noise are consistently lower than those obtained in quiet for both young and old adults, but older adults exhibit greater decrements even after adjusting for hearing loss (Wiley et al., 1998). Noise seems to exacerbate the speech-processing difficulties of older adults (Frisina & Frisina, 1997; Plomp & Mimpen, 1979; Tun, 1998; WGSUA, 1988; Wiley et al., 1998), as do other adverse listening conditions commonly found in the workplace, such as multiple sources of auditory distractions (Pichora-Fuller et al., 1995) and reverberation (for a review, see Gordon-Salant, 2005).

In one seminal investigation, Pichora-Fuller, Schneider, and Daneman (1995) examined sentence final word recognition among young and older listeners with normal-hearing and older presbycusic listeners across varying S/N ratios. The sentences provided either high or low contextual support to the listener. For example, a high contextual support sentence was, "The witness took a solemn oath" versus a low contextual sentence, "John hadn't discussed the oath" (Pichora-Fuller et al., 1995, p. 595). Contextual cues aided the word recognition performance of both older groups to a greater extent than for younger listeners. Context played a beneficial role at increasingly higher S/N ratios in the young, older-normal-hearing, and presbycusic listeners, respectively. That is, when background noise was present—even though the sentences were louder than the noise—older listeners made use of context to improve their word recognition score. Younger adults could achieve nearly 100% word recognition at negative S/N ratios (when the noise was louder than the sentences), and thus context could provide no further benefit. Context was particularly helpful for the presbycusic listeners, improving their word recognition performance across a much broader range of S/N ratios relative to the improvement exhibited by young and older-normal-hearing listeners.

In general, it seems that the greater the word recognition difficulties of an individual or group, the more detrimental the effects of noise. For example, Wiley et al. (1998) found that the impact of noise increased with age, and that noise was more detrimental to males—who initially had lower word recognition performance relative to females' performance.

In addition to these general effects of noise on speech comprehension, the specific effects of different types and levels of noise need to be examined. One important distinction is that between steady-state versus modulated noise. Several studies have compared the differential effects of these two types of noise.

Steady-State and Modulated Noise

Although noise in general impairs speech perception in the young and old alike, steady-state or constant noise is more disruptive than fluctuating or modulated noise, at least for listeners with normal hearing (Gifford, Bacon, & Williams, 2007; Smits & Houtgast, 2007). Young normal-hearing adults can take advantage of the short periods of quiet in fluctuating noise levels and thus are less disrupted in understanding speech compared to when they hear the same speech in constant noise. Older and hearing-impaired listeners are less able to benefit from the brief periods of quiet, particularly when they have gradually sloping hearing loss that is greatest at high frequencies (Eisenberg, Dirks, & Bell, 1995; Festen & Plomp, 1990). Whether this inability to benefit from the brief periods of quiet is due to reduced release from masking or from deficits in temporal processing is an open question.

Separating the noise source from the target signal benefits all listeners to some degree (Dubno, Ahlstrom, & Horwitz, 2002). But, much like modulating the noise, young listeners benefit from the spatial separation more than older listeners, and older listeners with good hearing benefit more than presbycusic listeners. Because all listeners tend to show some benefit, separating competing sound sources as much as possible is a highly recommended ergonomic practice.

The presence of noise can compound negative effects stemming from other adverse conditions, such as speeded or compressed speech (Jenstad & Souza, 2007). For example, Tun (1998) had young and old listeners recall sentences under varying conditions of speech rate and noise level. She observed that noise had more detrimental effects on sentence recall for older relative to younger listeners, particularly when the sentences were presented at fast speech rates. At the normal speech rate, older and younger listeners were equally efficient in recall at +6 dB S/N ratio. At the fastest speech rate, however, older listeners required S/N ratios of +18 dB to achieve word recall accuracies comparable to young listeners.

Reverberation

Reverberation is the persistence of sound in an enclosed space after the source of the sound has stopped. The reverberation or echoes caused by the original sound bouncing off the walls can cause distortion and lead to difficulties in perceiving and understanding the sound. Reverberation is measured in terms of the time it takes for a sound to decrease to 60 dB below its steady-state level. Long reverberation times and their processing challenges are characteristic of both large rooms with high ceilings and rooms with reflective materials such as glass (Gordon-Salant, 2005).

Hearing-impaired and older adults are particularly susceptible to the negative impact of reverberation. However, reverberation and the high noise levels that frequently accompany it in classroom settings are recognized as a problem for all ages (Picard & Bradley, 2001). Bilingual speakers fluent in a second language demonstrate greater word recognition disruption relative to native speakers in the presence of reverberation (Rogers, Lister, Febo, Besing, & Abrams, 2006). In general, reverberant environments are notably problematic for older listeners (Helfer & Wilber, 1990).

Moderate-to-large reverberation distortions—typically on the order of 0.4 and 0.6 reverberation times—affect the speech recognition performance of older listeners with good hearing (Gordon-Salant & Fitzgibbons, 1995a). That is, at these levels of distortion, age, independent of hearing loss, impairs recognition. The hearing loss among both young and old listeners results in significant decrements in speech recognition in the presence of reverberation even at much lower distortion levels. At reverberation times of 0.2 and 0.3, hearing-impaired listeners have lower recognition scores relative to their performance in a no-reverberation control condition (Gordon-Salant & Fitzgibbons, 1995a). In addition to these aspects of the listening environment that have an impact on the mental workload of speech processing, the quality of the speech signal also plays a role.

SPEECH SIGNAL QUALITY

Poor signal quality in the speech being processed will cause additional difficulties for older adults. Poor signal quality can stem from a number of factors, including low intensity (it simply is not loud enough) and fast speech (it either is spoken too quickly or is digitally speeded up).

Presentation Level or Intensity

One of the primary factors that will have an impact on the difficulty or ease of speech processing, particularly for older adults, is the presentation level or intensity of the speech signal. How loud the signal is, in relationship to the background noise, will have a significant impact on the effort required not only to recognize but also to understand and remember spoken material.

Results of several investigations provided evidence that even among young people, reducing the intensity of speech stimuli (while keeping it within an audible range) increased mental workload. The increased effort required to process lower-intensity speech stimuli impaired the listener's memory for the material (Baldwin & Ash, 2010; McCoy et al., 2005). In cognitively demanding situations (like while simultaneously driving and performing some other task), speech of lower intensity is more likely to be misunderstood and will take longer to respond to (Baldwin & Struckman-Johnson, 2002).

For example, decrements increase the attentional resource requirements of working memory, thus compromising performance in other resource-demanding cognitive tasks (Baldwin & Struckman-Johnson, 2002; Lindenberger, Scherer, & Baltes, 2001; Schneider & Pichora-Fuller, 2000). Empirical investigations with young non-hearing-impaired adults indicated that a reduction in presentation level of just 10 dB resulted in significant decrements in cognitive task performance (Baldwin & Struckman-Johnson, 2002). Young participants exhibited cognitive task performance decrements mirroring those commonly found in older adults when attentional demands were high and the presentation level of an auditory task was decreased from 55 to 45 dB. This 10 dB decrease in presentation level could be considered functionally equivalent to an elevation in threshold sensitivity that would be commonly experienced by an older listener. Note, however, that a 10 dB threshold increase would not result in a classification of even mild hearing impairment. Due to compromised hearing acuity, older adults expend greater attentional resources in auditory-processing tasks, thereby compromising their performance and safety in complex multitask environments. Understanding the interaction between sensory and cognitive processing can have great impact on our technological capability to augment performance among older adults.

Speech Rate

Speech processing under normal conditions requires the rapid translation of phonological segments into coherent semantic meanings. As discussed in Chapter 8, a normal conversational speech rate is approximately 140 words per minute (wpm; Stine et al., 1986). Older adults are more adversely affected by compressed or fast speech, particularly when time compression cooccurs with other adverse listening conditions, such as reverberation or noise (Gordon-Salant & Fitzgibbons, 1995b). Stine et al. (1986) examined proposition recall from spoken sentences of varying complexity among young and older listeners at speech rates of 200, 300, and 400 wpm. As illustrated in Figure 10.2, at the 200 wpm rate, no age differences were observed. However, at the faster 300 and 400 wpm rates, older adults recalled fewer propositions. Interestingly, although participants in general recalled fewer propositions as

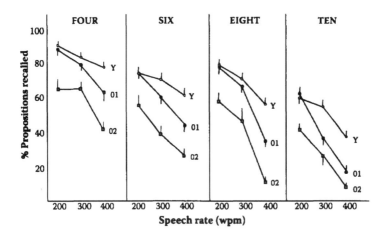

FIGURE 10.2   Proposition recall as a function of sentence complexity and speech rate for young and older listeners. (Data from Stine, E. L., Wingfield, A., & Poon, L. W. (1986). How much and how fast: Rapid processing of spoken language in later adulthood. *Psychology and Aging, 1*(4), 303–311.)

the sentences became more complex and as speech rate increased, the impact of sentence complexity was not greater for older adults than young adults. This finding suggests that it was the speech encoding that required more processing time rather than the processing of semantic complexities.

Note that for Stine et al.'s (1986) data, O1 indicates a sample of older adults with uncharacteristically high education and verbal ability scores relative to the younger sample. O2 represents a community-dwelling sample of older adults matched to the young sample on these two characteristics.

Verbatim recall of individual words in a spoken sentence was used as a measure of speech processing in another experiment examining the impact of speech rate on younger and older adults in full- and-divided attention conditions (Tun, Wingfield, Stine, & Mecsas, 1992). Verbatim recall of words decreased as speech rates increased across levels of 140, 182, and 280 wpm. Word recall was lower for older adults relative to younger adults at all speech rates, but recall among older adults was particularly disrupted by increased speech rate. However, similar to the Stine et al. (1986) results, recall was not differentially lower in older relative to younger adults under the more resource demanding dual-task conditions compared to the full-attention conditions. This finding is somewhat surprising given the observation that older adults are frequently (although not always) more disrupted by increased-complexity or divided-attention conditions. Speech processing is a well-learned task, and older adults appear to have developed compensatory mechanisms that allow them to maintain performance on communication tasks in many situations, particularly when listening conditions are ideal. As will be discussed further, differential dual-task costs (DTCs) among older adults relative to younger are not always observed. If the overall dual-task difficulty is not sufficiently great, perhaps because one or more of the tasks involved is highly practiced or automatized, then older adults may demonstrate preserved performance levels. In addition to differential age-related decrements observed in processing fast speech, the use of synthetic speech also tends to have increased negative consequences for older relative to younger adults.

Synthetic Speech

Older adults are differentially negatively affected by synthetic speech (Humes, Nelson, & Pisoni, 1991; Logan, Greene, & Pisoni, 1989; Roring, Hines, & Charness, 2007). For example, Roring et al. examined recognition performance among young and older adults for natural and synthetic speech. They examined both isolated words and words presented in supportive sentence contexts. Older adults had significantly lower recognition accuracy scores in all but the natural sentence condition. That is, the availability of sentential context seemed to assist older adults in achieving recognition scores comparable to those of younger listeners when natural speech was used, but the use of context was not sufficient when synthetic speech was used. Older adults' recognition performance was worst when attempting to recognize isolated words presented in synthetic voice. The differential decrements among older adults when processing synthetic speech are understandable considering the evidence that synthetic speech requires more cognitive effort to process than natural speech even among younger listeners (Koul, 2003).

UNDER DIVIDED ATTENTION

Older adults may often be able to compensate for their speech-processing abilities by exerting more effort into the listening task (McCoy et al., 2005; Tun et al., 2009; Wingfield, Tun, & McCoy, 2005). In dual-task situations, which may be frequently encountered in the workplace (Lindenberger et al., 2000; Schneider et al., 2000), competing demands make compensation more challenging. As Lindenberger et al. (2000) pointed out, older individuals appear to have greater difficulty performing more than one task at a time, relative to their younger counterparts. They found that even relatively well-learned tasks like walking seemed to require greater cognitive effort for older adults relative to younger adults. In their study, older adults were not able to successfully recall as many words when required to encode verbal lists while walking versus sitting. Younger adults were relatively unaffected by this dual-task demand. Lindenberger et al. concluded that basic sensory and motor tasks required greater cognitive control and effort for older relative to younger adults. Even after single-task performance decrements are accounted for, greater performance decrements are frequently observed in dual-task situations for older as compared to younger participants. This finding, often referred to as a DTC, is particularly evident under specific conditions.

As described by Lindenberger et al. (2000), the types of situations in which these DTCs become particularly evident are as follows: (a) when the two tasks share the same modality in either stimulus presentation or response modality; (b) when heavy demands are placed on working memory (i.e., when stimuli or task sequencing must be maintained and coordinated without the aid of external cues); and (c) when one or more of the tasks requires a high degree of cognitive control processes, such as focusing attention, divided-attention strategies (i.e., scheduling and planning), and coding contextual representations. The extent to which these dual-task costs have an impact on general speech-processing tasks remains an important area of investigation.

The extent to which DTCs exacerbate the speech-processing difficulties of older adults is of importance since listening to speech often occurs in conjunction with other tasks (i.e., having a conversation while driving or walking or while engaged in some activity at work). As previously discussed, Tun et al. (1992) observed that older adults recalled fewer words than younger adults in dual-task conditions, but this age-related recall difference was no greater in the dual-task condition than it was in the full-attention condition. While this observation would appear to conflict with the Lindenberger et al. (2000) results, the two investigations used substantially different measures of speech processing. Lindenberger et al. required memory for lists of unrelated words. Tun et al. (1992) required verbatim recall of 15 word sentences. Older adults are adept at using sentential context to improve speech-processing performance. Therefore, as discussed previously, older adults may be able to offset the additional processing requirements by placing greater reliance on top-down processing until some critical threshold is reached and the total processing resources of the dual-task demands exceed their available capacity (Pichora-Fuller, 2008). Using a variety of different concurrent task pairings in our lab, we frequently do not observe age differences in speech-processing performance until the resource demands of the concurrent task exceed some moderate level (Baldwin, 2001; Baldwin, Lewis, et al., 2006; Baldwin & Struckman-Johnson, 2002). In line with theories of mental workload discussed elsewhere in this book, older adults may still be expending greater effort to process the speech signal. It is likely that the greater effort required by older adults may not manifest in performance decrements until some critical demand threshold is reached and available resources are exceeded.

Further support for this idea is found in the link between working memory capacity and speech recognition. Lunner, Rudner, and Ronnberg (2009) discussed evidence that older individuals with greater working memory capacities are better able to cope with the distorting effects of hearing aids, particularly in challenging listening situations. In less-challenging situations, interindividual differences are less important. This suggests that having ample working memory capacity allows the older adult to rely more heavily on top-down processing to overcome the degraded bottom-up signal.

The question remains: How are older adults using top-down processes? Brain-imaging studies indicated that when performance levels are the same, older adults recruit more neural areas relative to their younger counterparts, typically prefrontal regions associated with executive control (Cabeza, 2002; Grady et al., 1994; Reuter-Lorenz & Cappell, 2008). Are these extracortical areas being activated because older adults are using different strategies to perform the same task, or are they relying on areas associated with higher-level cognitive processes (i.e., working memory) to compensate for reduced sensory activation—a decline compensation perspective? The issue is considered in the next section. But first, due the importance of age-related changes in speech processing, a summary is provided.

In sum, older adults frequently report difficulties in everyday auditory communication tasks (Pichora-Fuller, 1997; Schneider et al., 2000), particularly in the presence of noise (Pichora-Fuller et al., 1995; Plomp & Mimpen, 1979). In adverse listening situations (i.e., in the presence of noise), decrements in speech perception become evident in adults as early as age 50 (Bergman, 1980; Pichora-Fuller et al., 1995). Age-related changes in both peripheral and central auditory mechanisms as well as cognitive changes interactively contribute to the hearing difficulties of older adults.

Several possible reasons exist for why even mild, subclinical hearing loss can affect speech comprehension. Researchers and audiologists have long recognized the inadequacies and inherent difficulties involved in predicting everyday speech-processing abilities from clinical measures of pure-tone audiometry and speech recognition thresholds (American Speech and Hearing Association [ASHA], 1979, 1997, 2002). Conventional hearing assessments are based on detection of pure tones in the low frequencies (500, 1,000, and 2,000 Hz) and high frequencies (3,000 and 4,000 Hz) under ideal listening conditions (Lloyd & Kaplan, 1978). However, frequencies above 4,000 Hz play an essential role in speech perception for hearing-impaired listeners in both quiet and noisy environments and for persons with normal hearing in noisy environments. In addition, although higher levels are sometimes used, standard definitions indicated that pure-tone detection thresholds must be at least 26 dB to be classified as indicating even mild hearing impairment (Carhart, 1965; Miller et al., 1951; Silman & Silverman, 1991). Audiologists widely agree that speech recognition thresholds are dramatically affected by changes in presentation level as small as 2 dB (Pavlovic, 1993). Further, even pure-tone thresholds have been found to increase under conditions involving the imposition of a simultaneous dual task (Baldwin & Galinsky, 1999). In light of these findings, it comes as less of a surprise that older adults often experience speech-processing difficulties in everyday situations—for which listening conditions are likely to be less than ideal—and when people are often engaged in other simultaneous tasks (including such seemingly simple tasks as walking).

STRATEGY DIFFERENCES AND COMPENSATION

As discussed in the previous section, dual-task costs may be more evident in older adults due to their need to expend more effort during sensory processing stages to decipher degraded sensory stimuli. Using functional magnetic resonance imaging (fMRI), Wong et al. (2009) documented that during speech-processing tasks older adults, relative to young, showed reduced activation in the auditory cortex but greater activation in areas associated with higher-order processes such as working memory and attention. These cortical changes were particularly strong when older adults were listening to speech in a noise condition. These results suggest older adults are relying on working memory and other higher-order processes to compensate for a degraded sensory representation. Other investigations indicated that older adults may, in part, be using different strategies than young adults to perform the same task. Support for this came from investigations documenting reduced hemispheric asymmetry as a function of age. Patterns of asymmetry among older adults for tasks in which young adults demonstrated symmetrical processing have led researchers to question whether older individuals are experiencing dedifferentiation of neural mechanisms, using compensatory strategies, or simply using different strategies.

It is well known that the left hemisphere generally shows greater involvement in most speech-processing tasks for most individuals (Floel et al., 2004). The exception is the processing of emotion and prosodic cues, which rely heavily on right hemispheric mechanisms. However, older adults demonstrated more symmetrical patterns of activity during speech-processing tasks relative to their younger counterparts (Bellis et al., 2000; Cabeza, Daselaar, Dolcos, Budde, & Nyberg, 2004; Cabeza, McIntosh, Tulving, Nyberg, & Grady, 1997).

For example, Bellis et al. (2000) used ERPs to examine hemispheric response patterns as children, young adults, and older adults performed a phonological discrimination task. They observed symmetrical patterns of N1-P1 activity over the left and right temporal lobes in a group of older participants, while children and young adults exhibited characteristically greater left hemisphere activity when performing the same task. In addition, the same group of older adults demonstrated greater difficulty discriminating between speech syllables involving rapid spectrotemporal changes relative to the younger participants.

These patterns of asymmetry among older adults for tasks in which young adults demonstrated symmetrical processing have led researchers to question whether older individuals are experiencing dedifferentiation of neural mechanisms, using compensatory strategies, or simply different strategies (see reviews in Greenwood, 2000, 2007). It is of interest to note that, regardless of age, females also tended to demonstrate more symmetrical patterns of brain activation during language-processing tasks (Nowicka & Fersten, 2001) and music processing (Koelsch et al., 2003). Since females tend to demonstrate better verbal fluency across the life span and better word recognition abilities into advanced age (Gates et al., 1990), one must be reluctant to conclude that the patterns of symmetry mean older adults are using additional neural mechanisms in an attempt to compensate for what would otherwise be reduced performance. Whether older adults are recruiting additional brain mechanisms in an attempt to offset processing deficits or simply using different strategies to process the same information has important implications for mental workload. Compensatory mechanisms suggest that workload overall is higher for the same task. Conversely, strategy differences suggest that specific types of workload will increase.

STRATEGY DIFFERENCES

Evidence that older adults may use different strategies compared to the young to perform the same kinds of tasks comes from several studies. Older adults appear to rely more on subvocal rehearsal or phonological strategies than younger adults for performing a variety of tasks (Hartley et al., 2001; O'Hanlon, Kemper, & Wilcox, 2005). This increased reliance on verbal-phonological strategies occurs even when it hinders performance (such as when visually discriminable objects have phonologically similar names). For example, working memory is thought to consist of several separable systems for maintaining information: a phonological, a visual, and a spatial system (Cocchini et al., 2002; Della Sala, Gray, Baddeley, Allamano, & Wilson, 1999; Hartley et al., 2001). Younger adults showed greater dissociation between these systems when performing working memory tasks and therefore when they were performing a task such as visual recognition of objects with phonologically similar names; they were less susceptible to the phonological similarity effect (Hartley et al., 2001). Older adults appeared to use the confusing, phonologically similar names when it was not necessary and even when it tended to impair performance. Older adults in general were more susceptible to the phonological similarity effect (O'Hanlon et al., 2005).

Hartley et al. (2001) concluded that working memory processes remain dissociable in old age, but they suggested that older adults may rely more on acoustic strategies, even when they are not advantageous, and may in fact have difficulty inhibiting subvocal rehearsal. They suggested that older adults may have learned to rely on a subvocal rehearsal and naming strategy to aid memory systems associated with object identity and spatial location, a strategy available but not utilized in younger adults.

If adults do tend to rely on acoustic or verbal strategies more as they get older, a strategy shift of this sort would have important implications for use of displays and advanced technologies. For example, when navigating some people tend to form a map-like mental representation relying primarily on visuospatial working memory, while others tend to use a verbal-sequential list of directions relying primarily on verbal working memory (Baldwin & Reagan, in press; Garden, Cornoldi, & Logie, 2002). When young people rely on a verbal-sequential strategy, they are much more susceptible to interference from tasks that prevent articulatory rehearsal relative to tasks that require visuospatial processing (Baldwin & Reagan, in press; Garden et al., 2002). To the extent that older adults are relying on verbal strategies to aid performance, they could be expected to find verbal interference far more detrimental than visuospatial interference. Performance decrements under verbal interference conditions might be much greater than expected and present on a wider range of tasks in older adults relative to their younger counterparts.

The greater reliance on verbal strategies might concurrently help explain why auditory and verbal displays, in particular, seem to be more beneficial than visual displays to older adults in applied settings, such as during driving (Baldwin, 2002; Dingus et al., 1997; Liu, 2000; Reagan & Baldwin, 2006), and why older adults find distracting speech so much more problematic than do younger listeners (Beaman, 2005; Tun et al., 2002; Wingfield, Tun, O'Kane, & Peelle, 2005). Dingus and colleagues observed that older drivers made significantly more safety-related errors, particularly when attempting to use a navigational information system *without* voice guidance. Older drivers appeared to benefit from the advanced traveler information system as long as it contained the voice guidance display. Similarly, Liu (2001) observed that an auditory display either alone or in combination with a visual display reduced navigational errors in a driving simulation for all drivers, but particularly benefited older drivers. Providing information to the drivers in an auditory format (either with or without redundant visual information) also decreased steering wheel variability among older drivers. Variation in steering wheel movements has been used to assess the difficulty or attentional demand of a secondary task (Hoffmann & Macdonald, 1980). Since navigation is a challenging task for many older drivers (Burns, 1999; Dingus et al., Hulse et al., 1997), providing effective displays to aid in this task could substantially improve safety and mobility of this fast-growing segment of the driving population.

COMPENSATORY MECHANISMS

Compensatory mechanisms are an alternative explanation for the observation that older adults often demonstrate hemispheric asymmetry for tasks in which young adults demonstrate symmetrical processing. Older adults may be trying to compensate for what would otherwise be reduced performance (Cabeza, 2002; Dennis et al., 2008). Grady (1998) discussed several neuroimaging investigations that indicated that older individuals recruit different brain mechanisms than young to perform the same tasks to the same proficiency level. She referred to this observed phenomenon as "functional reorganization" and suggested that it probably plays a compensatory role in maintaining performance.

Observations that older adults make greater use of context (Abada et al., 2008) along with the commonly observed posterior-anterior processing shift in older adults (Dennis et al., 2008) suggests that older adults are using more executive control processes and associated neural mechanisms to compensate for reduced sensory-perceptual abilities. It may be that older adults exhibit both strategy shifts and compensatory processing in an effort to reduce the negative impact of aging sensory and cognitive mechanisms. An understanding of these age-related changes can facilitate design.

AGE-RELATED DESIGN

In this final section of this chapter, attention is turned to methods of facilitating the sensory-cognitive processing abilities of older adults through design. Suggestions are presented for constructing a supportive environment for listening for older adults that will serve to also benefit listeners of all ages.

Plomp and Mimpen (1979) suggested that to allow older listeners to communicate effectively, background noise levels in a room need to be on average 5 to 10 dB lower than those adequate for normal-hearing young listeners. In terms of S/N ratios, for older adults with normal hearing it is recommended that the signal be at least 8 dB above noise levels to ensure that speech with rich contextual cues is comprehended at levels that would be expected for young listeners (Pichora-Fuller et al., 1995). If contextual cues are low or absent or if designing for older adults with some degree of hearing loss, then effective S/N ratios will need to be much higher.

Whenever possible, when the distracting noise is irrelevant speech, separating the physical location of the noise from the signal—or at least its perceived physical location—can aid speech processing in both younger and older adults (Li et al., 2004). Providing contextual cues also aids the older listener.

Older listeners, particularly presbycusic listeners, benefit more than young listeners from supportive contextual cues (Pichora-Fuller et al., 1995). Older adults may be more adept at using contextual cues to compensate for reduced or noisy recognition processes (see discussion by Federmeier et al., 2003). Older adults consistently demonstrated larger priming effects in lexical decision tasks relative to their younger counterparts (see metanalysis by Laver & Burke, 1993). For example, Madden (1988) observed that older, relative to younger, adults' performance of a lexical decision task using visually presented words declined more when the words were degraded. Madden presented asterisks between letters. Presenting the degraded words in the context of a sentence benefited older participants more than the younger. These findings suggest that older adults have more difficulty with the feature-level extraction processes, and as Laver and Burke (1993) pointed out, these findings support process-specific models of cognitive aging rather than generalized slowing accounts.

For guidelines pertaining to the design of in-vehicle displays for older adults, see the work of Baldwin (2002). For example, auditory displays are preferable to visual displays for presenting both advance collision warning and navigational information. However, the presentation level of these auditory warnings should be on average 10 dB above a level adequate for younger drivers and ideally would be calibrated for the driver's individual hearing capabilities. Processing degraded or ignoring irrelevant auditory information requires mental effort—effort that must be diverted away from other tasks like maintaining vehicle control and hazard detection. Providing succinct auditory messages regarding time-critical events using standard terminology at acoustic levels clearly audible to older listeners can greatly enhance the safety and mobility of older drivers.

Additional information and methods aimed at designing more effective products, systems, and interfaces for older adults regardless of the presentation modality can be found in a book, *Designing for Older Adults* (Fisk, Rogers, Charness, Czaja, & Sharit, 2009).

SUMMARY

Age-related changes in hearing and cognition can independently and interactively compromise the information-processing capabilities of older adults. Both types of changes can alter temporal processing capabilities, resulting in a less-distinct signal that requires more cognitive effort to effectively distinguish. In this way, hearing impairments may masquerade as cognitive impairments, further complicating the communication process and contributing to frustration and decreased engagement levels on the part of both the older listeners and their social and work counterparts. Hearing impairment makes listening more effortful, thus compromising higher-order cognitive capabilities, such as short-term and long-term retention. At the same time that reduced hearing capability can be manifest as reduced working memory capability, larger working memory capability can offset some of the detriment caused by declining hearing. Thus, high-functioning older adults may be able to compensate for declining hearing abilities better or longer than their less-high-functioning counterparts. But, these compensatory efforts detract from the attentional effort a person might otherwise be able to devote to some other concurrent task. Attending to the human factors design approaches and guidelines for older adults set forth in the existing literature can greatly enhance the cognitive and performance capabilities of older adults.

CONCLUDING REMARKS

The focus of this chapter has been on age-related changes in hearing that have an impact on auditory cognition. After discussing the nature of some of these changes, their impact on specific aspects of auditory processing, such as understanding speech and dual-task performance, were discussed. In the final section, age-related design was discussed. Fortunately, generally the design approaches and guidelines that benefit older adults also benefit others. That is, the guidelines may be particularly important for ensuring that older adults are able to effectively benefit from a given device or display, but they generally improve performance in all age groups. In the next chapter, the focus is on auditory display design in general. In the same way that designing for older adults generally benefits everyone, the guidelines discussed in the next chapter that are meant as general guidelines for all adults will surely benefit older adults, perhaps even more than they do their younger counterparts.

11 Auditory Display Design

INTRODUCTION

The preceding 10 chapters in this book have discussed how aspects of different forms of auditory stimulation—speech, music, noise, nonverbal sounds, and so on—influence cognition and performance in a wide variety of tasks. A common theme that ran through many of the chapters is that while our phenomenal experience views auditory processing as seemingly effortless, in fact many aspects of auditory cognition pose significant demands on attentional resources and thus on mental workload, particularly in challenging listening environments. Accordingly, the design of auditory displays that minimize workload demands so that people can use them effectively and safely represents an important practical issue. This chapter examines the implications of the findings on auditory cognition discussed previously for the design of auditory displays.

Technological advances in recent years have provided engineers and designers the ability to display large amounts of information to human operators in many work environments. In some cases, the volume of data is greater than any human could possibly process at any given time. For example, the modern aircraft cockpit is replete with visual displays. In this environment of high visual demand, auditory and haptic displays are increasingly being used to provide information to the aircrew.

An auditory display is any device or interface in the environment that provides information to a human user in an auditory format. Defined this way, auditory displays can be discrete or continuous sounds that are either inherent or designed. They can also range from simple to complex. Inherent auditory displays are those that are an intrinsic part of some system and that sound automatically when the system is operational. The sounds of a computer operating, a printer in use, or the hum of a running engine in a car are examples of inherent displays. Changes in the auditory characteristics of the machines we interact with provide important cues regarding how the system is operating, such as when it is time to shift gears or when a machine may be malfunctioning. Recognizing the wealth of information these inherent displays can provide, designers sometimes artificially mimic them, such as implementing a key clicking sound in a touch screen display to provide feedback to the user. The focus of this chapter, however, is on designed rather than inherent auditory displays. Coverage of inherent displays is limited to the observation that they can inform ecological design.

Designed auditory displays include the more obvious sources, such as alarm clocks, telephones, and doorbells, as well as sophisticated sonifications or auditory graphs. They can be nonverbal, as the previous examples illustrate, or verbal, such as the in-vehicle routing and navigational directions informing us in our choice of voice (from a generic female avatar to the voice of the Mr. T. character) to "Turn right in two blocks on Ash Avenue." Such auditory displays vary tremendously in complexity and format, which largely influences the mental workload required to comprehend them.

Auditory displays can vary from simple beeps and buzzers to complex sonograms that require extensive training before they can be interpreted. Intermediate levels of complexity include auditory displays used in a wide variety of operational settings, including aviation, surface transportation, and medical facilities. Many operational environments utilize multiple auditory displays, including a range of inherent continuous and designed discrete categories as well as different complexity levels. For example, in the modern cockpit, continuous auditory information is available from running engines, while auditory alerts are used to signal changes in automation mode. Auditory warnings are used to signal system malfunctions, and speech interfaces are used for communication with air traffic controllers and other ground personnel.

Several important quality reviews of auditory displays have previously been compiled (Edworthy & Adams, 1996; Edworthy & Stanton, 1995; Kramer et al., 1999; Stanton & Edworthy, 1999b; Walker & Kramer, 2004). Their main conclusions are described, supplemented by a discussion of empirical findings of newer studies conducted since these reviews. Furthermore, this concluding chapter emphasizes the application of this existing body of knowledge for understanding and predicting the mental workload requirements of auditory displays in operational environments. First, some of the advantages and challenges with using auditory displays are presented, including guidelines for when they are preferred to other forms of displays. Next, key psychoacoustic factors are discussed, including such topics as how to ensure that auditory displays and warnings are detectable and distinguishable. The next section is devoted to auditory warnings, a specific type of auditory display. From there, auditory displays are discussed within the context of three specific application areas: aviation, surface transportation, and medicine. First, some important general considerations pertaining to auditory displays in work environments are discussed.

The next section provides answers to a number of key questions concerning the use of auditory displays in the workplace: When is it preferable to use the auditory rather than the visual channel to present information? How do the characteristics of auditory displays provide advantages for operational performance? How can this information be applied in different work settings?

ADVANTAGES OF AUDITORY DISPLAYS

Auditory displays date to at least the Industrial Revolution and even much earlier if you include natural auditory warnings like the human cry (see review in Haas & Edworthy, 2006). Many characteristics of the auditory modality make it particularly well suited for providing alerting and warning information in most settings. Because of this important function, auditory warnings are discussed more fully in a subsequent section. The auditory channel is also an effective means of conveying status or system-level changes and representational encodings of critical and sometimes multidimensional information. Their use in these more complex information representational forms is on the rise. Table 11.1 summarizes a number of circumstances when auditory displays are preferable to visual displays that were identified by Sanders and McCormick (1993).

TABLE 11.1
Circumstances When Auditory Displays Are Preferable to Visual Displays

Circumstance/Situation

When the origin of the signal itself is a sound
When the message is simple and short
When the message will not be referred to later
When the message deals with events in time
When warnings are sent or when the message calls for immediate action
When continuously changing information of some type is presented, such as aircraft, radio range, or flight path information
When the visual system is overburdened
When the speech channels are fully employed (in which case auditory signals such as tones should be clearly detectable from the speech)
When the illumination limits use of vision
When the receiver moves from one place to another
When a verbal response is required

Note: Data from Sanders, M. S., & McCormick, E. J. (1993). *Human factors in engineering and design* (7th ed.). New York: McGraw-Hill, p. 169.

AUDITORY SOURCE

The auditory modality is well suited when the information to be displayed is itself auditory. For example, the optimal means of conveying how something sounds is by demonstrating the sound. If you have ever had the experience of trying to imitate the unusual sound made by a car engine or other motor to a mechanic, you can appreciate this use of the auditory channel. Other instances include trying to convey how a professor's accent or mode of speech has an impact on comprehension or presenting a display of engine noise to a remote unmanned ground or aerial vehicle operator. The auditory modality is also well suited when the message is short.

BRIEF COMMUNICATIONS

Long or extensive messages that exceed working memory capacity are generally best presented in a visual format so that they can be referenced, thus increasing comprehension. But, when the message is brief, the auditory modality is generally preferred. To make sure that working memory limitations are not exceeded, auditory messages should be limited to two or three terse messages. For example, Barshi (1997) showed that the ability to execute procedural commands correctly deteriorates with increases in the number of commands contained in a set, and that performance bottoms out at three commands. Exceeding three commands overloads working memory capacity, particularly when the auditory display must be time-shared with a concurrent task (Scerbo et al., 2003). If more than three messages must be presented or if the

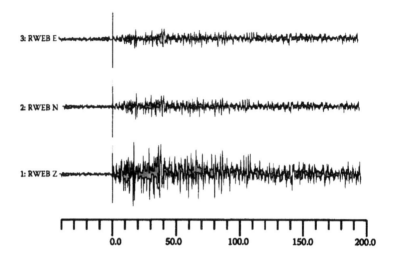

FIGURE 11.1   Seismic information.

individual messages are complex, then visual displays are advantageous (Dingus, Hulse, Mollenhauer, & Fleischman, 1997; Srinivasan & Jovanis, 1997a, 1997b).

A caveat must be noted regarding the length and complexity issue. Auditory displays are useful when messages are short and simple. However, auditory displays may also be used to present some types of information that are too complex to be easily interpreted visually. Consider seismic information like the sample illustrated in Figure 11.1. This sample illustrates activity from only a few recorders. An operator may be asked to monitor information from seismographs with 15 or more recording sites and to integrate the information across multiple seismographs. The complexity of this information is often difficult to understand in the visual domain (Walker & Kramer, 2004). Capitalizing on the human ability to perceive complex acoustic patterns, seismic information is much easier to interpret and categorize when it is presented in auditory rather than visual format. Translating complex information into auditory form is referred to as *audification*, and it is one of the many recognized categories of nonverbal auditory displays (Walker & Kramer, 2004). Others include auditory icons, earcons, and sonifications. Each is discussed in greater detail in subsequent sections. The main distinction is that in these instances the information is nonverbal and multidimensional. Our auditory system possesses unique pattern recognition capabilities that are well suited for processing this type of complexity, as long as the acquired information does not need to be stored for an extended period of time or referred to at a later point in time.

SHORT RETENTION INTERVALS

Visual displays may be preferable to the auditory channel when the acquired information must be retained over extended periods of time. For example, one of the advantages of a text-based data-link interface over traditional voice communications for providing air traffic control (ATC) information involves this issue. Text-based commands can be stored and referred to later. Therefore, they reduce working memory demands and decrease the chances of misinterpretation or forgetting. Conversely, pilots may attempt to write down spoken ATC commands, thus detracting from the benefit of eyes-free transmission. In fact, Helleberg and Wickens (2003) found that pilots spent more time with their head's down when using a voice ATC transmission system relative to a visual data-link system due to their attempt to write down spoken commands. Such head-down time disrupted both traffic monitoring and flight path tracking performance.

TIME-CRITICAL SIGNALING

Auditory displays are well suited for presenting time-critical information as well as information that requires temporal sequencing. Auditory signals can be superior to visual signals in terms of their attention-getting capabilities and often lead to faster response times (Liu, 2001). At one time, it was commonly believed that auditory signals resulted in about a 40-ms reaction time advantage over visual stimuli. This conclusion has subsequently been questioned, and the issue of modality superiority remains unresolved (see reviews in Kohfeld, 1971; Niemi & Naatanen, 1981). Equivocal results have been obtained and appear largely to do with different methods of equating stimulus intensity in each modality, as well as the duration and timing of stimulus intervals. Despite the debate, it is clear that auditory signals of sufficient intensity to be discerned attract attention even under conditions of distraction and therefore are well suited for presenting time-critical information. In the case of text-based controller-pilot communications discussed previously, for example, an auditory signal is likely to be necessary to alert the pilot to look at the text display. Because auditory signals are omnidirectional, they can be processed even when a person is not looking at them. For this reason, they may often be used in conjunction with visual displays to initiate and direct attention to time-critical information in a visual display (Chan & Chan, 2006).

An additional factor that influences how quickly people respond to auditory signals is the type of information presented and how it is portrayed to the listeners. Auditory signals, particularly speech, are processed slightly faster when presented to the right than to the left ear, a phenomenon referred to as the right ear advantage (REA) (Hiscock, Inch, & Kinsbourne, 1999). A REA is consistently observed for most verbal tasks (Hugdahl, Bodner, Weiss, & Benke, 2003; Shtyrov, Kujala, Lyytinen, Ilmoniemi, & Naatanen, 2000; Voyer & Boudreau, 2003), except when emotional or prosodic cues need to be processed (Sim & Martinez, 2005). As discussed in Chapter 3, the right hemisphere is specialized for processing prosodic cues. Hemispheric specialization for nonverbal tasks is not as consistent (Hiscock, Lin, & Kinsbourne, 1996). Nonverbal tasks such as melodic pattern recognition (Cohen, Levy, & McShane, 1989) or pitch identification (Itoh, Miyazaki, & Nakada, 2003) often show a left ear advantage. Providing auditory information to the ear best suited for processing that type of information could facilitate speed of processing.

CONTINUOUSLY CHANGING INFORMATION

The auditory modality is also ideal for presenting continuously changing information, such as flight path information in aviation (Lyons, Gillingham, Teas, & Ercoline, 1990; Simpson, Brungart, Dallman, Yasky, & Romigh, 2008; Veltman,

Oving, & Bronkhorst, 2004) or the status of a patient's vital functions to a nurse or physician (Watson & Sanderson, 2004). The auditory modality is ideal for presenting continuously changing information alone or to augment visual displays. Auditory presentation is particularly advantageous in environments where heavy demands are placed on the visual system.

UNDER HIGH VISUAL LOAD

Many activities and occupational settings require a heavy processing load on the visual system. For instance, it has been estimated that about 90% of resources required to drive are from the visual channel (Dingus et al., 1998). Clearly, there is little left to spare. Consequently, auditory displays can be used in these environments to avoid visual overload or provide additional information that would not otherwise be possible through the conventional visual channel. Heavy visual demands are also imposed on aircraft pilots during particular flight segments, such as taxiing, takeoff, final approach, and landing. Alerting pilots visually to potential hazards in the environment may add to such demands. Since many aircraft accidents and near misses occur on the airport runway rather than in the air, auditory alerts to hazards during taxiing or landing can help reduce the incidence of such "runway incursions." Squire et al. (2010) reported such an auditory alerting system design based on a cognitive task analysis of pilot activities during taxiing and landing.

Auditory displays have also been found to be particularly beneficial in other contexts, such as for older drivers (Baldwin, 2002; Dingus, Hulse, et al., 1997; Dingus et al., 1998). Older adults take longer to extract information from a visual display and longer to shift their attention from one visual area to another (Greenwood, Parasuraman, & Haxby, 1993; Parasuraman & Haxby, 1993). Providing information through the auditory modality can help offset some of the visual-processing load in these environments. Similarly, when visual processing cannot be accomplished or ensured, either due to poor illumination or because the operator is moving about and therefore unlikely to notice changes in a visual display, the auditory modality may be a suitable alternative choice for presentation of information.

DISPLAYS REQUIRING VERBAL RESPONSES

The auditory modality is also preferable in many situations for which a verbal or vocal response is required. Stimulus-response (S-R) compatibilities between pairings of visual-manual and auditory-vocal tasks are well documented (Hazeltine, Ruthruff, & Remington, 2006; Levy & Pashler, 2001; Proctor & Vu, 2006; Stelzel, Schumacher, Schubert, & D'Esposito, 2006). Visual-spatial tasks (i.e., indicating which direction an arrow is pointing) are carried out faster when a manual response mode is used. Auditory-verbal tasks (i.e., classifying tones as high or low) are carried out faster and more efficiently when a vocal response is used. Hazeltine et al. found that incompatible mappings (visual-vocal and auditory-manual) were particularly disruptive in dual-task trials. They computed response cost as time needed to perform in dual-task trials minus single-task performance time. Incompatible pairings in dual-task trials generally resulted in more than twice the response cost of compatible pairings.

AUDITORY DISPLAY CHALLENGES

Despite the many potential benefits that auditory displays can provide, they are not a panacea and pose some challenges in design. Auditory displays can be masked when background noise is high or if listeners have hearing impairment. They can be distracting and annoying (particularly if they are poorly designed or present information that is not useful). There is also a potential to overload the auditory channel, much like the visual overload that such displays may be designed to combat.

AUDITORY OVERLOAD

Numerous examples of the distraction that auditory signals impose in high-workload, high-stress situations have been cited in the literature. One of the best-known cases of this (at least in the human factors community) is probably the tragedy that occurred at the Chernobyl nuclear power plant in April 1986. It was not uncommon at the time for nuclear power plants to have as many as 100 or more auditory alarms that could potentially all go off simultaneously (Medvedev, 1990). It was such a cacophony of simultaneous alarms that Chernobyl plant workers faced while trying to diagnosis the crisis at hand.

Another vivid example of auditory overload was provided by Patterson (1990a). Some of the aircraft involved in accidents at the time had as many as 15 auditory warnings that potentially could all come on at the same time, some at intensities of over 100 dB. The number of auditory displays has increased dramatically since that time. According to Stanton and Edworthy (1999b), there were over 60 different auditory warnings simultaneously sounding during the Three Mile Island nuclear plant accident. In such situations, auditory warnings tend to distract attention and increase mental workload rather than reduce workload and guide effective action. This potential for auditory information to detract from visual or attentional processing rather than enhance it is closely related to another challenge for auditory displays briefly discussed in Chapter 9.

AUDITORY PREEMPTION

Cross-modal task pairings are generally preferred to within modality pairings. As previously discussed, this was predicted by Wickens' (1984, 2002) multiple resource theory (MRT) and a considerable body of evidence supporting it (Derrick, 1988; Horrey & Wickens, 2003; Klapp & Netick, 1988; Risser et al., 2006; Wickens & Liu, 1988). MRT leads to the prediction that, all other things being equal, a visual task will generally be time-shared more efficiently with an auditory task than with a visual task. However, as Wickens and Liu (1988) pointed out, there are exceptions to this general rule. Sometimes, a discrete auditory signal (particularly if it represents a task that is not extremely time critical) can disrupt a continuous visual task more than would a discrete visual task (Latorella, 1998; Wickens & Liu, 1988). Auditory stimuli have a tendency to capture or preempt attention. Providing a nonurgent secondary task in the visual modality can assist the operator in carrying out his or her own attention allocation strategy. Conversely, an auditory signal tends to divert attention and consequently may disrupt ongoing visual task performance at inopportune times.

Visual displays allow monitoring of the secondary task during periods deemed optimal by the operator rather than on the demand of an auditory signal.

In sum, complex patterns are frequently more easily discerned in auditory relative to visual format, and auditory information can be detected quickly from any direction—regardless of where the operator's visual attention is focused. Using auditory displays in operational environments that have high visual workload can facilitate performance provided the information is designed to match the capabilities of the human operator and auditory overload is avoided. Attention is now turned to the psychoacoustic characteristics that have an impact on the effectiveness of auditory displays.

PSYCHOACOUSTICS OF AUDITORY DISPLAYS

Effective auditory displays share some basic characteristics with displays in any other modality. To be effective, they must be detected and identified, allow discrimination, and sometimes be localized (Bonebright, Miner, Goldsmith, & Caudell, 2005; Sanders & McCormick, 1993; Walker & Kramer, 2004). Detection involves making sure that an operator or user is aware that a signal has been presented. Presuming the display is detected, it must also be identified and distinguished or discriminated from other sounds. For example, hearing a sound and knowing that it is an indicator that a machine has finished its series of cycles is important, but knowing which machine is finished is also important. Relative discrimination refers to differentiating between two or more signals presented close together. These first two aspects (detection and identification) are determined largely by the sensory characteristics of the signal, such as its intensity, frequency, and duration. Absolute identification involves identifying the particular class or categorization of the signal. Localization, as the function implies, involves determining where the signal is coming from. The process of locating the display or its referent may or may not be essential in all situations and is thus left out of some categorization schemes. Similar stages or processes have been discussed in reference to a particular class of auditory displays: warnings (Stanton & Edworthy, 1999a). Auditory warnings simply add an element of time urgency to the equation that may not be present with other forms of auditory display. Factors that have an impact on each of these key aspects are discussed.

DETECTION

Any auditory display that cannot be heard, or for which important patterns of fluctuation cannot be perceived, can be considered completely ineffective. Designing auditory displays that are detectable without being disruptive or annoying is an ongoing ergonomic challenge (Edworthy & Adams, 1996). Early implementation of auditory alarms often took an approach of "better safe than sorry" (Patterson, 1990a). Detection was ensured by creating alarms with high intensities. If alarms or displays are made loud enough, detection is virtually guaranteed. However, as Patterson pointed out, high intensities also cause a startle response and can prevent communication in time-critical situations.

Startle responses can be elicited from sudden acoustic stimuli of at least 85 dBA (Blumenthal, 1996). In laboratory settings, the startle response is most frequently associated with the eye-blink response as measured by electromyographic (EMG) activity. However, in tasks outside the laboratory, such as during driving, a startle effect may result in not only an eye-blink response but also the prepotent response of grabbing the steering wheel tightly and quickly applying the brakes. This is obviously not always the optimum response, and any time spent recovering from the startle effect is time lost for responding to a critical event.

Detection and perception of the display depend on a complex interaction between the acoustic characteristics of the sound, the listening environment, and the hearing capabilities of the listener (Walker & Kramer, 2004). As Edworthy and Adams (1996) pointed out, in quiet environments an auditory signal can easily be too loud, while in noisy environments signals can be inaudible and therefore missed. Therefore, the key to whether an auditory display can be perceived depends on its masked threshold rather than its intensity (Stanton & Edworthy, 1999a).

Recall from Chapter 3 that any background noise or unwanted sound can obscure or mask a target sound. Because of upward spread of masking, the impact of background noise will tend to be greatest on frequencies at and above the spectral makeup of the noise. Since engine noise (i.e., jet engines) tends to have a high concentration of low-frequency noise their potential to mask target sounds is particularly high.

Compounding the situation is that typical work environments often have unpredictable levels of noise; therefore, the optimum appropriate amplitude level for an auditory display or warning may vary from moment to moment. (See Edworthy & Adams, 1996, for a discussion of remedies for fluctuating background noise levels.) Patterson (1982) developed a set of guidelines for auditory warning systems in aircraft that have had widespread application in a number or occupational environments, including aviation, surface transportation (Campbell, Richman, Carney, & Lee, 2004), and medical facilities (Edworthy & Hellier, 2006a; Mondor & Finley, 2003; Sanderson, 2006). The aim of his guidelines was to ensure detection and recognition of warning signals while reducing disruption to cognitive performance and flight crew communication. A computerized model called Detectsound has also been developed (Laroche, Quoc, Hetu, & McDuff, 1991). An additional feature of the Detectsound program is that it takes into account the reduced hearing sensitivity common among older adults. Improving detection while minimizing distraction and annoyance remains an ongoing challenge in many operational environments.

To make sure that auditory displays (and warnings in particular) are audible, Patterson (1982, 1990a) recommended that they contain four or more spectral components that are at least 15 dB above the ambient background spectral frequencies. Figure 11.2 provides an example of the application of Patterson's (1982) method of assessing background frequencies in 0.01-kHz increments—using the spectral components of a Boeing 727 in flight—to assess the optimal range for auditory warnings.

Additional standards and design guidelines for ensuring alarm detection include international standard ISO (International Organization for Standardization) 7731, "Danger Signals for Work Places—Auditory Danger Signals," (ANSI, 2003) military standard 1472C, "Human Engineering Design Criteria for Military System, Equipment, and Facilities" (DOD, 1981), and IEC (International Electrotechnical Commission) 60601-1-8, 2006, which provides guidelines for alarm systems in medical equipment. In general, these guidelines recommend that auditory warnings

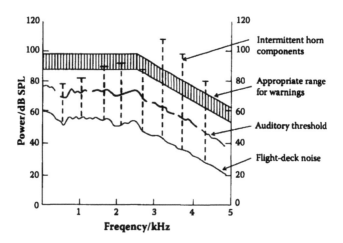

FIGURE 11.2   Patterson detectability.

should be 15–20 dB(A) above ambient background noise levels, with ISO 7731 considering the spectral content of the background noise and recommending levels of 13 dB or more above the masked threshold in one or more octave bands (see Begault, Godfroy, Sandor, & Holden, 2007). These guidelines, although primarily aimed at auditory warnings, can be used to ensure that other forms of auditory displays are audible as well.

A report compiled by the Federal Highway Administration (FHWA) has also composed guidelines for ensuring that auditory displays are both detectable and identifiable when used in in-vehicle automotive displays (Campbell et al., 2004). The interested reader is referred specifically to Chapter 6 of the report, "The Auditory Presentation of In-Vehicle Information." Recommendations found in this report are based largely on empirical results with detectability recommendations from Patterson (1982).

In summary, to promote detection, critical display sounds should contain four or more spectral components at least 15 dB above the background auditory levels while avoiding sounds louder than 90 dBA.

IDENTIFICATION AND DISCRIMINATION

When an operational environment includes numerous auditory displays or alarms, identifying what any given sound means can be onerous, if not impossible. Time spent trying to recognize what a sound is designed to represent detracts from effectiveness by adding to the time necessary to respond to the sound. Therefore, the number of time-critical displays or alarms should be limited to a small, finite set. In fact, the potential advantage of signifying different states with multiple unique alarms may be offset by the increased time spent in identifying what each alarm is meant to represent. For example, master alarms (one alarm signifying a number of different critical states) may be preferable to multiple individual alarms in some settings

(Cummings, Kilgore, Wang, Tijerina, & Kochhar, 2007). One method of combating the recognition/identification issue that is increasingly being considered is mapping the display or alarm sound to something that represents the natural or inherent sound that the situation or failure is designed to represent. This method, involving the use of auditory icons such as screeching brakes or breaking glass to represent the sound of an impending crash, has been shown to improve alarm response in a number of settings. For example, such auditory icons improve crash avoidance when used in in-vehicle collision avoidance systems (Belz et al., 1999; Graham, 1999).

In an aviation context, the use of auditory icons (i.e., such as coughing to indicate carbon monoxide exposure) resulted in faster learning of alarm associations and resulted in greater identification accuracy during a test phase in both high- and low-workload situations relative to abstract alarm sounds (Perry, Stevens, Wiggins, & Howell, 2007). Earcons—melodic sequences that must be learned—may not be beneficial, particularly if they signal infrequent events or if many different states must be discriminated.

Attempts to use earcons to aid learning and recognition of alarms among nurses in a medical context met with relatively little success (Wee & Sanderson, 2008). Recognizing numerous multiple alarms by learning a specific melodic pattern resulted in error-prone and slow performance, particularly when tested under divided-attention conditions.

Patterson (1990b) pointed out that immediate action warnings should be kept to a maximum of about six in any given work environment. These immediate action warnings are ones such as fire alarms that will gain attention through "sheer brute force." They should each be distinct in both melody and pattern so that confusion does not ensue. Prior to the establishment of Patterson's guidelines for auditory warnings, pilots apparently complained—and rightfully so according to Patterson—that there were too many existing alarms, they were too loud (as high as 100 dB or more), and too confusing. Further, there seemed to be no coherent ordering of the sounds used, such that two or more could come on simultaneously, with the combined sound preventing identification of either individual warning. Patterson noted that a similar situation plagued operating rooms at the time. Often, patients could be linked to as many as 10 different systems, each with its own auditory warning sounds. The result was a discordant symphony of sound that, as discussed further in this chapter, nurses often cannot discriminate.

LOCALIZATION

Auditory displays can be used to assist operators with tasks requiring localization of targets or other objects or to keep track of where things are in operational space. Tremendous progress has been made in understanding how the use of spatial and three-dimensional (3-D) audio displays can have an impact on performance. The auditory system is intrinsically suited for localizing and perceptually grouping objects in the environment. This ability can be capitalized on in designed auditory displays to present location information pertaining to critical and potentially interacting or conflicting elements. Auditory displays providing location information for real-world sound sources can be of benefit in a wide range of occupational settings, including ATC, traffic collision avoidance for pilots and drivers, telerobotics in hazardous environments, and any other environment where visual cues are limited or overly taxed (Wenzel, Wightman, & Foster, 1988). In addition to location information, auditory spatial cues enhance the feeling of presence in virtual reality (VR) displays. Auditory cues may even enhance neurobiological presence—such as increased hippocampal activity (previously shown to be related to both spatial navigation and learning in general) when viewing scenes of moving through space and locating objects relative to viewing the scenes without auditory information (Andreano et al., 2009).

Spatial Audio Cues

Spatial audio or 3-D audio cues can aid localization in a number of different situations (Begault, 1991; Begault et al., 2007; Begault & Pittman, 1996; Bolia, 2004; MacDonald, Balakrishnan, Orosz, & Karplus, 2002). For example, 3-D audio significantly reduces the time needed to acquire visual targets (Begault & Pittman, 1996; Veltman et al., 2004), improves speech intelligibility in noise, and enhances realism in virtual environments (Begault, 1993). Spatial audio makes use of the sound localization cues discussed in Chapter 3. Namely, these are interaural time differences (ITDs) and interaural level differences (ILDs) as well as cues provided by the outer ear (called the pinna) and the head. These direction-dependent acoustic cues can be used to compute a head-related transfer function (HRTF). HRTFs simulate the acoustic effects of the listener's head, pinnae, and shoulders, and they can be used to synthesize location information through speakers or headphones. Generally, 3-D communication systems make use of air conduction (sound waves traveling through air), but bone conduction spatial audio transmission has also been considered since it has some advantages in certain environments (MacDonald, Henry, & Letowski, 2006). For example, wearing extensive hearing protection (i.e., insert devices in combination with earmuffs) makes sound localization poor (Brungart, Kordik, Simpson, & McKinley, 2003). Bone conduction transmission localization interfaces can be used in extremely noisy environments and when hearing protection devices are being worn, conditions often found in military field operations (MacDonald et al., 2006). However, air transmission is still more prevalent, and therefore unless specified otherwise, that is the form referred to here.

The ability to localize real versus virtual sounds was explicitly studied by Bronkhorst (1995). Although localization accuracy is better for real versus virtual sounds, accuracy improves if individualized HRTFs are used and for left-right judgments relative to up-and-down discriminations. Allowing listeners to move their heads and making sure that the sound source is of sufficient duration also increases the localization of virtual sound sources (Bronkhorst, 1995).

Several experiments investigated 3-D audio as a means of improving collision avoidance and other target detection tasks in visually demanding flight tasks. For example, Veltman et al. (2004) found that providing 3-D audio cues assisted pilots with an aircraft pursuit task that involved following the flight path of a target jet while maintaining a specified distance. 3-D audio cues can improve accuracy and response time in aircraft target detection tasks and can decrease reliance on head-down displays (Oving, Veltman, & Bronkhorst, 2004; Parker, Smith, Stephan,

Martin, & McAnally, 2004), resulting in decreased mental workload and improved situation awareness. Oving and colleagues observed that the combination of semantically informative (e.g., "up") 3-D spatial cues resulted in particularly dramatic improvements in detection times relative to a visual-only display and either semantically informative or 3-D cues alone.

Spatial Auditory Alerts

Spatial auditory alerts have met with less success as an efficacious method of decreasing crash rates in in-vehicle warning systems (Bliss & Acton, 2003; Ho & Spence, 2005). It is reasonable to think that spatially predictive alerts might guide visual attention, which could potentially aid appropriate crash avoidance responses. However, Ho and Spence (2005) observed no benefit in either response time or accuracy for making a collision avoidance response from a spatially predictive nonverbal audio cue in a driving simulation task. Conversely, nonspatial but semantically informative (e.g., front, back) verbal warnings did produce significantly faster response times than the spatial audio cues. Further, similar to the results of Oving et al. (2004), Ho and Spence observed additive benefits by providing the semantically informative verbal warning from a spatially predictive location. Verbal directional words presented from a spatially congruent direction produced even faster response times than nonspatial verbal directional warnings.

Bliss and Acton (2003) observed that reliable spatially predictive cues increased the likelihood that drivers would swerve in the correct direction but appeared to have no effect on their ability to avoid having a collision. However, in their simulated driving paradigm, the spatial predictability of the cue may have been confounded with overall alarm reliability. They manipulated alarm reliability for both a spatial and a nonspatial alarm. In two experiments, they observed the best collision avoidance behavior when the alarm was reliable only 50% of the time, compared to 75% or 100% reliability. This surprising result may have stemmed from the fact that participants were told ahead of time the reliability of the alarm system and experienced either 6, 9, or 12 collision events (in the 50%, 75%, and 100% reliability conditions, respectively) over the course of a 20-min drive. The frequency of the collision events (participants had to look in a rearview mirror and determine which direction to swerve for a car approaching from the rear) is rather unrealistically high and may partially explain why participants in the 100% reliable alarm conditions (regardless of whether the alarm provided spatial information) experienced more collisions than those in the 50% reliable condition. In the 100% reliable condition, participants would have been experiencing a potential collision event approximately once every 100 s relative to once every 200 s in the 50% reliable condition.

Alarm reliability has been shown to have a significant impact on driver collision . avoidance response (Cummings et al., 2007). However, Cummings et al. did not find spatially predictive audio warnings to result in improved accuracy or response time over a nonspatially predictive auditory warning in their simulations. So, to date research regarding the potential benefits of spatially predictive auditory collision avoidance alarms has been equivocal at best, with the strongest support at present being for nonspatial but semantically informative cues. Barrow and Baldwin (2009) also found semantically informative directional cues to aid response time.

Importantly, though, since warning systems will not always be reliable, when the semantic directional word guided attention in an incorrect direction (e.g., the word "left" was presented when a response from stimuli to the right was required), the incongruity led to a significant increase in response time relative to a congruent semantic word or a neutral word. The incongruent semantic word resulted in greater response time detriment than an incongruent nonverbal spatial cue. This observation was replicated for a majority of participants in a subsequent study. However, Barrow and Baldwin (in preparation) noted strong individual differences in the impact of incongruity in the two conditions. That is, some individuals experienced differentially greater disruption from incongruent semantic information, while others were more disrupted by incongruent spatial information. This pattern of individual differences could be predicted based on a wayfinding strategy questionnaire developed by Kato and Takeuchi (2003). Further research is needed to determine whether spatially predictive semantic cues of imperfect reliability would improve or hinder effective collision avoidance behaviors or if such systems should ideally be adaptable based on individual differences in spatial orientation.

Spatial audio can also be used to assist in monitoring the activity of several different people in real or virtual space. For example, separating voices along the azimuth improves both detection and recognition of the speech signal relative to diotic (binaural) listening conditions or when all the signals are presented from the same location (Brungart & Simpson, 2002, 2005; McAnally & Martin, 2007). Brungart and Simpson (2005), for instance, found that listeners were over three times as likely to accurately identify which of seven different virtual speakers had presented a verbal instruction when the virtual speakers were presented in 3-D audio rather than nonspatialized audio.

In sum, auditory displays have the advantage of reducing the frequency with which visual attention must be diverted from more primary tasks (such as monitoring the position of an aircraft). In addition, sound naturally captures attention, and auditory displays in general have the advantage of being able to be presented simultaneously in different directions, freeing the operator from remaining in a stationary position to extract the signal. Auditory displays have strong potential for reducing visual and mental workload and for improving situation awareness if well designed. So far, we have been discussing all forms of auditory displays, although much of the discussion has included examples of auditory alerts and alarms. Now, attention is directed specifically to this important category of auditory display.

AUDITORY WARNINGS

Auditory alerts and warnings are frequently used to present time-critical information during periods of high workload and stress. The alerting capabilities inherent in the auditory modality as well as the omnipresent (vision-free) nature of audition make it well suited for these types of time-critical situations. Unfortunately, however, many past applications have added auditory alerts and warnings one at a time as needed, with little attention paid to the effective integration of auditory warnings and alerts into the existing system. Lack of integration can result in auditory displays having the exact opposite effect for which they are intended, thus increasing workload by causing startle responses, distracting the operator from concentrating on time-critical tasks in the event of an emergency situation, and simply being too numerous for the operator to effectively interpret in a time-pressured situation.

An examination of key principles of auditory warnings design is presented in the following section. First, a general introduction to warning design theory is presented. This is followed by an examination of both verbal and nonverbal auditory alerts and warnings in complex operational environments, such as aviation and medicine.

DESIGN OF WARNINGS

Warnings can be viewed as an extension of human sensory systems, providing an additional venue for obtaining system state information (McDonald, Gilson, Mouloua, Dorman, & Fouts, 1999). In this sense, warnings enhance or supplement sensory input. Warnings have the greatest potential benefit in high-workload, high-stress environments in which immediate response is critical, such as on the flight decks of aircraft and in surgical operating rooms. In complex multitask environments such as these, the efficacy of warnings and alerts extends beyond mere detectability. In these environments, operators are required to process and respond to simultaneous competing sources of information from several sensory modalities. The level of technological complexity present in many modern operational environments results in the need for warning systems that extend beyond error detection aids toward integrated displays that support error recovery, error management, and team/crew decision making (Noyes & Starr, 2000).

The warning process can be said to involve detection, encoding, comprehension, and compliance (Rogers, Lamson, & Rousseau, 2000). In more specific terms, effective warnings are compatible with human sensory, attentional, and cognitive capacities and are mapped to represent an appropriate level of urgency for the system states they represent. At a minimum, effective warnings are detectable and understandable, convey appropriate levels of urgency, and induce appropriate levels of compliance while minimizing attentional resource demands.

Ineffective warnings are inappropriately intrusive or not easily detectable, are difficult to interpret, or are improperly mapped to represent an appropriate level of system urgency. Further, ineffective warnings may require excessive attentional resources or invoke an unnecessary state of physiological stress that may interfere with decision making and corrective response. Alarms that are excessively redundant, have high false-alarm rates, or are perceived as representing an unrealistic state of urgency will promote noncompliance by increasing the likelihood that the operator will ignore the alarm, disable the alarm system, or in the event of a system state emergency, expend critical time and attentional resources to disable the warning system prior to responding to critical system components. Current warning design theory cautions against the overuse of redundant or unnecessary alarms (Rogers, et al., 2000).

Numerous investigations have established empirical support for the implementation of human factors guidelines for the design of auditory warnings, particularly for operational environments such as aviation and more recently in the area of in-vehicle systems. Unfortunately, a wide gap persists between scientific theory and application in this area. Acoustic characteristics such as intensity, frequency, pulse rate, and onset-offset time affect important warning parameters, such as perceived urgency, alerting effectiveness, and noise penetration (Baldwin, 2011; Baldwin & May, 2011; Edworthy et al., 1991; Haas & Edworthy, 1996; Hellier et al., 1993; Patterson & Datta, 1999). However, many auditory alerts and warnings are still installed in a piecemeal fashion with little attention to how the sounds will affect operator detectability, intelligibility, and workload. This can lead to numerous problems, contributing to distraction, alarm mistrust, and general ineffectiveness. Well-designed auditory warnings are capable of telling what the problem is, where the problem is, and when to expect the problem—the urgency of the situation (Catchpole, McKeown, & Withington, 2004). Ensuring detection, identification, and discrimination and potentially localization were discussed in this chapter, and these issues are particularly relevant to auditory warnings.

As discussed previously in this chapter in reference to auditory displays in general, ensuring that auditory warnings contain several spectral components (at least four) within a range of 15–25 dB above background noise levels (Patterson, 1982) is critical to ensure detection. Intensity levels lower than this may result in the warning not being heard, while intensity levels higher than this are annoying and irritating and run the risk of being disabled. Warnings need to include periods of silence to promote communication and because periodic sounds are more attention getting than continuous sounds. Pulses of sound lasting 200–500 ms with onset and offset times of 20–30 s will reduce startle while preventing unduly long sound periods. Pulses of sound of different frequencies can be combined to produce recognizable patterns of sounds in repetitive "bursts" lasting 1–2 s. Bursts of sound can be repeated, perhaps increasing somewhat in intensity if not attended for as long as necessary. Adhering to these psychoacoustic design principles can ensure detection of auditory alarms. However, once detected the sound should be recognized and, as discussed in the next section, matched to the hazard level of the situation it is designed to represent. Verbal warnings convey relative hazard levels and can easily be made recognizable but can be more difficult to make detectable in environments with high ambient noise levels. Ensuring that warnings are designed to convey an appropriate level of urgency for the situation they represent cannot be overemphasized. Fortunately, a considerable body of literature has been constructed that can inform this aspect of design.

Urgency Mapping

Warning design research has emphasized the need for consideration of urgency mapping between the perceived urgency of the warning and the situational urgency of the condition represented by the warning (Edworthy et al., 1991). Considerable research has been conducted regarding the signal characteristics associated with levels of perceived urgency, particularly for design of auditory warnings. For example, the parameters of pitch, speed, repetition rate, inharmonicity, duration, and loudness have been shown to affect the perceived urgency of auditory warnings (Edworthy et al., 1991; Haas & Casali, 1995; Haas & Edworthy, 1996; Hellier & Edworthy, 1999a; Hellier et al., 1993; Momtahan, 1990; Wiese & Lee, 2004). Other things being equal, sounds that are higher in pitch, faster in rate, louder, and contain more random mismatch in harmonics are perceived as more urgent. Hellier and Edworthy presented a compelling case that existing research pertaining to these parameters and others can be used to design sounds with varying degrees of urgency for multiple-priority alerting systems. In addition to providing a solid design example, they discussed the results of two investigations that validated the use of guidelines from the existing literature (Edworthy et al., 1991; Patterson, 1982).

Hellier and Edworthy (1999a) constructed a set of nine auditory sounds, which they labeled "attensons." The set was constructed to be similar, but not identical, to warning sounds in actual use in work environments. Using the design guidelines, they constructed three attensons at each of three urgency levels, representing the most critical (Priority 1), moderate urgency (Priority 2), and least urgent (Priority 3). They then validated the newly constructed attensons by obtaining urgency ratings for each. Results indicated that attensons ranged in urgency according to the priority levels they were designed to represent, providing strong support for the use of the guidelines in achieving appropriate urgency mappings.

Edworthy and Adams (1996) pointed out that urgency mapping is particularly critical in high-workload environments in which a multitude of alarms may be present. With appropriate mapping, the apparent urgency of the situation may be assessed instantaneously, thus providing a decisional aid for determining how quickly the operator must divert attention to the malfunctioning system.

In general, a relationship between perceived urgency and response time has been observed such that as perceived urgency increases, response time decreases (Haas & Casali, 1995).

However, the evidence for this relationship is far from unequivocal. For example, when auditory warnings of different designed urgency level were presented in combination with automated or manual tracking, Burt, Bartolome-Rull, Burdette, and Comstock (1999) found no differences in reaction time between different levels of warning urgency. Interestingly, subjective ratings of perceived urgency obtained from participants before the experiment followed the expected pattern, with higher urgency ratings given for warnings designed to signify higher priority and correspondingly lower ratings for moderate- and low-priority levels. However, following participation in the dual-task warning response and tracking trials, no differences in urgency ratings were obtained across the three priority levels. Participants appeared to reassign urgency ratings based on task demands. This suggests that urgency levels achieved in work environments after exposure to contextual factors may not always match those obtained in laboratory rating tasks.

Other investigations have found the expected relationship between urgency ratings and response time. For example, decreasing the number of pulses or bursts per second increases the perceived urgency of a sound (Edworthy et al., 1991; Hass & Edworthy, 1996). Suied, Susini, and McAdams (2008) found that this same parameter (the interval between two pulses of sound) decreased response time to the sounds. Further, the association between urgency level and response time was more pronounced in divided-attention conditions.

In addition to burst rate, a number of other acoustic parameters have been shown to affect perceived urgency. Fundamental frequency and pitch range are two important parameters (Edworthy et al., 1991). Additional psychoacoustic properties that have an impact on perceptions of the urgency of a sound include its harmonic series, amplitude envelope, and temporal and melodic parameters, such as speed, rhythm, pitch range, and melodic structure (Edworthy et al., 1991). These psychoacoustic parameters can be combined to construct nonverbal auditory warnings with predictable levels of urgency. Less attention has been systematically placed on the influence of alarm intensity, perceived as loudness, on perceptions of urgency. In one notable exception, Momtahan (1990) systematically manipulated intensity in conjunction with interpulse interval (IPI), number of harmonics, spectral shape, fundamental frequency, and frequency glide. While IPI, spectral shape, frequency glide, and the number of harmonics all had an impact on urgency, loudness independently influenced perceived urgency. Specifically, sounds presented at 90 dB were rated as significantly more urgent than sounds presented at 75 dB, regardless of how the other sound characteristics were manipulated. Loudness also interacted with a number of other sound characteristics, a result that has been supported in other recent investigations (Baldwin, 2011; Baldwin & May, 2011). Although there is some evidence indicating that loudness has an impact on the perceived urgency and annoyance of an auditory alarm (Haas & Edworthy, 1996; Momtahan, 1990), loudness is generally examined in conjunction with other parameters (i.e., loudness and pulse rate are both varied simultaneously) rather than being systematically varied while holding other parameters (i.e., pulse rate) constant.

The interaction between individual acoustic parameters (i.e., frequency and pulse rate) is one of the many issues that make designing sounds so that their perceived urgency matches the actual hazard level a very complex process (see reviews in Edworthy & Adams, 1996; Haas & Edworthy, 2006; Stanton & Edworthy, 1999b). An additional challenge that has received considerably less attention is the potential influence of contextual factors on perceptions of urgency. Research has established that when an auditory warning is presented in the absence of appropriate context and hazard level, highly urgent sounds are also highly annoying (Marshall, Lee, & Austria, 2007; Wiese & Lee, 2004).

Urgency in Context

An example of this contextually based urgency-annoyance trade-off is provided by a study by Marshall et al. (2007). They obtained ratings of the urgency and annoyance of warnings as a function of their auditory characteristics after providing listeners with a description of a driving context. Listeners were asked to imagine that sounds were coming from one of three different in-vehicle systems ranging in urgency: (a) a collision avoidance system, (b) a navigational system; or (c) an e-mail system. Marshall and colleagues examined numerous acoustic parameters (i.e., pulse duration, IPI, alert onset and offset, sound type) and found a relatively consistent pattern indicating that as sounds were rated as more urgent, they were also rated as more annoying. At the same time, listeners utilized the contextual descriptions indicating that highly urgent sounds were more appropriate in highly urgent driving scenarios, and they provided higher annoyance ratings for urgent sounds in the low-urgency driving scenario. The psychoacoustic parameters of pulse duration, IPI, alert duty cycle, and sound type influenced perceived urgency more than annoyance. Nevertheless, they found a strong relationship between ratings of urgency and appropriateness in the collision avoidance scenario and between ratings of annoyance and appropriateness in the low-urgency e-mail scenario. These results confirmed that both psychoacoustic properties and context will have an impact on perceptions of auditory warnings.

Baldwin and May (2011) examined the impact of performing a contextually related task on ratings of the perceived urgency and annoyance of verbal collision avoidance warnings. A low-fidelity simulated driving task was used to provide a limited situational context. The task provided limited situational context because the warnings were not actually related to events within the driving scene, but participants were encouraged to prioritize the driving task at all times and to respond and rate warnings only when they could do so without disrupting their driving performance. Within this context, louder warnings (+10 dB signal/noise [S/N] ratio vs. -2 dB S/N ratio) were rated as more urgent and were responded to faster but were also rated as more annoying, thus confirming previous results obtained with ratings only.

Baldwin and May (2011) used a high-fidelity driving simulator to examine the impact of acoustic and semantic warning parameters on collision avoidance behavior. The signal word *notice* or *danger* was presented at one of two loudness levels (70 or 85 dBA) immediately prior to the onset of an event with a high probability of a crash. An interaction between acoustic and semantic characteristics was observed. The two combinations designed to be of intermediate urgency (danger at 70 dB and notice at 85 dB) resulted in significant reductions in crash probability. Neither the low-urgency (notice at 70 dB) nor the most urgent (danger at 85 dB) warning reduced crash probability. Results supported the importance of both appropriate hazard matching and the need to avoid startling operators in contextual settings. Cognitive issues, including contextual factors, also play a significant role in the learning and recognition of auditory warnings.

Auditory warnings come in many types. The type implemented will have ramifications for cognitive factors such as how easily the warnings are learned (e.g., see a review in Edworthy & Hellier, 2006b) as well as recognized after initially being learned (Petocz, Keller, & Stevens, 2008). One broad classification that will be used here involves distinguishing between nonverbal and verbal auditory warnings. Nonverbal warnings can then be further divided into categories such as auditory icons versus abstract sounds.

WARNING TYPES

Nonverbal warnings can be classified in several ways other than the verbal-nonverbal distinction. As with any auditory display, a warning sound may be intentional or incidental. Incidental sounds that accompany a critical system state may not be considered warnings at all in some classifications (Edworthy & Adams, 1996; Haas & Edworthy, 2006). Intentional warnings are those that are designed. They consist of sound added to signify critical states or changes in state to a system or piece of equipment (Haas & Edworthy, 2006). Intentional warnings are the focus of the present discussion. Auditory warnings, like visual warnings, may be abstract or representational (Blattner, Sumikawa, & Greenberg, 1989; and see discussion in Stanton & Edworthy, 1999a). Traditionally, warnings have tended to be abstract sounds (i.e., bells, buzzers, sirens, and klaxons). However, representative sounds are gaining increasing attention, if not implementation.

Warning sounds can further be classified into symbolic, nomic, and metaphoric (Gaver, 1986). Symbolic sounds are abstract, that is, they are arbitrary learned mappings (abstract symbols). Nomic and metaphoric are categories of representational sounds. Nomic sounds are derived from the physics of the object they represent, much like an auditory photograph. Metaphoric sounds are somewhere in between the symbolic and nomic categories. These types of sound refer to metaphorical similarities of the physics of the object, rather than completely arbitrary mappings.

Another classification system includes four main categories: (a) conventional nonverbal sounds; (b) earcons, which are sounds having musical qualities and learned associations (Blattner et al., 1989; Brewster, Wright, & Edwards, 1995); (c) auditory icons, which are environmental or representational sounds (Belz et al., 1999; Graham, 1999); and (d) speech (Edworthy & Hellier, 2006b; Hellier et al., 2002). See the work of Edworthy and Hellier (2006b) and Petocz et al. (2008) for reviews of several other classification schemes. Edworthy and Hellier pointed out that the lack of an established taxonomy of auditory warnings is an indication of the relative infancy, compared to visual cognition, of the field of auditory cognition. More research and further refinement of classification taxonomies and methods are needed. However, as discussed in the following sections, a number of strides in this direction are currently under way.

Auditory icons and speech warnings are learned, recognized, and retained more easily than conventional nonverbal sounds and earcons (Keller & Stevens, 2004; Petocz et al., 2008). The more direct the relationship between the warning sound and its referent, the more easily the warning is both learned and retained. Additional characteristics of both nonverbal and verbal warnings are provided in the following sections.

Nonverbal Warnings

To date, the overwhelming majority of auditory warnings have been nonverbal. Traditionally, these tended to be bells, buzzers, and sirens. More recently, new classes, including designed attensons, auditory icons, and earcons, have received considerable attention, research, and application.

Attensons

Edworthy and Hellier (1999) have used the term *attenson* to refer to short, attention-grabbing sounds. A well-established body of literature now exists for guiding design of attensons (Edworthy & Adams, 1996; Edworthy et al., 1991; Patterson, 1982, 1990b). As discussed, varying nonverbal parameters such as the fundamental frequency of a sound, the overall loudness (often measured in root-mean-square [RMS] intensity), and the burst rate. Edworthy and Hellier recommended using a dominant frequency range of 500–3,000 Hz to capitalize on human hearing capabilities (we are most sensitive in this range) and to decrease the probability that the noise will be startling or annoying. As discussed in several reviews (Edworthy & Hellier, 2006b; Hellier & Edworthy, 1999a), attensons can be effectively designed to be detectable and recognizable and to convey an appropriate level of urgency in a wide variety of work and leisure settings.

Auditory icons, the second type of nonverbal warning to be discussed, have received considerably less attention until quite recently.

Auditory Icons

Auditory icons consist of representational nonverbal sounds, or those that sound like the object, machine, or event they represent. Examples include screeching tires and breaking glass for in-vehicle collision warning systems (CWS) or a cash register sound on a Web site to indicate that an electronic payment has been processed. Auditory icons were first systematically examined for use in computer interfaces by Gaver (1986, 1989). Since that time, they have received considerable attention for use as in-vehicle auditory warnings (Belz et al., 1999; Graham, 1999). Auditory icons use representations of natural sounds to represent different types of objects or actions (Gaver, 1986). As pointed out by Graham (1999), they differ from earcons in several key ways.

Earcons

Earcons are abstract representations of synthetic tones. Earcons have a musical connotation not associated with a defined meaning. Therefore, the relationship between earcons and their referents must be learned (Graham, 1999). There are several notable sounds that might be associated with either complex attensons or earcons. For example, the French police siren, known to anyone who has either visited France or watched the movie of Dan Brown's book, *The da Vinci Code*, makes a repeating two-chord sound that certainly grabs attention but has nearly a musical quality to it. More complex earcons are not in common use as warnings presently, but they have potential, particularly in medical environments, where they might be used to convey time-critical information to trained medical staff without alarming patients and family members.

The final category of warnings discussed here is verbal warnings.

Verbal Warnings

Verbal warnings have the advantage of being able to both alert and inform without imposing a heavy memory load on the part of the receiver. Further, as foreshadowed in science fiction films such as *2001: A Space Odyssey* and *The Andromeda Strain*, verbal warnings can use semantics rather than acoustics to convey the urgency of the warning. That is, synthesized voices can provide time-critical information aimed at avoiding impending disaster in a calm, monotone voice rather than in high-pitched, high-rate tones that are likely to startle the listener (Edworthy & Adams, 1996). However, there is a paucity of research regarding the use of verbal warnings in operational environments. This may be due in large part to technical challenges.

Voice synthesis technology has only recently begun to achieve acceptable design-level standards. Natural speech presented in digitized form has several important methodological issues that must be considered for it to be utilized in operational environments. Edworthy and Adams (1996) noted that, relative to synthetic speech, natural speech is more likely to be masked by other concurrent communications present in the environment, and issues of intelligibility and detectability need further research.

Masking of Speech Signal

As previously discussed, for an auditory display or warning to be audible in noisy operational environments, it should contain multiple spectral components with intensity levels in the range between 15 and 25 dB above ambient background noise levels. The dominant frequencies for speech recognition (~400–5,000 Hz) include the frequency range at which humans hear best. This frequency range is also one that is the least susceptible to age-related threshold increases. The positive side of this feature is that, in favorable listening conditions (i.e., low-noise conditions), the auditory signal is presented within a listeners' most sensitive range, but the downside is that design flexibility is consequently reduced. A speech signal cannot be altered significantly to contain additional frequencies outside the normal range to accommodate a noisy environment without significant distortion to the signal, potentially rendering it unrecognizable.

Older individuals and people with hearing loss are particularly susceptible to the difficulties of perceiving speech in the presence of noise. Frequencies at the high end of the speech range (>~3,500 Hz) are critical for distinguishing between consonants, a task that is critical to recognition (Botwinick, 1984; Villaume et al., 1994). Recall that due to upward spread of masking, frequencies slightly to moderately above the frequency range of the noise are more susceptible to masking. This makes recognition of consonant sounds—which are critical to speech recognition—more susceptible to masking than lower-frequency sounds. Not surprisingly, people with high-frequency hearing loss have more difficulty understanding speech in the presence of noise (Amos & Humes, 2007; Helfer & Wilber, 1990).

Depending on the type and level of background noise, some potential masking issues can be countered by using synthetic speech in a slightly different frequency range. However, there are limits to what can be achieved since altering the frequency range too much would unduly distort the speech signal. In addition, there is a potential trade-off in recognition accuracy.

Synthesized and Digitized Speech

Digitized natural speech is easier to recognize and comprehend than synthesized speech (Simpson & Marchionda-Frost, 1984; Simpson & Williams, 1980). The trade-off is that synthesized speech requires less data storage than digitized speech and may be generated on an as-needed basis. System capabilities and requirements will thus dictate the use of each. For verbal displays requiring an extensive vocabulary (i.e., in-vehicle route guidance systems [RGSs]), synthesized speech displays will be required. Use of a limited vocabulary set, which is more indicative of verbal warnings, may benefit from the use of digitized natural speech. Limited use of key signal words such as *danger, caution,* and *notice* or terse advisory commands in aviation such as *pull up* to signal critical low-altitude states are examples of where it may be possible to use digitized rather than synthesized speech.

Both digitized natural and synthetic verbal warnings have the advantage of being able not only to alert an operator but also to convey information. Verbal warnings reduce learning time relative to nonverbal warnings, and an extensive body of literature now exists to facilitate designing verbal warnings that vary in perceived urgency level.

Verbal Urgency Mapping

The importance of urgency mapping was discussed previously in this chapter. We return to this important topic and discuss specific empirical findings and guidelines relevant to verbal urgency mapping. One of the first systematic examinations of hazard level and verbal warning parameters was conducted by Wogalter and colleagues (Wogalter, Kalsher, Frederick, Magurno, & Brewster, 1998). Wogalter had listeners judge the connoted hazard level of various signal words presented under one of three voice styles (monotone, emotional, and whisper) at one of two presentation levels (60 and 90 dBA) spoken by a male or female. The signal word, *danger* received a higher hazard rating than *warning* and *caution*, which did not differ. *Notice* received the lowest hazard rating. The hazard level (also termed the perceived urgency) of spoken signal words has been confirmed by subsequent research (Baldwin, 2011; Hellier et al., 2002; Hellier, Wright, Edworthy, & Newstead, 2000).

Hellier et al. (2002) examined the relationship between semantic and acoustic parameters of spoken words. Male and female actors spoke signal words in an *urgent, nonurgent,* and *monotone* style and listeners rated the perceived urgency of each of 10 warning signal words, including *deadly, danger, warning, caution, risky, no, hazard, attention, beware,* and *note.* Hellier found that speaking style, signal word, and gender of the speaker all significantly affected ratings of perceived urgency. Words spoken in the urgent style were rated as more urgent than those spoken in the monotone style. In rank order, *deadly* was perceived as more urgent than *danger,* followed by *warning.* No differences were found between the perceived urgency of the signal words *warning* and *caution.* Further, words spoken by female speakers were rated as more urgent relative to words spoken by males. These results were consistent across both male and female listeners.

Hellier et al. (2002) did not specifically manipulate the presentation level in their investigation. However, they did implement a careful procedural control during the recording of stimuli to ensure that the presentation levels, expressed in peak SPL (sound pressure level) were consistent across male and female speakers. The mean presentation level for both male and female actors was 80 dBA in the urgent style. Slight differences in presentation level were observed for male and female speakers in the other two conditions, for which presentation levels of 62 dBA for the male and 64 dBA for the female were recorded in the nonurgent style, and presentation levels of 58 dBA for the male and 64 dBA for the female were recorded in the monotone style. It is possible that the slightly louder and higher-frequency characteristics of the female voice contributed to higher ratings of perceived urgency. However, from these data alone one cannot rule out the possibility that some other factor (i.e., learned associations) accounted for the different urgency ratings of male and female voices.

In sum, ample empirical evidence from psychoacoustic investigations now exists to guide the design of both nonverbal and verbal warnings, so that they are audible and can be distinguished from each other. After ensuring that auditory warnings are audible, it is essential to consider whether the warnings are meaningful to their intended recipients. That is, do such warnings convey meaningful information to allow their recipients to understand what they are designed to represent, and are they presented at an appropriate urgency level—one that matches the actual hazard level of the situation? Considerable information is currently available on the interaction of acoustic and semantic factors that have an impact on ratings of urgency. Considerably less is known regarding how different types of sounds (i.e., artificial, environmental, and verbal) will have an impact on signal-referent relationships and perceptions of urgency and appropriateness in different operational contexts. This last area is awaiting additional research.

Next, a discussion of an emerging area of auditory research is presented: sonography or data sonification.

DATA SONIFICATION

Representing continuous information (such as system state) or data relations through sound is referred to as *sonification*. Data sonification is particularly useful for presenting information to the visually impaired, when visual attention must be devoted somewhere else, or when representing multiple complex relationships that are not easily amenable to visual graphical depiction (Flowers, Buhman, & Turnage, 2005). The "science of turning data into sound" (Edworthy, Hellier, Aldrich, & Loxley, 2004, p. 203) is a burgeoning area of applied auditory research efforts. Data sonification has strong potential for application in both medicine and aviation.

One of the most important factors in sonification involves parameter mapping, or determining which specific data dimensions should be mapped to an acoustic characteristic (Edworthy & Hellier, 2006b). An example would be identifying which sound parameters (i.e., pitch, loudness, tempo) should be used to represent specific data dimensions (i.e., temperature, pressure, rate, etc.).

Effective parameter mapping involves finding out which sound parameters are best for representing different types of dimensions. This area has proven more challenging than was originally thought. Intuitive guesses for parameter mapping do not generally lead to the most effective performance outcomes (Walker, 2002). A related and equally important issue is determining the most effective polarity for the mapped parameters. That is, should increases in the sound parameter be associated with increases in the dimension (i.e., increasing pitch to represent increased temperature). This positive mapping (Walker, 2002) can be contrasted with a negative mapping in which decreases in the acoustic dimension are associated with increases in the perceptual dimension it is designed to represent. For example, a decrease in pitch might be used to represent increases in size, which would have natural representational mappings in many contexts.

Psychophysical scaling is a third important issue to consider in the parameter-mapping relationship (Walker, 2002). This involves determining how much the acoustic parameter should change to connote a level of change in the data dimension. Magnitude estimation has become the accepted standard for determining psychophysical scaling for data sonification (Edworthy et al., 2004; Walker, 2002).

Considerable work is currently under way in the area of data sonification (e.g., Anderson & Sanderson, 2009; Harding & Souleyrette, 2010; Pauletto & Hunt, 2009). This promises to be a significant new area of application in areas as diverse as determining suitable highway locations (Harding & Souleyrette, 2010), physical therapy and rehabilitation (Pauletto & Hunt, 2009), as well as aviation and medicine. Several reviews discussing the state-of-the-art work in this area have been published (see Flowers et al., 2005; Kramer et al., 1999; Walker & Kramer, 2005).

In the remainder of this chapter, applications of auditory displays are discussed within the context of three environments: aviation, surface transportation, and medicine.

AUDITORY DISPLAYS IN COMPLEX ENVIRONMENTS

Aviation, surface transportation, and medical operations are three complex environments that have been the focus of considerable human factors work. The issues pertaining to auditory displays that have been discussed so far in this chapter are applicable, for the most part, in each of these environments. Yet, each has some unique characteristics along with the commonalities, and each environment has been the focus of considerable research and application in the area of auditory cognition.

AVIATION

Dramatic changes in cockpit displays and warning systems have taken place over the last several decades. Advances in avionics capabilities have transformed both the display format and the quantity of information provided to today's flight crew. Considerable research has been conducted regarding the efficacy of the warning and alerting systems currently in existence. Human performance capabilities and the mental workload requirements of these systems have been examined both in the laboratory and in the field.

Advances in technology and the advent of the "glass cockpit" have led to increasingly complex interacting systems in the cockpits of modern aircraft (Figure 11.3). Cockpit automation requires pilots to divide attention between traditional flight tasks and monitoring the system states of automated functions, often with the aid of complex multimodal displays. Numerous visual and auditory displays and warnings provide invaluable information regarding system states, automation mode changes, and traffic management. Careful consideration of the attentional workload involved in utilizing auditory displays is critical to their integration into existing cockpit designs. At present, new visual and auditory cockpit displays are often added in piecemeal fashion with inadequate emphasis placed on the overall attentional impact of their integration with other system components. What follows is a discussion of some examples of flight deck warnings that highlight some of the key issues. It is by no means a comprehensive review of the literature on auditory displays in aviation environments.

FIGURE 11.3 Modern glass cockpit.

Flight Deck Warnings

With the increasing sophistication and technological advances in display technologies, the number of flight deck warnings has also risen sharply. As Noyes et al. (1995) pointed out, "Over the last several decades, civil aircraft warnings systems have gradually evolved from little more than a fire bell and a few lights to highly sophisticated visual warning displays accompanied by a cacophony of aural signals" (p. 2432). Noyes and colleagues also reiterated that, unlike machines, which have the potential to have their capabilities increased through technological advances, humans are limited in the extent to which their human information-processing capabilities can be enhanced.

Auditory alerts and warnings are used extensively in the modern cockpit to present time-critical information to pilots. For example, traffic collision avoidance systems (TCASs) and cockpit displays of traffic information (CDTIs) both supplement a visual display with an auditory warning if time-critical traffic situations are detected. Wickens and Colcombe (2007) demonstrated that even when these types of systems were prone to false alarms (presenting alerts when no real danger was present), they tended to improve overall flight control performance. The alerts decreased the mental workload of the traffic-monitoring tasks so that pilots could devote more resources to primary flight control tasks.

Another key area in which auditory cognition plays a critical role is in communication between the flight crew and ATC personnel. Considerable attention has been placed on reducing mental workload and errors in this important area.

ATC Communications

Communicating with and remembering ATC commands is a cognitively demanding task for both the pilot and the air traffic controller. The ATC communication task can be particularly challenging either in high-workload conditions or for older pilots (Morrow, Wickens, Rantanen, Chang, & Marcus, 2008). The importance of understanding various signal quality dynamics and their effects on speech processing and comprehension is illustrated by a tragedy in the Canary Islands that cost the lives of 583 individuals. The accident resulted from poor verbal communication between the pilot and the air traffic controllers (see discussion in Wickens, 1992). Rerouted, tired, and trying to navigate a crowded airport through dense fog, a KLM pilot misinterpreted ATC instructions. Believing it was safe to initiate takeoff, the KLM initiated takeoff and had just reached lift speed when it clipped a Pan American plane that was crossing the runway. Of the 249 people aboard the KLM flight, there were no survivors, and the crash is one of the deadliest on record to this date.

Despite an extensive review of ATC phraseology and new standards for restricting the use of certain terms such as *clear/clearance* and *takeoff*, auditory communication remains a pervasive safety issue in aviation settings (Hawkins, 1987). According to Hawkins (1987), an analysis in 1986 of more than 50,000 aviation accidents stored in the Aviation Safety Reporting System (ASRS) databank revealed that about 70% involved some kind of oral communication problem related to the operation of the aircraft. Despite potential for error, the auditory modality has several advantages, and speech communication continues to be an integral and perhaps irreplaceable part of many complex aviation environments.

Efforts to replace speech-based ATC communications have met with a number of hurdles. Even as technological advances are made through efforts such as next-generation (NexGen) aviation operations that make alternative forms of communication possible, speech-based communication within the cockpit will continue to play an important role in aviation safety. Speech-based communications will be used between personnel on the flight deck as well as to confirm nonroutine and poorly understood communication with other aircraft and air traffic controllers.

In an effort to understand some of the many challenges to speech communication in the cockpit, speech intelligibility has been investigated using a variety of methods of degradation. For example, the impact of intermittent speech disrupted by a chopping circuit on speech intelligibility has been examined (Payne et al., 1994), as well as comparisons of natural versus synthetic speech (Simpson & Williams, 1980), fast versus time-compressed speech (Adank & Janse, 2009), and after sustained periods of time on task and background noise (Abel, 2009). These are just some of the many factors that threaten aviation communications.

The NexGen air transportation system is the strategy of the Federal Aviation Administration to maintain safety and efficiency through 2025 as the national air space (NAS) becomes increasingly congested. Several new cockpit technologies have been designed in accordance with this aim. Controller-pilot data-link communications (CPDLC or data link/data com) and the CDTI are aimed at reducing congestion of overburdened radio-frequency channels, reducing communication errors, and offsetting pilot working memory load.

Pilots must often perform multiple tasks that require competing attentional demands, such as monitoring and interpreting displays, monitoring traffic, as well as interpreting information from ATC. Aviation prioritization strategies emphasize the need for pilots to aviate, navigate, and communicate (Helleberg & Wickens, 2003; Jonsson & Ricks, 1995; Schutte & Trujillo, 1996). Thus, while communication is an essential component of the piloting task, it is not the highest-priority task. Information communicated to pilots from ATC, regardless of modality, may be subjected to interference from multiple sources within the flight deck, potentially affecting the recall and execution of ATC commands.

CPDLC or data-link text communication was developed in an effort to reduce some of the errors associated with current voice communications, to reduce radio-frequency congestion, and as an initial step toward greater pilot autonomy in support of free flight. However, research to date indicates that rather than reducing communication errors, data link appears to introduce errors of a different type (Dunbar, McGann, Mackintosh, & Lozito, 2001; Issac & Ruitenberg, 1999; Kerns, 1999; McGann, Morrow, Rodvold, & Mackintosh, 1998).

Data-Link and Voice Communications

Data link, as an alternative to traditional voice communication, has been in existence for at least three decades and has been implemented in several air systems, particularly for transoceanic flights (Kerns, 1999). Data link is a predominantly visually based communications format that allows messages to be typed for visual display between pilots and controllers.

Essential to understanding the implications of redistributing communication workload is an examination of the interaction between voice and visual communications (or mixed). Currently, implementations of the data-link system result in pilots and controllers using both systems for different communications rather than exclusively relying on the visual-manual system. Even when visual displays are the dominant form of ATC-to-pilot communication, an auditory alert is used to signal the presence of a new visual message, and radio/speech channels continue to be used to supplement and clarify visually presented information. Such "mixed" communication environments appear to place even greater demand on attentional resources than either format alone demands.

Another complex environment that shares a number of the same communication challenges as aviation is surface transportation. While the focus of the following discussion is on one component—automobile driving—it is important to keep in mind that there are other important areas of surface transportation, including railways and runways. For the purposes of brevity, the coverage here focuses on emerging in-vehicle displays that capitalize on the auditory modality.

SURFACE TRANSPORTATION

Driving is primarily a visual-motor task. Due to the heavy visual demands placed on the driver, use of the auditory modality for presenting time-critical or supplementary information may be advantageous. Advanced in-vehicle information systems (AVISs) are rapidly making their way into the modern automobile. Many of these systems provide information through both visual and auditory displays. Three systems that rely on auditory displays as a major form of presentation are CWSs, RGSs, and lane departure and fatigue detection systems. Each of these emerging areas is discussed in turn.

Collision Warning Systems

Many automobile models currently come equipped with some form of auditory CWS. Commercially available systems currently most often utilize a visual alert in combination with an auditory alert, although previous research indicated that drivers may have lower levels of reliance on multimodal systems relative to single-modality systems (Maltz & Shinar, 2004). In simulation studies, CWSs have shown great potential for decreasing both the severity and the rate of occurrence of motor vehicle collisions (Brown, Lee, & McGehee, 2001; Graham, 1999; Maltz & Shinar, 2004). As discussed, a wide variety of auditory warnings has been investigated in this context, ranging from nonverbal attensons (Brown et al., 2001; Lee et al., 2002; Siuru, 2001) to auditory icons such as breaking glass and screeching tires (Belz et al., 1999; Graham, 1999). Research is currently under way to determine if these systems will actually reduce roadway collisions outside the laboratory. But, preliminary work is promising, particularly for certain categories of drivers, such as older drivers (Dingus, McGehee, Manakkal, & Jahns, 1997; May, Baldwin, & Parasuraman, 2006). At present, there is still great concern, and some evidence (Maltz & Shinar, 2007), that some younger drivers may misuse CWSs by overrelying on them and maintaining closer vehicle headways without sufficient visual attention. On the other hand, there is some evidence that providing an auditory alert to middle-aged drivers can be used to teach safe driving behaviors. Shinar and Schechtman (2002) provided an auditory alert to drivers whenever their temporal headway reduced to unsafe levels. Drivers learned to maintain safer following distances, and the beneficial effects observed during the initial experimental session remained at a session administered 6 months later.

Another emerging in-vehicle system that can be a great transportation aid to all, and particularly to older drivers, are the RGSs. Although they nearly always have both a visual and an auditory display, it is the auditory guidance that is particularly helpful to older drivers and, in fact, all drivers in high-workload situations. The auditory interface of these systems is the focus here.

In-Vehicle Route Guidance Systems

In-vehicle RGSs are one of the forms of complex verbal displays rapidly increasing in prevalence. Several high-end automobile manufacturers include RGSs as part of their standard option package. Optional RGSs are available in virtually all vehicles, if not from the factory, then from one of the off-the-shelf stand-alone systems widely commercially available.

These systems use global positioning satellite (GPS) technologies to track the location of the automobile and then provide navigational guidance in the form of visual maps and terse auditory commands. Drivers have demonstrated a preference for auditory commands relative to visual maps for navigation tasks in a number of investigations (Dingus, Hulse, et al., 1997; Streeter et al., 1985). Auditory guidance instructions have the advantage of allowing drivers to keep their eyes on the road. Switching visual attention back and forth from the roadway to a visual map incurs increased executive processing requirements to control task switching. The increased demand for executive processing results in switching costs normally manifested by increased response time and errors (DiGirolamo et al., 2001). Using functional magnetic resonance imaging (fMRI), DiGirolamo and colleagues have shown that task switching involves greater recruitment of the medial and dorsolateral frontal cortex. Interestingly, unlike their younger counterparts, older individuals demonstrated increased recruitment of these areas continuously during dual-task performance rather than just when switching between tasks. This increased use of executive processes during concurrent performance may be particularly problematic for older drivers, who require more time for task switching (Parasuraman & Haxby, 1993).

Navigational Message Complexity

Empirical evidence indicates that auditory navigational commands must be kept short to avoid overloading the driver (Srinivasan & Jovanis, 1997a, 1997b). If an auditory message is too long or too complex, it threatens to exceed the working memory capacity of the driver and not be retained (for a review see Reagan & Baldwin, 2006). Several investigations indicated that auditory route guidance messages should contain no more than three or four informational units or propositional phrases and should take no longer than 5–7 s to present (Barshi, 1997; Green, 1992; Kimura, Marunaka, & Sugiura, 1997; Walker et al., 1990). When procedural or navigational commands contain more than three propositional phrases, execution errors increase sharply (Scerbo et al., 2003), drivers make more navigational errors (Walker et al., 1990), and driving performance becomes less stable.

Srinivasan and Jovanis (1997a) suggested that to prevent excessive load on a driver's information-processing resources, auditory directional information be provided in the form of terse commands, such as, "Turn left in 2 blocks onto Park Avenue." In the systems currently available on the market, these types of terse commands are generally followed by even shorter commands just prior to the turn, such as, "Turn left." Reagan and Baldwin (2006) found that including one additional piece of information, either a salient landmark or cardinal direction, could be accomplished without noticeable negative impact on simulated driving performance or memory for the instructions. In fact, including a salient landmark in the guidance statement improved navigational performance and decreased ratings of subjective mental workload relative to the standard guidance command without this information.

The FHWA guidelines, based largely on empirical research, have been established to aid designers in determining the appropriate complexity level of in-vehicle displays (Campbell et al., 2004). According to these guidelines, complexity can be thought of as falling along a continuum from high to low. For auditory displays, high-complexity displays contain over nine information units and take over 5 s to process. Conversely, low-complexity displays contain three to five information units and require less than 5 s to process.

Fatigue and Lane Departure Detection Systems

In addition to forward CWSs and navigational route guidance systems, a number of other driver assistance devices are making their way into the modern automobile. Detection systems have also been devised to detect potentially dangerous driver states, like inattention, distraction, and fatigue. A number of fatigue alerts designed to warn drivers of safety-critical sleep episodes are commercially available. Figure 11.4 illustrates an example of one of these devices. Note that it can be mounted in the vehicle. These devices may monitor eye closure (or the percentage of time the eye is closed, called *perclos*), changes in lane position variability, or even the driver's rate of physical movements.

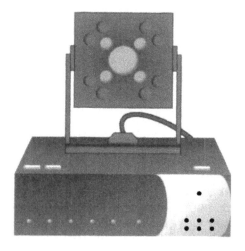

FIGURE 11.4   Artistic illustration of the Eye Alert™ Fatigue Warning System 2011.

Auditory warnings that alert drivers to inadvertent lane departures may benefit both fatigued and distracted drivers. Auditory warnings simulating the sound of rumble strips have been shown to decrease the duration of lane departure events and speed corrective steering behaviors when departures occur (Navarro et al., 2007). It should be noted that Navarro et al. found that steering wheel movements designed to prime appropriate motor behaviors were even more effective than auditory warnings alone.

The number of in-vehicle driver assistance and information systems is on the rise, and further research into how best to provide drivers with this information will continue to be of importance for many years to come (Baldwin, 2006; Baldwin & Reagan, 2009; Cummings et al., 2007; Maltz & Shinar, 2007; Reagan & Baldwin, 2006; Shinar, 2008). Another environment replete with auditory alarms is the modern medical center.

MEDICAL ENVIRONMENTS

Medical environments are the third specific work domain that are considered with respect to auditory displays. Auditory alerts and displays are common in many health care systems. As in other high-workload environments, such as the aviation cockpit and the modern automobile, auditory interfaces allow surgeons, anesthesiologists, and other medical personnel to monitor critical information (patient physiology) while performing other tasks (Watson & Sanderson, 2004). Medical patient-monitoring equipment generally provides patient status information to medical personnel in both visual and auditory formats. Auditory displays may provide continuous information regarding patient status or intermittent alerts to signal that operators should focus attention on the visual displays.

Watson, Sanderson, and Russell (2004) described three categories of auditory information found in the surgical operating room and discussed the impact of each on attention, mental workload, working memory, and expectations. The first category is the continuous auditory display; Watson et al. (2004) provided the pulse oximeter as an example of this category. The pulse oximeter provides continuous information on a patient's status, generally in the form of a two-dimensional auditory display, with one dimension (pulse rate) mapped to heart rate and another (pitch level) mapped to oxygen saturation.

A second category, the informative alarm signal, is designed to capture or direct the operator's attention to a discrete event, such as an unexpected change in patient status or equipment functioning. As described in the model (Watson et al., 2004), the presence of informative alarms will decrease attentional costs and mental workload (presumably because they aid the operator in patient monitoring) but increase working memory demands since they interrupt ongoing activities. However, since they are often used to direct attention to life-threatening patient states, the added working memory demands can be viewed as worth the cost. In addition, Watson and colleagues pointed out that the auditory alarms on various medical devices are sometimes tailored by changing system settings so that they can be used as reminders to check patient status information and offset the workload of continuous monitoring.

The final category in the model of Watson and colleagues (2004) is called "alarm noise and other auditory distractions" (p. 277). Included in this category are false alarms or nuisance alarms that do not convey important patient or equipment information. Auditory information in this category has a negative impact on attention, workload, and working memory without any operator benefit. They pointed out that this category includes auditory distractions unrelated to patient status or current operating situation. Any auditory information, including continuous displays and informational displays, can be distracting during some procedures and to some individuals or work teams. Turning off all auditory alarms continues to be the most common form of alarm tailoring in the operating room (Watson et al., 2004). As is discussed in the next section, because the prevalence of nuisance alarms is so high, it is not surprising that health care workers might be tempted to turn off auditory warnings altogether.

Nuisance Alarms

Nuisance alarms continue to be pervasive in medical environments. Edworthy and Hellier (2005) illustrated this with the example of a workstation used in anesthesia. An average anesthetic workstation might have several pieces of equipment, each capable of producing 20 or more different alarm sounds. Now, multiply this by the number of pieces of equipment in an intensive care unit, and the resulting number of alarms is staggering. Alarmingly, up to 90% of these are false or nuisance alarms (Imhoff & Kuhls, 2006).

Gorges, Markewitz, and Westenskow (2009) provided a sobering real-life example of this phenomenon. They conducted an observational study of alarm sounds and behavioral responses in an intensive care unit. During their 200-h observational study, 1,214 alarms went off. This is an average of 6.07 alarms per hour directly observed in this one unit. They observed that only 23% of these alarms were effective. Thirty-six percent were ineffective (meaning no response was taken by medical personnel), while the remaining 41% were consciously ignored or disabled.

To make matters worse, Patterson (1990a) pointed out that for reasons of economy, auditory alarms too often consist of high frequencies and vary only in intensity and frequency. High frequencies are much more difficult to localize than lower speech frequencies, and high intensities are associated with greater annoyance and stress. The poor design of some auditory warnings only increases the negative consequences of nuisance alarms.

Watson and Sanderson (2004) noted that the most commonly cited problem with auditory alarms in the medical environment is that they too often present less-than-useful information at inappropriate times. Notably, they pointed out that alarms present critical state information rather than trend information, and this information generally comes at an inopportune time (i.e., high-workload times occurring as a result of multiple problems or failures). Then, to make matters worse, the anesthesiologist must waste valuable time shutting off the auditory alarm when he or she could be attending to the patient. For these reasons in particular, Watson and Sanderson (2004) suggested that a continuous auditory display presenting patient status trend data (i.e., a sonograph) may be a more effective means of presenting patient physiological data, one that allows medical personnel to maintain eye contact with the patient and perform other tasks without losing awareness of patient status. Sonographic displays, discussed in this chapter, have many applications in medical environments.

SUMMARY

Auditory displays are preferable to other types of displays in a number of circumstances. For example, when the information to be conveyed is auditory in nature, when the operational environment places heavy visual demands on the operator, and when the receiver may not be able to see a visual display either because he or she is moving around or because of low visibility, auditory displays are preferable to visual displays. Despite these advantages, including too many auditory displays can result in auditory overload: Critical information is masked, and the soundscape becomes so annoying that attention and communication are disrupted. Auditory information also has the tendency to preempt visual processing—a double-edged sword. On the positive side, this auditory characteristic makes it well suited for presenting time-critical alerting information. On the negative side, presenting auditory information may disrupt performance of an ongoing visual task at inopportune times. Therefore, caution should be exercised in both the type and the timing of auditory displays. Alerting an individual or operator to time-critical information is often best carried out through an auditory display. Appropriate hazard matching between the acoustic and semantic characteristics of the sound and the threat level of the situation it represents is essential to the design of effective auditory warnings. Auditory displays are increasingly being used in such arenas as aviation, surface transportation, and medical facilities. Each of these environments not only shares some commonalities but also exhibits some unique aspects that warrant careful attention to the design and implementation of new auditory displays.

CONCLUDING REMARKS

The design and implementation of auditory displays across domains represent the culmination of a considerable body of literature in the burgeoning field of applied auditory cognition. In this book, I attempted to discuss not only the importance of continued research aimed at uncovering the mysteries of human auditory processing but also what can be understood from the existing literature.

Many auditory tasks are commonly used to investigate cognitive and neuropsychological functioning, in both healthy and neurological populations. The wrong conclusions can be drawn from the results of their use unless careful attention is paid to both the stimulus quality of the material presented and the hearing capabilities of the listener. Older adults, for example, may experience communication and even cognitive difficulties that can be attributed at least partly to declining hearing capabilities. Degraded listening conditions and hearing impairment make speech understanding and other complex auditory processing tasks more effortful, which can in turn compromise performance on other important tasks, such as remembering, driving an automobile, or performing a complex surgical procedure.

Presenting redundant information through the visual channel can offset some of the load associated with processing auditory information. However, in many of the complex operational environments in existence today, rather than presenting redundant information, visual and auditory displays are used simultaneously to present independent streams of information. In complex domains such as aviation, surface transportation, and medicine, understanding principles of auditory cognition and their implications for auditory display design is essential for facilitating and maintaining effective, efficient human performance capabilities.

Sounds are everywhere in the modern world. The complex soundscape that we live in influences where we direct our attention, how we communicate with each other, and how we interact with the technological systems prevalent in modern life. As I hope the material presented in this book has illustrated, we know increasingly more of the mechanisms of auditory cognition, knowledge that can be put to good use to better design our auditory world. Yet, at the same time, much remains to be learned about auditory cognition. The future promises to reveal more discoveries concerning this essential human faculty.

References

Abada, S. H., Baum, S. R., & Titone, D. (2008). The effects of contextual strength on phonetic identification in younger and older listeners. *Experimental Aging Research, 34*(3), 232–250.

Abel, S. M. (2009). Stimulus complexity and dual tasking effects on sustained auditory attention in noise. *Aviation, Space, and Environmental Medicine, 80*(5), 448–453.

Ackerman, D. (1990). *A natural history of the senses.* New York: Vintage Books, Random House.

ANSI. (2003). Ergonomics – Danger signals for public and work areas: Auditory danger signals (No. ISO 7731:2003).

Adank, P., & Janse, E. (2009). Perceptual learning of time-compressed and natural fast speech. *The Journal of the Acoustical Society of America, 126*(5), 2649–2659.

Alain, C., Achim, A., & Woods, D. L. (1999). Separate memory-related processing for auditory frequency and patterns. *Psychophysiology, 36,* 737–744.

Alain, C., Arnott, S. R., & Picton, T. W. (2001). Bottom-up and top-down influences on auditory scene analysis: Evidence from event-related brain potentials. *Journal of Experimental Psychology: Human Perception and Performance, 27*(5), 1072–1089.

Alain, C., & Izenberg, A. (2003). Effects of attentional load on auditory scene analysis. *Journal of Cognitive Neuroscience, 15*(7), 1063–1073.

Aldrich, K. M., Hellier, E. J., & Edworthy, J. (2009). What determines auditory similarity? The effect of stimulus group and methodology. *The Quarterly Journal of Experimental Psychology, 62*(1), 63–83.

Alho, K., Woods, D. L., Algazi, A., Knight, R. T., & Naatanen, R. (1994). Lesions of frontal cortex diminish the auditory mismatch negativity. *Electroencephalography and Clinical Neurophysiology, 91*(5), 353–362.

Allen, D. N., Goldstein, G., & Aldarondo, F. (1999). Neurocognitive dysfunction in patients diagnosed with schizophrenia and alcoholism. *Neuropsychology, 13*(1), 62–68.

Altmann, G. T. M. (1990). Cognitive models of speech processing: An introduction. In G. T. M. Altmann (Ed.), *Cognitive models of speech processing: Psycholinguistic and comutational perspectives,* 1–23. Cambridge, MA: MIT Press.

American Speech and Hearing Association. (1979). Guidelines for determining the threshold level for speech (American Speech and Hearing Association). *ASHA, 21,* 353–356.

American Speech and Hearing Association. (1997). Noise and hearing loss. Retrieved March 2, 2008, from http://www.asha.org/public/hearing/disorders/noise.htm.

American Speech and Hearing Association. (2002). *Guidelines for audiology service provision in and for schools.* Retrieved from http://www.asha.org/policy.

Amos, N. E., & Humes, L. E. (2007). Contribution of high frequencies to speech recognition in quiet and noise in listeners with varying degrees of high-frequency sensorineural hearing loss. *Journal of Speech, Language, and Hearing Research, 50*(4), 819–834.

Anderer, P., Saletu, B., Saletu-Zyhlarz, G., Gruber, D., Metka, M., Huber, J., et al. (2004). Brain regions activated during an auditory discrimination task in insomniac postmenopausal patients before and after hormone replacement therapy: Low resolution brain electromagnetic tomography applied to event-related potentials. *Neuropsychobiology, 49*(3), 134–153.

Anderer, P., Semlitsch, H. V., Saletu, B., Saletu-Zyhlarz, G., Gruber, D., Metka, M., et al. (2003). Effects of hormone replacement therapy on perceptual and cognitive event-related potentials in menopausal insomnia. *Psychoneuroendocrinology, 28*(3), 419–445.

Anderson, J. E., & Sanderson, P. (2009). Sonification design for complex work domains: Dimensions and distractors. *Journal of Experimental Psychology: Applied, 15*(3), 183–198.

Anderson, J. R. (2000). *Cognitive psychology and its implications* (5th ed.). New York: Worth.

Anderson, J. R., & Bower, G. H. (1972). Configural properties in sentence memory. *Journal of Verbal Learning and Verbal Behavior, 11*(5), 594–605.

Anderson, N. D., Craik, F. I. M., & Naveh-Benjamin, M. (1998). The attentional demands of encoding and retrieval in younger and older adults: I. Evidence from divided attention costs. *Psychology and Aging, 13*(3), 405–423.

Andersson, M., Reinvang, I., Wehling, E., Hugdahl, K., & Lundervold, A. J. (2008). A dichotic listening study of attention control in older adults. *Scandinavian Journal of Psychology, 49*(4), 299–304.

Andreano, J., Liang, K., Kong, L. J., Hubbard, D., Wiederhold, B. K., & Wiederhold, M. D. (2009). Auditory cues increase the hippocampal response to unimodal virtual reality. *Cyberpsychology and Behavior, 12*(3), 309–313.

Andrianopoulos, M. V., Darrow, K., & Chen, J. (2001). Multimodal standardization of voice among four multicultural populations formant structures. *Journal of Voice, 15*(1), 61–77.

Anshel, M. H., & Marisi, D. Q. (1978). Effect of music and rhythm on physical performance. *Research Quarterly, 49*(2), 109–113.

Arieh, Y., & Marks, L. E. (2003). Recalibrating the auditory system: A speed-accuracy analysis of intensity perception. *Journal of Experimental Psychology: Human Perception and Performance, 29*(3), 523–536.

Arlinger, S., Lunner, T., Lyxell, B., & Pichora-Fuller, M. K. (2009). Background and basic processes: The emergence of cognitive hearing science. *Scandinavian Journal of Psychology, 50*(5), 371–384.

Arnold, P., & Hill, F. (2001). Bisensory augmentation: A speechreading advantage when speech is clearly audible and intact. *British Journal of Psychology, 92*(2), 339–355.

Atienza, M., Cantero, J. L., & Gomez, C. M. (2000). Decay time of the auditory sensory memory trace during wakefulness and REM sleep. *Psychophysiology, 37*(4), 485–493.

Atkinson, R. C., & Shiffrin, R. M. (1968). Human memory: A proposed system and its control processes. In K. Spence & J. Spence (Eds.), *The psychology of learning and motivation* (Vol. 2), 742–775. New York: Academic Press.

Atkinson, R. C., & Shiffrin, R. M. (1971). The control of short-term memory. *Scientific American, 225*(2), 82–90.

Ayres, T. J., & Hughes, P. (1986). Visual acuity with noise and music at 107 dBA. *Journal of Auditory Research, 26*(1), 65–74.

Backs, R. W. (1997). Psychophysiological aspects of selective and divided attention during continuous manual tracking. *Acta Psychologica, 96*(3), 167–191.

Baddeley, A. (2003). Working memory: Looking back and looking forward. *Nature Reviews Neuroscience, 4*(10), 829–839.

Baddeley, A., Chincotta, D., Stafford, L., & Turk, D. (2002). Is the word length effect in STM entirely attributable to output delay? Evidence from serial recognition. *Quarterly Journal of Experimental Psychology A, 2*, 353–369.

Baddeley, A., Gathercole, S., & Papagno, C. (1998). The phonological loop as a language learning device. *Psychological Review, 105*(1), 158–173.

Baddeley, A. D. (1968). How does acoustic similarity influence short-term memory? *Quarterly Journal of Experimental Psychology A, 20*(3), 249–264.

Baddeley, A. D. (1992). Working memory. *Science, 255*, 556–559.

Baddeley, A. D. (1997). *Human memory: Theory and practice* (rev. ed.). Upper Saddle River, NJ: Allyn & Bacon.

Baddeley, A. D. (2002). Is working memory still working? *European Psychologist, 7*(2), 85–97.

Baddeley, A. D., & Hitch, G. (1974). Working memory. In G. H. Bower (Ed.), *The psychology of learning and motivation* (Vol. 8, 47–89). Orlando, FL: Academic Press.

Baddeley, A. D., & Hitch, G. J. (1994). Developments in the concept of working memory. *Neuropsychology, 8,* 4485–4493.

Baddeley, A. D. (1998). *Human memory: Theory and practice.* Needham Heights, MA: Allyn and Bacon.

Baddeley, A. D., Lewis, V., & Vallar, G. (1984). Exploring the articulatory loop. *The Quarterly Journal of Experimental Psychology A: Human Experimental Psychology, 36*(2), 233–252.

Baddeley, A. D., & Logie, R. H. (1999). Working memory: The multiple-component model. In A. Miyake & P. Shah (Eds.), *Models of working memory,* 28–61. New York: Cambridge University Press.

Bagley, W. C. (1900–1901). The apperception of the spoken sentence: A study in the psychology of language. *American Journal of Psychology, 12,* 80–130.

Baldwin, C. L. (October, 2001). *Impact of age-related hearing impairment on cognitive task performance.* Paper presented at the 45th Annual Conference of the Human Factors and Ergonomics Society, Minneapolis, MN.

Baldwin, C. L. (2002). Designing in-vehicle technologies for older drivers: Application of sensory-cognitive interaction theory. *Theoretical Issues in Ergonomics Science, 3*(4), 307–329.

Baldwin, C. L. (2003). Neuroergonomics of mental workload: New insights from the convergence of brain and behavior in ergonomics research [Commentary]. *Theoretical Issues in Ergonomics Science, 4,* 132–141.

Baldwin, C. L. (2006). User-centered design of in-vehicle route guidance systems. In D. de Waard, K. Brookhuis, & A. Toffetti (Eds.), *Developments in human factors in transportation, design, and evaluation,* 43–49. Maastricht, The Netherlands: Shaker.

Baldwin, C. L. (2007). Cognitive implications of facilitating echoic persistence. *Memory & Cognition, 35*(4), 774–780.

Baldwin, C. L. (2011). Verbal collision avoidance messages during simulated driving: Perceived urgency, annoyance, and alerting effectiveness. *Ergonomics, 54*(4), 328.

Baldwin, C. L., & Ash, I. (2010). Sensory factors influence complex span scores of verbal working memory in older listeners. *Psychology and Aging, 25*(3)85–91.

Baldwin, C. L., & Coyne, J. T. (2003, July). *Mental workload as a function of traffic density in an urban environment: Convergence of physiological, behavioral, and subjective indices.* Paper presented at the Human Factors in Driving Assessment 2003 Symposium, Park City, UT.

Baldwin, C. L., & Coyne, J. T. (2005). Dissociable aspects of mental workload: Examinations of the P300 ERP component and performance assessments. *Psychologia, 48,* 102–119.

Baldwin, C. L., Freeman, F., & Coyne, J. T. (2004, September). *Mental workload as a function of road type and visibility: Comparison of neurphysiological, behavioral, and subjective indices.* Paper presented at the Human Factors and Ergonomics Society, New Orleans, LA.

Baldwin, C. L., & Galinsky, A. M. (1999). Pure-tone threshold shifts during moderate workload conditions: Human performance implications for automated environments. In M. S. Scerbo & M. Mouloua (Eds.), *Automation technology and human performance: Current research trends,* 296–300. Mahwah, NJ: Erlbaum.

Baldwin, C. L., Lewis, K. R., & Morris, A. E. (2006, July 10–14). *Sensory-cognitive interactions in young and older listeners.* Paper presented at the 16th World Congress on Ergonomics, Maastricht, the Netherlands.

Baldwin, C. L., & May, J. F. (2005, June). *Verbal collision avoidance messages of varying perceived urgency reduce crashes in high risk scenarios.* Paper presented at the Driving Assessment 2005, Rockport, ME.

Baldwin, C. L., & May, J. F. (2011). Loudness interacts with semantics in auditory warnings to impact rear-end collisions. *Transportation Research Part F, 14*(1), 36–42.

Baldwin, C. L., May, J. F., & Reagan, I. (2006, October). *Auditory in-vehicle messages and older drivers*. Paper presented at the Human Factors and Ergonomics Society, San Francisco.

Baldwin, C. L., & Reagan, I. (2009). Individual differences in route-learning strategy and associated working memory resources. *Human Factors, 51*, 368–377.

Baldwin, C. L., & Schieber, F. (1995). *Age Differences in Mental Workload with Implications for Driving*. Paper presented at the 39th Annual Conference of the Human Factors and Ergonomics Society, San Diego, CA.

Baldwin, C. L., & Struckman-Johnson, D. (2002). Impact of speech presentation level on cognitive task performance: Implications for auditory display design. *Ergonomics, 45*(1), 61–74.

Ball, K. (1997). Attentional problems and older drivers. *Alzheimer Disease and Associated Disorders, 11*(Suppl. 1), 42–47.

Ballas, J. A. (1993). Common factors in the identification of an assortment of brief everyday sounds. *Journal of Experimental Psychology: Human Perception and Performance, 19*(2), 250–267.

Ballas, J. A. (1999). The interpretation of natural sound in the cockpit. In N. A. Stanton & J. Edworthy (Eds.), *Human factors in auditory warnings*, 93–112. Aldershot, UK: Ashgate.

Banich, M. T. (2004). *Cognitive neuroscience and neuropsychology* (2nd ed.). Boston: Houghton Mifflin.

Banich, M. T., & Mack, M. (Eds.). (2003). *Mind, brain, and language*. Mahwah, NJ: Erlbaum.

Baños, J. H., Elliott, T. R., & Schmitt, M. (2005). Factor structure of the Rey Auditory Verbal Learning Test in adults with spinal cord injury. *Rehabilitation Psychology, 50*(4), 375–380.

Barr, R. A., & Giambra, L. M. (1990). Age-related decrement in auditory selective attention. *Psychology and Aging, 5*(4), 597–599.

Barrett, L. F., Tugade, M. M., & Engle, R. W. (2004). Individual differences in working memory capacity and dual-process theories of the mind. *Psychological Bulletin, 130*(4), 553–573.

Barrouillet, P. (1996). Transitive inferences from set-inclusion relations and working memory. *Journal of Experimental Psychology: Learning, Memory, & Cognition, 22*, 1408–1422.

Barrow, J. H., & Baldwin, C. L. (2009). *Verbal-spatial cue conflict: Implications for the design of collision avoidance warning systems*. Paper presented at the Driving Assessment, Big Sky, MT.

Barrow, J., & Baldwin, C. L. (in preparation). Verbal-spatial facilitation and conflict in rapid spatial orientation tasks.

Barshi, I. (1997). *Effects of linguistic properties and message length on misunderstandings in aviation communication*. Unpublished doctoral dissertation, University of Colorado, Boulder.

Bates, E., Dick, F., & Wulfeck, B. (1999). Not so fast: Domain-general factors can account for selective deficits in grammatical processing [Commentary to Caplan & Waters, 1999]. *Behavioral and Brain Sciences, 22*, 77–126.

Bavelier, D., Corina, D., Jezzard, P., Clark, V., Karni, A., Lalwani, A., et al. (1998). Hemispheric specialization for English and ASL: Left invariance-right variability. *NeuroReport: For Rapid Communication of Neuroscience Research, 9*(7), 1537–1542.

Bayliss, D. M., Jarrold, C., Gunn, D. M., & Baddeley, A. D. (2003). The complexities of complex span: Explaining individual differences in working memory in children and adults. *Journal of Experimental Psychology: General, 132*(1), 71–92.

Beaman, C. P. (2005). Irrelevant sound effects amongst younger and older adults: Objective findings and subjective insights. *European Journal of Cognitive Psychology, 17*(2), 241–265.

Beaty, D. (1995). *The naked pilot: The human factor in aircraft accidents*. Shrewsbury, UK: Airlife.

Begault, D. R. (1991). Challenges to the successful implementation of 3-D sound. *Journal of the Audio Engineering Society, 39*(11), 864–870.

Begault, D. R. (1993). Head-up auditory displays for traffic collision avoidance system advisories: A preliminary investigation. *Human Factors, 35*(4), 707–717.

Begault, D. R., Godfroy, M., Sandor, A., & Holden, K. (2007, June 26–29). *Auditory alarm design for NASA CEV applications.* Paper presented at the 13th International Conference on Auditory Displays, Montreal, Canada.

Begault, D. R., & Pittman, M. T. (1996). Three-dimensional audio versus head-down Traffic Alert and Collision Avoidance System displays. *International Journal of Aviation Psychology, 6*(1), 79–93.

Beh, H. C., & Hirst, R. (1999). Performance on driving-related tasks during music. *Ergonomics, 42*(8), 1087–1098.

Bekesy, G. V. (1960). *Experiments in hearing.* New York: McGraw-Hill.

Belin, P., & Zatorre, R. J. (2000). "What," "where" and "how" in auditory cortex. *Nature Neuroscience, 3*(10), 965–966.

Belin, P., Zatorre, R. J., Lafaille, P., & Ahad, P. (2000). Voice-selective areas in the human auditory cortex. *Nature, 403,* 309–312.

Bell, P. A. (1978). Effects of noise and heat stress on primary and subsidiary task performance. *Human Factors, 20*(6), 749–752.

Bellis, T. J. (2002). *When the brain can't hear: Unraveling the mystery of auditory processing disorder.* New York: Simon & Schuster.

Bellis, T. J., Nicol, T., & Kraus, N. (2000). Aging affects hemispheric asymmetry in the neural representation of speech sounds. *Journal of Neuroscience, 20*(2), 791–797.

Bellis, T. J., & Wilber, L. A. (2001). Effects of aging and gender in interhemispheric function. *Journal of Speech, Language, & Hearing Research, 44*(2), 246–263.

Belz, S. M., Robinson, G. S., & Casali, J. G. (1999). A new class of auditory warning signals for complex systems: Auditory icons. *Human Factors, 41*(4), 608–618.

Bentin, S., Kutas, M., & Hillyard, S. A. (1995). Semantic processing and memory for attended and unattended words in dichotic listening: Behavioral and electrophysiological evidence. *Journal of Experimental Psychology: Human Perception and Performance, 21*(1), 54–67.

Benton, D., & Jarvis, M. (2007). The role of breakfast and a mid-morning snack on the ability of children to concentrate at school. *Physiology and Behavior, 90*(2–3), 382–385.

Ben-Yaacov, A., Maltz, M., & Shinar, D. (2002). Effects of an in-vehicle collision avoidance warning system on short- and long-term driving performance. *Human Factors, 44*(2), 335–342.

Berckmoes, C., & Vingerhoets, G. (2004). Neural foundations of emotional speech processing. *Current Directions in Psychological Science, 13*(5), 182–185.

Bergman, M. (1971). Changes in hearing with age. *Gerontologist, 11*(2), 148–151.

Bergman, M. (1980). *Aging and the perception of speech.* Baltimore: University Park Press.

Beringer, D., Harris, H. C., & Joseph, K. M. (1998, October 5–9). *Hearing thresholds among pilots and non-pilots: Implications for auditory warning design.* Paper presented at the 42nd Annual Meeting of the Human Factors and Ergonomics Society, Chicago.

Bernardi, L., Porta, C., & Sleight, P. (2006). Cardiovascular, cerebrovascular, and respiratory changes induced by different types of music in musicians and non-musicians: The importance of silence. *Heart, 92*(4), 445–452.

Bernstein, I. H. (1970). Can we see and hear at the same time? Some recent studies of intersensory facilitation of reaction time. *Acta Psychologica, Amsterdam, 33,* 21–35.

Bernstein, L. E., Auer, E. T., & Moore, J. K. (2004). Audiovisual speech binding: Convergence or association? In G. A. Calvert, C. Spence, & B. E. Stein (Eds.), *The handbook of multisensory processes,* 203–223. Cambridge, MA: MIT Press.

Bigand, E. (1993). Contributions of music to research on human auditory cognition. In S. McAdams & E. Bigand (Eds.), *Thinking in sound: The cognitive psychology of human audition*, 231–277. Oxford, UK: Clarendon Press.

Blattner, M. M., Sumikawa, D. A., & Greenberg, R. M. (1989). Earcons and icons: Their structure and common design principles. *Human-Computer Interaction, 4*(1), 11–44.

Blesser, B., & Salter, L.-R. (2007). *Spaces speak, are you listening?* Cambridge, MA: MIT Press.

Bliss, J. P., & Acton, S. A. (2003). Alarm mistrust in automobiles: How collision alarm reliability affects driving. *Applied Ergonomics, 34*(6), 499–509.

Blosser, J. L., Weidner, W. E., & Dinero, T. (1976). The effect of rate-controlled speech on the auditory receptive scores of children with normal and disordered language abilities. *The Journal of Special Education, 10*(3), 291–298.

Blumenthal, J. A., & Madden, D. J. (1988). Effects of aerobic exercise training, age, and physical fitness on memory-search performance. *Psychology and Aging, 3*(3), 280–285.

Blumenthal, T. D. (1996). Inhibition of the human startle response is affected by both prepulse intensity and eliciting stimulus intensity. *Biological Psychology, 44*(2), 85–104.

Blumstein, S. E., Myers, E. B., & Rissman, J. (2005). The perception of voice onset time: An fMRI investigation of phonetic category structure. *Journal of Cognitive Neuroscience, 17*(9), 1353–1366.

Blumstein, S. E., & Stevens, K. N. (1981). Phonetic features and acoustic invariance in speech. *Cognition, 10*(1, Suppl. 3), 25–32.

Bolia, R. S. (2004). Special issue: Spatial audio displays for military aviation. *International Journal of Aviation Psychology, 14*(3), 233–238.

Bonebright, T. L., Miner, N. E., Goldsmith, T. E., & Caudell, T. P. (2005). Data collection and analysis techniques for evaluating the perceptual qualities of auditory stimuli. *ACM Transactions on Applied Perception, 2*(4), 505–516.

Botwinick, J. (1984). *Aging and behavior* (3rd ed.). New York: Springer.

Bowles, R. P., & Salthouse, T. A. (2003). Assessing the age-related effects of proactive interference on working memory tasks using the Rasch model. *Psychology and Aging, 18*(3), 608–615.

Braitman, K. A., McCartt, A. T., Zuby, D. S., & Singer, J. (2010). Volvo and Infiniti drivers' experiences with select crash avoidance technologies. *Traffic Injury Prevention, 11*(3), 270–278.

Brant, L. J., & Fozard, J. L. (1990). Age changes in pure-tone hearing thresholds in a longitudinal study of normal human aging. *Journal of the Acoustical Society of America, 88*(2), 813–820.

Brattico, E., Winkler, I., Naatanen, R., Paavilainen, P., & Tervaniemi, M. (2002). Simultaneous storage of two complex temporal sound patterns in auditory sensory memory. *NeuroReport: For Rapid Communication of Neuroscience Research, 13*(14), 1747–1751.

Bregman, A. S. (1990). *Auditory scene analysis: The perceptual organization of sound.* Cambridge, MA: MIT Press.

Bregman, A. S. (1993). Auditory scene analysis: hearing in complex environments. In S. McAdams & E. Bigand (Eds.), *Thinking in sound: The cognitive psychology of human audition*, 10–36. Oxford, UK: Clarendon Press.

Bregman, A. S., & Campbell, J. (1971). Primary auditory stream segregation and perception of order in rapid sequences of tones. *Journal of Experimental Psychology, 89*(244–249).

Brenner, M., Doherty, E. T., & Shipp, T. (1994). Speech measures indicating workload demand. *Aviation, Space, and Environmental Medicine, 65*(1), 21–26.

Brewster, S. (2002). Visualization tools for blind people using multiple modalities. *Disability and Rehabilitation: An International Multidisciplinary Journal, 24*(11–12), 613–621.

Brewster, S., Wright, P. C., & Edwards, A. D. N. (1995). Parallel earcons: Reducing the length of audio messages. *International Journal of Human-Computer Studies, 43*(2), 153–175.

Brisson, B., & Jolicoeur, P. (2007). Cross-modal multitasking processing deficits prior to the central bottleneck revealed by event-related potentials. *Neuropsychologia, 45*(13), 3038–3053.

Brisson, B., Leblanc, E., & Jolicoeur, P. (2009). Contingent capture of visual-spatial attention depends on capacity-limited central mechanisms: Evidence from human electrophysiology and the psychological refractory period. *Biological Psychology, 80*(2), 218–225.

Broadbent, D. E. (1958). *Perception and communication.* New York: Pergamon.

Broadbent, D. E. (1971). *Decision and stress.* New York: Academic Press.

Broadbent, D. E. (1978). The current state of noise research: Reply to Poulton. *Psychological Bulletin, 85*(5), 1052–1067.

Broadbent, D. E. (1979). Human performance and noise. In C. S. Harris (Ed.), *Handbook of noise control,* 2066–2085. New York: McGraw-Hill.

Broadbent, D. E. (1982). Task combination and selective intake of information. *Acta Psychologica, 50*(3), 253–290.

Brodsky, W. (2001). The effects of music tempo on simulated driving performance and vehicular control. *Transportation Research Part F: Traffic Psychology and Behaviour, 4*(4), 219–241.

Bronkhorst, A. W. (1995). Localization of real and virtual sound sources. *Journal of the Acoustical Society of America, 98*(5), 2542–2553.

Brookhuis, K. A., & de Waard, D. (2001). Assessment of drivers' workload: Performance and subjective and physiological indexes. In P. A. Hancock & P. A. Desmond (Eds.), *Stress, workload, and fatigue,* 321–333. Mahwah, NJ: Erlbaum.

Brooks, L. R. (1968). Spatial and verbal components of the act of recall. *Canadian Journal of Psychology, 22*(5), 349–368.

Brown, B. L., Strong, W. J., & Rencher, A. C. (1974). Fifty-four voices from two: The effects of simultaneous manipulations of rate, mean fundamental frequency, and variance of fundamental frequency on ratings of personality from speech. *Journal of the Acoustical Society of America, 55*(2), 313–318.

Brown, I. D. (1965). A comparison of two subsidiary tasks used to measure fatigue in car drivers. *Ergonomics, 8*(4), 467–473.

Brown, I. D., & Poulton, E. C. (1961). Measuring the spare "mental capacity" of car drivers by a subsidiary task. *Ergonomics, 4,* 35–40.

Brown, S. W., & Boltz, M. G. (2002). Attentional processes in time perception: Effects of mental workload and event structure. *Journal of Experimental Psychology: Human Perception and Performance, 28*(3), 600–615.

Brown, T. L., Lee, J. D., & Hoffman, J. (2001). *The effect of rear-end collision warnings on on-going response.* Paper presented at the Human Factors and Ergonomics Society 45th Annual Meeting, Minneapolis, MN (October).

Brown, T. L., Lee, J. D., & McGehee, D. V. (2001). Human performance models and rear-end collision avoidance algorithms. *Human Factors, 43*(3), 462–482.

Brumback, C. R., Low, K. A., Gratton, G., & Fabiani, M. (2004). Sensory ERPs predict differences in working memory span and fluid intelligence. *NeuroReport: For Rapid Communication of Neuroscience Research, 15*(2), 373–376.

Brungart, D. S., & Simpson, B. D. (2002). The effects of spatial separation in distance on the informational and energetic masking of a nearby speech signal. *Journal of the Acoustical Society of America, 112*(2), 664–676.

Brungart, D. S., & Simpson, B. D. (2005). Optimizing the spatial configuration of a seven-talker speech display. *ACM Transactions on Applied Perception, 2*(4), 430–436.

Brungart, D. S., Kordik, A. J., Simpson, B. D., & McKinley, R. L. (2003). Auditory localization in the horizontal plane with single and double hearing protection. *Aviation, Space, and Environmental Medicine, 74*(9), 937–946.

Bryant, G. A., & Barrett, H. C. (2007). Recognizing intentions in infant-directed speech: Evidence for universals. *Psychological Science, 18*(8), 746–751.

Buchwald, J. S. (1990). Animal models of cognitive event-related potentials. In J. W. Rohrbaugh, R. Parasuraman, & R. Johnson, Jr. (Eds.), *Event-related brain potentials: Basic issues and applications*, 57–75. New York: Oxford University Press.

Burke, D. M., & Shafto, M. A. (2008). Language and aging. In F. I. M. Craik & T. A. Salthouse (Eds.), *The handbook of aging and cognition* (3rd ed., 373–443). New York: Psychology Press.

Burnett, G. (2000a). "Turn right at the traffic lights": The requirement for landmarks in vehicle navigation systems. *Journal of Navigation, 53*(3), 499–510.

Burnett, G. (2000b, June 29–30). *Usable vehicle navigation systems: Are we there yet?* Paper presented at the Vehicle Electronic Systems 2000—European conference and exhibition. Stratford-upon-Avon, UK.

Burns, P. C. (1999). Navigation and the mobility of older drivers. *Journals of Gerontology Series B: Psychological Sciences and Social Sciences, 54*(1), S49–S55.

Burt, J. L., Bartolome-Rull, D. S., Burdette, D. W., & Comstock, J. R. (1999). A psychophysical evaluation of the perceived urgency of auditory warning signals. In N. A. Stanton & J. Edworthy (Eds.), *Human factors in auditory warnings*, 151–169. Brookfield, VT: Ashgate.

Byrne, E. A., & Parasuraman, R. (1996). Psychophysiology and adaptive automation. *Biological Psychology, 42*(3), 249–268.

Cabeza, R. (2002). Hemispheric asymmetry reduction in older adults: The HAROLD model. *Psychology and Aging, 17*(1), 85–100.

Cabeza, R., Daselaar, S. M., Dolcos, F., Budde, M., & Nyberg, L. (2004). Task-independent and task-specific age effects on brain activity during working memory, visual attention and episodic retrieval. *Cerebral Cortex, 14*(4), 364–375.

Cabeza, R., McIntosh, A. R., Tulving, E., Nyberg, L., & Grady, C. L. (1997). Age-related differences in effective neural connectivity during encoding and recall. *NeuroReport: For Rapid Communication of Neuroscience Research, 8*(16), 3479–3483.

Caird, J. K., Willness, C. R., Steel, P., & Scialfa, C. (2008). A meta-analysis of the effects of cell phones on driver performance. *Accident Analysis and Prevention, 40*(4), 1282–1293.

Calvert, G. A., Brammer, M. J., Bullmore, E. T., Campbell, R., Iversen, S. D., & David, A. S. (1999). Response amplification in sensory-specific cortices during crossmodal binding. *NeuroReport: For Rapid Communication of Neuroscience Research, 10*(12), 2619–2623.

Calvert, G. A., Bullmore, E. T., Brammer, M. J., & Campbell, R. (1997). Activation of auditory cortex during silent lipreading. *Science, 276*(5312), 593–596.

Calvert, G. A., & Campbell, R. (2003). Reading speech from still and moving faces: The neural substrates of visible speech. *Journal of Cognitive Neuroscience, 15*(1), 57–70.

Campbell, J. L., Richman, J. B., Carney, C., & Lee, J. D. (2004). *In-vehicle display icons and other information elements volume I: Guidelines* (No. FHWA-RD-03–065). McLean, VA: Office of Safety Research and Development: Federal Highway Administration.

Campbell, R. (1986). The lateralization of lip-read sounds: A first look. *Brain and Cognition, 5*(1), 1–21.

Cansino, S., Williamson, S. J., & Karron, D. (1994). Tonotopic organization of human auditory association cortex. *Brain Research, 663*(1), 38–50.

Caplan, D., Alpert, N., & Waters, G. (1998). Effects of syntactic structure and propositional number on patterns of regional cerebral blood flow. *Journal of Cognitive Neuroscience, 10*(4), 541–552.

Caplan, D., & Waters, G. S. (1999). Verbal working memory and sentence comprehension. *Behavioral and Brain Sciences, 22*(1), 77–126.

Caporael, L. R. (1981). The paralanguage of caregiving: Baby talk to the institutionalized aged. *Journal of Personality and Social Psychology, 40*(5), 876–884.

Carhart, R. (1965). Problems in the measurement of speech discrimination. *Archives of Otolaryngology, 82*, 253–260.

Carlyon, R. P., Cusack, R., Foxton, J. M., & Robertson, I. H. (2001). Effects of attention and unilateral neglect on auditory stream segregation. *Journal of Experimental Psychology: Human Perception and Performance, 27*(1), 115–127.

Carrier, L. M., & Pashler, H. (1995). Attentional limits in memory retrieval. *Journal of Experimental Psychology: Learning, Memory, and Cognition, 21*(5), 1339–1348.

Carroll, D. W. (1994). *Psychology of language* (2nd ed.). Pacific Grove: CA: Brooks/Cole.

Carroll, D. W. (2004). *Psychology of language* (4th ed.). Belmont, CA: Thomson Wadsworth.

Casali, J. G., & Wierwille, W. W. (1983). A comparison of rating scale, secondary-task, physiological, and primary-task workload estimation techniques in a simulated flight task emphasizing communications load. *Human Factors, 25*(6), 623–641.

Casali, J. G., & Wierwille, W. W. (1984). On the measurement of pilot perceptual workload: A comparison of assessment techniques addressing sensitivity and intrusion issues. *Ergonomics, 27*(10), 1033–1050.

Casey, M. A. (1998). *Auditory group theory with applications to statistical basis methods for structured audio.* Unpublished doctoral dissertation, Massachusetts Institute of Technology Media Laboratory, Boston.

Cassity, H. D., Henley, T. B., & Markley, R. P. (2007). The Mozart effect: Musical phenomenon or musical preference? A more ecologically valid reconsideration. *Journal of Instructional Psychology, 34*(1), 13–17.

Catchpole, K. R., McKeown, J. D., & Withington, D. J. (2004). Localizable auditory warning pulses. *Ergonomics, 47*(7), 748–771.

Cerella, J. (1985). Information processing rates in the elderly. *Psychological Bulletin, 98*(1), 67–83.

Cerella, J. (1991). Age effects may be global, not local: Comment on Fisk and Rogers (1991). *Journal of Experimental Psychology: General, 120*(2), 215–223.

Chabris, C. F. (1999). Prelude or requiem for the "Mozart effect"? *Nature, 400*(6747), 826–827.

Chambers, C. G., & Smyth, R. (1998). Structural parallelism and discourse coherence: A test of centering theory. *Journal of Memory and Language, 39*(4), 593–608.

Chamorro-Premuzic, T., & Furnham, A. (2007). Personality and music: Can traits explain how people use music in everyday life? *British Journal of Psychology, 98*(2), 175–185.

Chamorro-Premuzic, T., Goma -i-Freixanet, M., Furnham, A., & Muro, A. (2009). Personality, self-estimated intelligence, and uses of music: A Spanish replication and extension using structural equation modeling. *Psychology of Aesthetics, Creativity, and the Arts, 3*(3), 149–155.

Chan, A. H. S., & Chan, K. W. L. (2006). Synchronous and asynchronous presentations of auditory and visual signals: Implications for control console design. *Applied Ergonomics, 37*(2), 131–140.

Cherry, C. (1953). Some experiments on the reception of speech with one and with two ears. *Journal of the Acoustical Society of America, 25*, 975–979.

Cherry, C. (1957). On human communication; a review, a survey, and a criticism.

Chiu, C. Y. P., Coen-Cummings, M., Schmithorst, V. J., Holland, S. K., Keith, R., Nabors, L., et al. (2005). Sound blending in the brain: A functional magnetic resonance imaging investigation. *NeuroReport: For Rapid Communication of Neuroscience Research, 16*(9), 883–886.

Chomsky, N. (1984). *Modular approaches to the study of the mind.* San Diego, CA: San Diego State University Press.

Chomsky, N. (1995). *The minimalist program.* Cambridge, MA: MIT Press.

Chou, C.-D., Madhavan, D., & Funk, K. (1996). Studies of cockpit task management errors. *International Journal of Aviation Psychology, 6*(4), 307–320.

Christensen, N., D'Souza, M., Zhu, X., & Frisina, R. D. (2009). Age-related hearing loss: Aquaporin 4 gene expression changes in the mouse cochlea and auditory midbrain. *Brain Research, 1253*, 27–34.

Christianson, K., Hollingworth, A., Halliwell, J. F., & Ferreira, F. (2001). Thematic roles assigned along the garden path linger. *Cognitive Psychology, 42*(4), 368–407.

Clark, C., & Stansfeld, S. A. (2007). The effect of transportation noise on health and cognitive development: A review of recent evidence. *International Journal of Comparative Psychology, 20*(2/3), 145–158.

Clarke, S., & Thiran, A. B. (2004). Auditory neglect: What and where in auditory space. *Cortex, 40*(2), 291–300.

Cnossen, F., Rothengatter, T., & Meijman, T. (2000). Strategic changes in task performance in simulated car driving as an adaptive response to task demands. *Transportation Research Part F: Traffic Psychology and Behaviour, 3*(3), 123–140.

Cocchini, G., Logie, R. H., Della Sala, S., MacPherson, S. E., & Baddeley, A. D. (2002). Concurrent performance of two memory tasks: Evidence for domain-specific working memory systems. *Memory & Cognition, 30*(7), 1086–1095.

Cohen, H., Levy, J. J., & McShane, D. (1989). Hemispheric specialization for speech and non-verbal stimuli in Chinese and French Canadian subjects. *Neuropsychologia, 27*(2), 241–245.

Cohen, S., Evans, G. W., Krantz, D. S., & Stokols, D. (1980). Physiological, motivational, and cognitive effects of aircraft noise on children: Moving from the laboratory to the field. *American Psychologist, 35*(3), 231–243.

Cole, R. A., & Jakimik, J. (1980). A model of speech perception. In R. A. Cole (Ed.), *Perception and production of fluent speech*, 130–163. Hillsdale, NJ: Erlbaum.

Cole, R. A., & Scott, B. (1974). The phantom in the phoneme: Invariant cues for stop consonants. *Perception & Psychophysics, 15*(1), 101–107.

Coles, M. G., & Rugg, M. D. (1995). Event-related brain potentials: An introduction. In M. D. Rugg & M. G. Coles (Eds.), *Electrophysiology of mind*, 1–26. Oxford: Oxford University Press.

Colle, H. A., & Reid, G. B. (1997). A framework for mental workload research and applications using formal measurement theory. *International Journal of Cognitive Ergonomics, 1*(4), 303–313.

Colle, H. A., & Welsh, A. (1976). Acoustic masking in primary memory. *Journal of Verbal Learning and Verbal Behavior, 15*(1), 17–31.

Collins, A. M., & Quillian, M. R. (1969). Retrieval time from semantic memory. *Journal of Verbal Learning and Verbal Behavior, 8*(2), 240–247.

Colquhoun, W. P. (1975). Evaluation of auditory, visual, and dual-mode displays for prolonged sonar monitoring in repeated sessions. *Human Factors, 17*(5), 425–437.

Coltheart, M. (2004). Are there lexicons? *Quarterly Journal of Experimental Psychology A, 57*(7), 1153–1171.

Conrad, R. (1964). Acoustic confusions in immediate memory. *British Journal of Psychology, 55*, 75–84.

Conrad, R., Baddeley, A. D., & Hull, A. J. (1966). Rate of presentation and the acoustic similarity effect in short-term memory. *Psychonomic Science, 5*(6), 233–234.

Conrad, R., & Hull, A. J. (1964). Information, acoustic confusion and memory span. *British Journal of Psychology, 55*(4), 429–432.

Conrad, R., & Hull, A. J. (1968). Input modality and the serial position curve in short-term memory. *Psychonomic Science, 10*(4), 135–136.

Conway, A. R. A., Cowan, N., Bunting, M. F., Therriault, D. J., & Minkoff, S. R. B. (2002). A latent variable analysis of working memory capacity, short-term memory capacity, processing speed, and general fluid intelligence. *Intelligence, 30*(2), 163–184.

Conway, A. R. A., & Kane, M. J. (2001). Capacity, control and conflict: An individual differences perspective on attentional capture. In C. L. Folk & B. S. Gibson (Eds.), *Attraction, distraction and action: Multiple perspectives on attentional capture. Advances in psychology, 133*, 349–372. New York: Elsevier Science.

Conway, A. R. A., Kane, M. J., Bunting, M. F., Hambrick, D. Z., Wilhelm, O., & Engle, R. W. (2005). Working memory span tasks: A methodological review and user's guide. *Psychonomic Bulletin & Review, 12*(5), 769–786.

Cooper, J. C., Jr., & Gates, G. A. (1991). Hearing in the elderly—The Framingham cohort, 1983–1985: Part II. Prevalence of central auditory processing disorders. *Ear and Hearing, 12*(5), 304–311.

Cooper, J. M., & Strayer, D. L. (2008). Effects of simulator practice and real-world experience on cell-phone related driver distraction. *Human Factors, 50*(6), 893–902.

Corballis, M. C. (2000). How laterality will survive the millennium bug, *Brain & Cognition, 42*, 160–162.

Corballis, M. C. (1966). Rehearsal and decay in immediate recall of visually and aurally presented items. *Canadian Journal of Psychology/Revue canadienne de psychologie, 20*(1), 43–51.

Coren, S., Ward, L. M., & Enns, J. T. (1999). *Sensation and perception* (5th ed.). Fort Worth, TX: Harcourt College.

Corso, J. F. (1963a). Aging and auditory thresholds in men and women. *Archives of Environmental Health, 6*, 350–356.

Corso, J. F. (1963b). Age and sex differences in pure-tone thresholds. *Archives of Otolaryngology, 77*, 385–405.

Corso, J. F. (1981). *Aging, sensory systems and perception.* New York: Praeger.

Courtney, S. M., & Ungerleider, L. G. (1997). What fMRI has taught us about human vision. *Current Opinion in Neurobiology, 7*(4), 554–561.

Cowan, N. (1984). On short and long auditory stores. *Psychological Bulletin, 96*(2), 341–370.

Cowan, N., Winkler, I., Teder, W., & Naatanen, R. (1993). Memory prerequisites of mismatch negativity in the auditory event-related potential (ERP). *Journal of Experimental Psychology: Learning, Memory, and Cognition, 19*(4), 909–921.

Coyne, J. T., & Baldwin, C. L. (April, 2003). *ERP indices of mental workload for traditional and text-based ATC commands during a simulated flight task.* Paper presented at the 12th International Symposium on Aviation Psychology, Dayton, OH.

Craik, F. I. M. (1977). Age difference in human memory. In J. E. Birren & K. W. Schaie (Eds.), *Handbook of the psychology of aging,* 384–420. New York: Van Nostrand Reinhold.

Craik, F. I. M., Govoni, R., Naveh-Benjamin, M., & Anderson, N. D. (1996). The effects of divided attention on encoding and retrieval processes in human memory. *Journal of Experimental Psychology: General, 125*(2), 159–180.

Craik, F. I. M., Naveh-Benjamin, M., Ishaik, G., & Anderson, N. D. (2000). Divided attention during encoding and retrieval: Differential control effects? *Journal of Experimental Psychology: Learning, Memory, and Cognition November, 26*(6), 1744–1749.

Crawford, H. J., & Strapp, C. M. (1994). Effects of vocal and instrumental music on visuospatial and verbal performance as moderated by studying preference and personality. *Personality and Individual Differences, 16*(2), 237–245.

Crowder, R. G. (1970). The role of one's own voice in immediate memory. *Cognitive Psychology, 1*(2), 157–178.

Crowder, R. G. (1972). Visual and auditory memory. In J. F. Kavanagh & I. G. Mattingly (Eds.), *Language by ear and by eye: The relationships between speech and reading,* 251–275. Cambridge, MA: MIT Press.

Crowder, R. G. (1978). Mechanisms of auditory backward masking in the stimulus suffix effect. *Psychological Review, 85*(6), 502–524.

CTIA. (2009). *Annualized wireless industry survey results—June 1985 to June 2009. Semi-annual wireless industry survey.* Retrieved February 18, 2010, from http://www.ctia.org/advocacy/research/index.cfm/AID/10538

. CTIA. (2010). *U.S. wireless quick facts*. Retrieved April 15, 2011, from http://www.ctia.org/advocacy/research/index.cfm/aid/10323

Cummings, M. L., Kilgore, R. M., Wang, E., Tijerina, L., & Kochhar, D. S. (2007). Effects of single versus multiple warnings on driver performance. *Human Factors, 49*(6), 1097–1106.

Curran, T., Tucker, D. M., Kutas, M., & Posner, M. I. (1993). Topography of the N400: Brain electrical activity reflecting semantic expectancy. *Electroencephalography and Clinical Neurophysiology: Evoked Potentials, 88*(3), 188–209.

Cutler, A. (1995). Spoken word recognition and production. In J. L. Miller & P. D. Eimas (Eds.), *Speech, language, and communication* (2nd ed.), 115–123. San Diego, CA: Academic.

Damasio, A. R. (1989). The brain binds entities and events by multiregional activation from convergence zones. *Neural Computation, 1*(1), 123–132.

Daneman, M., & Carpenter, P. A. (1980). Individual differences in working memory and reading. *Journal of Verbal Learning and Verbal Behavior, 19*(4), 450–466.

Daneman, M., & Merikle, P. M. (1996). Working memory and language comprehension: A meta-analysis. *Psychonomic Bulletin & Review, 3*(4), 422–433.

Darwin, C. J., Turvey, M. T., & Crowder, R. C. (1972). An auditory analogue of the Sperling partial report procedure: Evidence for brief auditory storage. *Cognitive Psychology, 3*(2), 255–267.

Davidson, R. J., Chapman, J. P., Chapman, L. J., & Henriques, J. B. (1990). Asymmetrical brain electrical activity discriminates between psychometrically-matched verbal and spatial cognitive tasks. *Psychophysiology, 27*(5), 528–543.

Davies, D. R., & Jones, D. M. (1975). The effects of noise and incentives upon attention in short-term memory. *British Journal of Psychology, 66*(1), 61–68.

Davies, D. R., Lang, L., & Shackleton, V. J. (1973). The effects of music and task difficulty on performance at a visual vigilance task. *British Journal of Psychology, 64*(3), 383–389.

Davies, D. R., & Tune, G. S. (1970). *Human vigilance performance*. London: Staples Press.

Davis, M. H., Marslen-Wilson, W. D., & Gaskell, M. G. (2002). Leading up the lexical garden path: Segmentation and ambiguity in spoken word recognition. *Journal of Experimental Psychology: Human Perception and Performance, 28*(1), 218–244.

Day, R. F., Lin, C. H., Huang, W. H., & Chuang, S. H. (2009). Effects of music tempo and task difficulty on multi-attribute decision-making: An eye-tracking approach. *Computers in Human Behavior, 25*(1), 130–143.

Dean, P., Redgrave, P., Sahjbzada, N., & Tsuji, K. (1986). Head and body movements produced by electrical stimulation of superior colliculus in rats: Effects of interruption of crossed tectoreticulospinal pathway. *Neuroscience, 19*(2), 367–380.

Deatherage, B. H. (1972). Auditory and other sensory forms of information presentation. In H. P. Van Cott & J. P. Kincaid (Eds.), *Human engineering guide to equipment design*, 123–160. Washington, DC: Government Printing Office.

de Beauport, E., & Diaz, A. S. (1996). *The three faces of mind. Developing your mental, emotional, and behavioral intelligences*. Wheaton, IL: Theosophical Publishing House.

de Boer, E., & Dreschler, W. A. (1987). Auditory psychophysics: Spectrotemporal representation of signals. *Annual Review of Psychology, 38*, 181–202.

DeDe, G., Caplan, D., Kemtes, K., & Waters, G. (2004). The relationship between age, verbal working memory, and language comprehension. *Psychology and Aging, 19*(4), 601–616.

Dehaene-Lambertz, G. (2000). Cerebral specialization for speech and non-speech stimuli in infants. *Journal of Cognitive Neuroscience, 12*(3), 449–460.

Dehaene-Lambertz, G., & Baillet, S. (1998). A phonological representation in the infant brain. *NeuroReport: For Rapid Communication of Neuroscience Research, 9*(8), 1885–1888.

Dehaene-Lambertz, G., & Pena, M. (2001). Electrophysiological evidence for automatic phonetic processing in neonates. *NeuroReport: For Rapid Communication of Neuroscience Research, 12*(14), 3155–3158.

Della Sala, S., Gray, C., Baddeley, A. D., Allamano, N., & Wilson, L. (1999). Pattern span: A tool for unwelding visuo-spatial memory. *Neuropsychologia, 37*, 1189–1199.

De Lucia, M., Camen, C., Clarke, S., & Murray, M. M. (2009). The role of actions in auditory object discrimination. *NeuroImage, 48*(2), 475–485.

Dennis, N. A., Hayes, S. M., Prince, S. E., Madden, D. J., Huettel, S. A., & Cabeza, R. (2008). Effects of aging on the neural correlates of successful item and source memory encoding. *Journal of Experimental Psychology: Learning, Memory, and Cognition, 34*(4), 791–808.

Derrick, W. L. (1988). Dimensions of operator workload. *Human Factors, 30*(1), 95–110.

Deutsch, D. (1970). Tones and numbers: Specificity of interference in immediate memory. *Science, 168*(3939), 1604–1605.

Deutsch, D. (Ed.). (1999). *The psychology of music*. San Diego, CA: Academic Press.

Deutsch, J., & Deutsch, D. (1963). Attention: Some theoretical considerations. *Psychological Review, 70*(1), 51–61.

Deyzac, E., Logie, R. H., & Denis, M. (2006). Visuospatial working memory and the processing of spatial descriptions. *British Journal of Psychology, 97*(2), 217–243.

Diamond, A. (2005). Attention-deficit disorder (attention-deficit/hyperactivity disorder without hyperactivity): A neurobiologically and behaviorally distinct disorder from attention-deficit/hyperactivity disorder (with hyperactivity). *Development and Psychopathology, 17*(3), 807–825.

Dick, F., Bates, E., Wulfeck, B., Utman, J. A., Dronkers, N., & Gernsbacher, M. A. (2001). Language deficits, localization, and grammar: Evidence for a distributive model of language breakdown in aphasic patients and neurologically intact individuals. *Psychological Review, 108*(4), 759–788.

Dick, F., Saygin, A. P., Galati, G., Pitzalis, S., Bentrovato, S., D'Amico, S., et al. (2007). What is involved and what is necessary for complex linguistic and nonlinguistic auditory processing: Evidence from functional magnetic resonance imaging and lesion data. *Journal of Cognitive Neuroscience, 19*(5), 799–816.

DiGirolamo, G. J., Kramer, A. F., Barad, V., Cepeda, N. J., Weissman, D. H., Milham, M. P., et al. (2001). General and task-specific frontal lobe recruitment in older adults during executive processes: A fMRI investigation of task-switching. *NeuroReport: For Rapid Communication of Neuroscience Research, 12*(9), 2065–2071.

Di Lollo, V., & Bischof, W. F. (1995). Inverse-intensity effect in duration of visible persistence. *Psychological Bulletin, 118*(2), 223–237.

Dingus, T. A., Hulse, M. C., & Barfield, W. (1998). Human-system interface issues in the design and use of Advanced Traveler Information Systems. In W. Barfield & T. A. Dingus (Eds.), *Human factors in intelligent transportation systems*, 359–395. Mahwah, NJ: Erlbaum.

Dingus, T. A., Hulse, M. C., Mollenhauer, M. A., & Fleischman, R. N. (1997). Effects of age, system experience, and navigation technique on driving with an Advanced Traveler Information System. *Human Factors, 39*(2), 177–199.

Dingus, T. A., McGehee, D. V., Manakkal, N., & Jahns, S. K. (1997). Human factors field evaluation of automotive headway maintenance/collision warning devices. *Human Factors, 39*(2), 216–229.

Dirks, D. D., Takayanagi, S., & Moshfegh, A. (2001). Effects of lexical factors on word recognition among normal-hearing and hearing-impaired listeners. *Journal of the American Academy of Audiology, 12*(5), 233–244.

DOD. (1981). *Military Standard: Human engineering design criteria for military systems, equipment and facilities* (No. MIL-STD 1472C): US Department of Defense.

Doll, T. J., & Folds, D. J. (1986). Auditory signals in military aircraft: Ergonomics principles versus practice. *Applied Ergonomics, 17*(4), 257–264.

Donchin, Y. M. D., & Seagull, F. J. P. (2002). The hostile environment of the intensive care unit. *Current Opinion in Critical Care August, 8*(4), 316–320.

Doupe, A. J., & Kuhl, P. K. (1999). Birdsong and human speech: Common themes and mechanisms. *Annual Review of Neuroscience, 22,* 567–631.

Drager, B., Jansen, A., Bruchmann, S., Forster, A. F., Pleger, B., Zwitserlood, P., et al. (2004). How does the brain accommodate to increased task difficulty in word finding? A functional MRI study. *NeuroImage, 23*(3), 1152–1160.

Driver, J., & Spence, C. J. (1994). Spatial synergies between auditory and visual attention. In C. Umilta & M. Moscovitch (Eds.), *Attention and performance 15: Conscious and nonconscious information processing,* 311–331. Cambridge, MA: MIT Press.

Dubno, J. R., Ahlstrom, J. B., & Horwitz, A. R. (2002). Spectral contributions to the benefit from spatial separation of speech and noise. *Journal of Speech, Language, and Hearing Research, 45*(6), 1297–1310.

Dubno, J. R., Lee, F. S., Matthews, L. J., Ahlstrom, J. B., Horwitz, A. R., & Mills, J. H. (2008). Longitudinal changes in speech recognition in older persons. *The Journal of the Acoustical Society of America, 123*(1), 462–475.

Duka, T., Tasker, R., & McGowan, J. F. (2000). The effects of 3-week estrogen hormone replacement on cognition in elderly healthy females. *Psychopharmacology, 149*(2), 129–139.

Dunbar, M., McGann, A., Mackintosh, M., & Lozito, S. (2001). *Re-examination of mixed media communication: The impact of voice, data-link, and mixed air traffic control environments on the flight deck.* (No. NASA/TM-2001–210919). Moffet Field, CA: NASA Ames Research Center.

Duncan, J. (1984). Selective attention and the organization of visual information. *Journal of Experimental Psychology: General, 113*(4), 501–517.

Dyson, B. J., & Ishfaq, F. (2008). Auditory memory can be object based. *Psychonomic Bulletin & Review, 15*(2), 409–412.

Ebert, T., Pantev, C., Wienbruch, C., Rockstroth, B., & Taub, E. (1995). Increased cortical representation of the fingers of the left hand in string players. *Science, 270*(5234), 305–307.

Eby, D. W., & Kostyniuk, L. P. (1999). An on-the-road comparison of in-vehicle navigation assistance systems. *Human Factors, 41*(2), 295–311.

Edworthy, J. (1985). Melodic contour and musical structure. In P. Howell, I. Cross, & R. West (Eds.), *Musicical structure and cognition,* 169–188. New York: Academic Press.

Edworthy, J., & Adams, A. (1996). *Warning design: A research prospective.* London: Taylor & Francis.

Edworthy, J., & Hellier, E. (2005). Fewer but better auditory alarms will improve patient safety. *Quality & Safety in Health Care, 14*(3), 212–215.

Edworthy, J., & Hellier, E. (2006a). Alarms and human behaviour: Implications for medical alarms. *British Journal of Anaesthesia, 97*(1), 12–17.

Edworthy, J., & Hellier, E. (2006b). Complex nonverbal auditory signals and speech warnings. In M. S. Wogalter (Ed.), *Handbook of warnings,* 199–220. Mahwah, NJ: Erlbaum.

Edworthy, J., Hellier, E., Aldrich, K., & Loxley, S. (2004). Designing trend-monitoring sounds for helicopters: Methodological issues and an application. *Journal of Experimental Psychology-Applied, 10*(4), 203–218.

Edworthy, J., Loxley, S., & Dennis, I. (1991). Improving auditory warning design: Relationship between warning sound parameters and perceived urgency. *Human Factors, 33*(2), 205–231.

Edworthy, J., & Stanton, N. (1995). A user-centred approach to the design and evaluation of auditory warning signals: I. Methodology. *Ergonomics, 38*(11), 2262–2280.

Edworthy, J., & Waring, H. (2006). The effects of music tempo and loudness level on treadmill exercise. *Ergonomics, 49*(15), 1597–1610.

Egan, J. P., Carterette, E. C., & Thwing, E. J. (1954). Some factors affecting multi-channel listening. *Journal of the Acoustical Society of America, 26,* 774–782.

Eggemeier, F. (1988). Properties of workload assessment techniques. In P. A. Hancock & N. Meshkati (Eds.), *Human mental workload. Advances in psychology, 52,* 41–62. Oxford, UK: North-Holland.

Eggemeier, F. T., Wilson, G. F., Kramer, A. F., & Damos, D. L. (1991). Workload assessment in multi-task environments. In D. L. Damos (Ed.), *Multiple-task performance,* 207–216. London: Taylor & Francis.

Eisenberg, L. S., Dirks, D. D., & Bell, T. S. (1995). Speech recognition in amplitude-modulated noise of listeners with normal and listeners with impaired hearing. *Journal of Speech and Hearing Research, 38*(1), 222–233.

Elliott, D., Carr, S., & Orme, D. (2005). The effect of motivational music on sub-maximal exercise. *European Journal of Sport Science, 5*(2), 97–106.

Elliott, D., Carr, S., & Savage, D. (2004). Effects of motivational music on work output and affective responses during sub-maximal cycling of a standardized perceived intensity. *Journal of Sport Behavior, 27*(2), 134–147.

Engle, R. W. (1974). The modality effect: Is precategorical acoustic storage responsible? *Journal of Experimental Psychology, 102*(5), 824–829.

Engle, R. W. (2001). What is working memory capacity? In H. L. Roediger III, J. S. Nairne, et al. (Eds.), *The nature of remembering: Essays in honor of Robert G. Crowder science conference series,* 297–314. Washington, DC: American Psychological Association.

Engle, R. W., Kane, M. J., & Tuholski, S. W. (1999). Individual differences in working memory capacity and what they tell us about controlled attention, general fluid intelligence, and functions of the prefrontal cortex. In A. Miyake & P. Shah (Eds.), *Models of working memory: Mechanisms of active maintenance and executive control,* 102–134. New York: Cambridge University Press.

Engle, R. W., Tuholski, S. W., Laughlin, J. E., & Conway, A. R. A. (1999). Working memory, short-term memory, and general fluid intelligence: A latent-variable approach. *Journal of Experimental Psychology: General, 128*(3), 309–331.

Erber, N. P. (1969). Interaction of audition and vision in the recognition of oral speech stimuli. *Journal of Speech and Hearing Research, 12*(2), 423–425.

Erskine, J. M., & Seymour, P. H. K. (2005). Proximal analysis of developmental dyslexia in adulthood: The cognitive mosaic model. *Journal of Educational Psychology, 97*(3), 406–424.

Etholm, B., & Belal, A. (1974). Senile changes in the middle ear joints. *Annals of Otolaryngology, 83,* 49–54.

Evans, G. W., Bullinger, M., & Hygge, S. (1998). Chronic noise exposure and physiological response: A prospective study of children living under environmental stress. *Psychological Science, 9*(1), 75–77.

Farrag, A.-k. F., Khedr, E. M., Abdel-Aleem, H., & Rageh, T. A. (2002). Effect of surgical menopause on cognitive functions. *Dementia and Geriatric Cognitive Disorders, 13*(3), 193–198.

Faustmann, A., Murdoch, B. E., Finnigan, S., & Copland, D. A. (2007). Effects of advanced age on the processing of semantic anomalies in adults: Evidence from even-related brain potentials. *Experimental Aging Research, 33*(4), 439–460.

Federmeier, K. D., Mclennan, D. B., De Ochoa, E., & Kutas, M. (2002). The impact of semantic memory organization and sentence context information on spoken language processing by younger and older adults: An ERP study. *Psychophysiology, 39*(2), 133–146.

Federmeier, K. D., van Petten, C., Schwartz, T. J., & Kutas, M. (2003). Sounds, words, sentences: Age-related changes across levels of language processing. *Psychology and Aging, 18*(4), 858–872.

Fedorenko, E., Gibson, E., & Rohde, D. (2006). The nature of working memory capacity in sentence comprehension: Evidence against domain-specific working memory resources. *Journal of Memory and Language, 54*(4), 541–553.

Feinberg, T. E., & Farah, M. J. (2006). A historical perspective on cognitive neuroscience. In M. J. Farah & T. E. Feinberg (Eds.), *Patient-based approaches to cognitive neuroscience* (2nd ed., 3–20). Cambridge, MA: MIT Press.

Ferguson, K. (2008). *The music of Pythagoras.* New York: Walker.

Fernald, A. (1989). Intonation and communicative intent in mothers' speech to infants: Is the melody the message? *Child Development, 60*(6), 1497–1510.

Festen, J. M., & Plomp, R. (1990). Effects of fluctuating noise and interfering speech on the speech-reception threshold for impaired and normal hearing. *Journal of the Acoustical Society of America, 88*(4), 1725–1736.

Filion, D. L., & Poje, A. B. (2003). Selective and nonselective attention effects on prepulse inhibition of startle: A comparison of task and no-task protocols. *Biological Psychology, 64*(3), 283–296.

Finger, S. (1994). *Origins of neuroscience: A history of explorations into brain function.* New York: Oxford University Press.

Finkelman, J. M., & Glass, D. C. (1970). Reappraisal of the relationship between noise and human performance by means of a subsidiary task measure. *Journal of Applied Psychology, 54*(3), 211–213.

Fischler, I. (1998). Attention and language. In R. Parasuraman (Ed.), *The attentive brain,* 381–399. Cambridge, MA: MIT Press.

Fishbach, A., Yeshurun, Y., & Nelken, I. (2003). Neural model for physiological responses to frequency and amplitude transitions uncovers topographical order in the auditory cortex. *Journal of Neurophysiology, 90*(6), 3663–3678.

Fisk, A. D., Rogers, W. A., Charness, N., Czaja, S. J., & Sharit, J. (Eds.). (2009). *Designing for older adults: Principles and creative human factors approaches* (2nd ed.). Boca Raton, FL: CRC Press.

Flach, J. M., & Kuperman, G. (2001). The human capacity for work: A (biased) historical perspective. In P. A. Hancock & P. A. Desmond (Eds.), *Stress, Workload, and Fatigue.* Mahwah, NJ: Lawrence Erlbaum.

Fletcher, H. (1953). *Speech and hearing in communication* (2nd ed.). New York: Van Nostrand.

Floel, A., Poeppel, D., Buffalo, E. A., Braun, A., Wu, C. W.-H., Seo, H.-J., et al. (2004). Prefrontal cortex asymmetry for memory encoding of words and abstract shapes. *Cerebral Cortex, 14*(4), 404–409.

Flowers, J. H., Buhman, D. C., & Turnage, K. D. (1997). Cross-modal equivalence of visual and auditory scatterplots for exploring bivariate data samples. *Human Factors, 39*(3), 341–351.

Flowers, J. F., Buhman, D., C., & Turnage, K., D. (2005). Data sonification from the desktop: Should sound be part of standard data analysis software? *ACM Transactions on Applied Perception, 2*(4), 467–472.

Flowers, J. H., & Hauer, T. A. (1995). Musical versus visual graphs: Cross-modal equivalence in perception of time series data. *Human Factors, 37*(3), 553–569.

Fodor, J. A. (1983). *The modularity of mind: An essay on faculty psychology.* Cambridge, MA: MIT Press.

Fowler, B. (1994). P300 as a measure of workload during a simulated aircraft landing task. *Human Factors, 36*(4), 670–683.

Fox, J. G. (1971). Background music and industrial efficiency—A review. *Applied Ergonomics, 2*(2), 70–73.

Fozard, J. L. (1990). Vision and hearing in aging. In J. E. Birren & K. W. Schaie (Eds.), *Handbook of the psychology of aging* (3rd ed., 150–170). San Diego, CA: Academic Press.

Fozard, J. L., & Gordon-Salant, S. (2001). Changes in vision and hearing with aging. In J. E. Birren (Ed.), *Handbook of the psychology of aging*, 241–266. San Diego, CA: Academic Press.

Francis, A. L., & Nusbaum, H. C. (2009). Effects of intelligibility on working memory demand for speech perception. *Attention, Perception, & Psychophysics, 71*(6), 1360–1374.

Friederici, A. D. (1999). The neurobiology of language comprehension. In *Language comprehension: A biological perspective* (2nd ed.). Berlin: Springer.

Friederici, A. D., & Alter, K. (2004). Lateralization of auditory language functions: A dynamic dual pathway model. *Brain and Language Language, 89*(2), 267–276.

Friedmann, N., & Gvion, A. (2003). Sentence comprehension and working memory limitation in aphasia: A dissociation between semantic-syntatic and phonological reactivation. *Brain and Language, 86*(1), 23–39.

Frisina, D. R., & Frisina, R. D. (1997). Speech recognition in noise and presbycusis: Relations to possible neural mechanisms. *Hearing Research, 106*(1–2), 95–104.

Fucci, D., Reynolds, M. E., Bettagere, R., & Gonzales, M. D. (1995). Synthetic speech intelligibility under several experimental conditions. *AAC: Augmentative and Alternative Communication, 11*(2), 113–117.

Furnham, A., & Bradley, A. (1997). Music while you work: The differential distraction of background music on the cognitive test performance of introverts and extraverts. *Applied Cognitive Psychology, 11*(5), 445–455.

Furnham, A., & Stanley, A. (2003). The influence of vocal and instrumental background music on the cognitive performance of introverts and extraverts. In S. P. Shohov (Ed.), *Topics in cognitive psychology*, 151–167. Hauppauge, NY: NOVA Science.

Furnham, A., & Stephenson, R. (2007). Musical distracters, personality type and cognitive performance in school children. *Psychology of Music, 35*(3), 403–420.

Furnham, A., & Strbac, L. (2002). Music is as distracting as noise: The differential distraction of background music and noise on the cognitive test performance of introverts and extraverts. *Ergonomics, 45*(3), 203–217.

Furukawa, H., Baldwin, C. L., & Carpenter, E. M. (2004). Supporting drivers' area-learning task with visual geo-centered and auditory ego-centered guidance: Interference or improved performance? In D. A. Vincenzi, M. Mouloua, & P. A. Hancock (Eds.), *Human performance, situation awareness and automation: Current research and trends, HPSAA II*, 124–129. Mahwah, NJ: Erlbaum.

Gage, N., & Hickok, G. (2005). Multiregional cell assemblies, temporal binding and the representation of conceptual knowledge in cortex: A modern theory by a 'classical' neurologist, Carl Wernicke. *Cortex, 41*(6), 823–832.

Ganong, W. F. (1980). Phonetic categorization in auditory word perception. *Journal of Experimental Psychology: Human Perception and Performance, 6*(1), 110–125.

Garden, S., Cornoldi, C., & Logie, R. H. (2002). Visuo-spatial working memory in navigation. *Applied Cognitive Psychology, 16*(1), 35–50.

Garvey, W. D., & Knowles, W. B. (1954). Response time patterns associated with various display-control relationships. *Journal of Experimental Psychology, 47*, 315–322.

Gates, G. A., Cooper, J. C., Kannel, W. B., & Miller, N. J. (1990). Hearing in the elderly: The Framingham Cohort, 1983–1985. Part 1. Basic audiometric test results. *Ear and Hearing, 11*(4), 247–256.

Gathercole, S. E. (1994). Neuropsychology and working memory: A review. *Neuropsychology, 8*(4), 494–505.

Gaver, W. W. (1986). Auditory icons: Using sound in computer interfaces. *Human-Computer Interaction, 2*(2), 167–177.

Gaver, W. W. (1989). The SonicFinder: An interface that uses auditory icons. *Human-Computer Interaction, 4*(1), 67–94.

Gaver, W. W. (1993). How do we hear in the world? Explorations in ecological acoustics. *Ecological Psychology, 5*(4), 285–313.

Gawron, V. J. (1982). Performance effects of noise intensity, psychological set, and task type and complexity. *Human Factors, 24*(2), 225–242.

Gay, T. (1978). Effect of speaking rate on vowel formant movements. *Journal of the Acoustical Society of America, 63*(1), 223–230.

Gazzaniga, M. S. (2000). Cerebral specialization and interhemispheric communication: Does the corpus callosum enable the human condition? *Brain: A Journal of Neurology, 123*(7), 1293–1326.

Gazzaniga, M. S., & Sperry, R. W. (1967). Language after section of the cerebral commissures. *Brain: A Journal of Neurology, 90*(1), 131–148.

Geen, R. G., McCown, E. J., & Broyles, J. W. (1985). Effects of noise on sensitivity of introverts and extraverts to signals in a vigilance task. *Personality and Individual Differences, 6*(2), 237–241.

Gevins, A., Leong, H., Du, R., Smith, M. E., Le, J., DuRousseau, D., et al. (1995). Towards measurement of brain function in operational environments. *Biological Psychology, 40*(1–2), 169–186.

Giard, M. H., Lavikainen, J., Reinikainen, K., Perrin, F., & Naatanen, R. (1995). Separate representation of stimulus frequency, intensity, and duration in auditory sensory memory: An event-related potential and dipole-model analysis. *Journal of Cognitive Neuroscience, 7*(2), 133–143.

Gibson, J. J. (1966). *The senses considered as perceptual systems.* Oxford, UK: Houghton Mifflin.

Gifford, R. H., Bacon, S. P., & Williams, E. J. (2007). An examination of speech recognition in a modulated background and of forward masking in younger and older listeners. *Journal of Speech, Language, and Hearing Research, 50*(4), 857–864.

Glanzer, M., & Cunitz, A. R. (1966). Two storage mechanisms in free recall. *Journal of Verbal Learning & Verbal Behavior 5*(4), 351–360. UK: Elsevier Science.

Glassbrenner, D. (2005). *Driver cell phone use in 2005—Overall results* (No. DOT HS 809 967). Washington, DC: National Highway Traffic Safety Administraion, Department of Transportation.

Golding, M., Carter, N., Mitchell, P., & Hood, L. J. (2004). Prevalence of central auditory processing (CAP) abnormality in an older Australian population: The Blue Mountains Hearing Study. *Journal of the American Academy of Audiology, 15*, 633–642.

Goldman-Eisler, F. (1956). The determinants of the rate of speech output and their mutual relations. *Journal of Psychosomatic Research, 1*, 137–143.

Goldstein, E. B. (2002). *Sensation and perception* (6th ed.). Pacific Grove, CA: Wadsworth.

Goldstein, J. L. (1973). An optimum processor theory for the central formation of the pitch of complex tones. *Journal of the Acoustical Society of America, 54*(6), 1496–1516.

Gonzalez-Crussi, F. (1989). *The five senses.* San Diego, CA: Harcourt Brace Jovanovich.

Goodman, M. J., Bents, F. D., Tijerina, L., Wierwille, W. W., Lerner, N., & Benel, D. (1997). *An investigation of the safety implications of wireless communications in vehicles* (No. DOT HS 808-635). Washington, DC: U.S. Department of Transportation, National Highway Traffic Safety Administration (NHTSA).

Gopher, D., & Donchin, E. (1986). Workload—An examination of the concept. In K. R. Boff, L. Kaufman, & J. P. Thomas (Eds.), *Handbook of perception and human performance. Vol. 2. Cognitive processes and performance,* 41-41–41-49. New York: Wiley.

Gordon-Salant, S. (1986). Effects of aging on response criteria in speech-recognition tasks. *Journal of Speech and Hearing Research, 29*(2), 155–162.

Gordon-Salant, S. (2005). Hearing loss and aging: New research findings and clinical applications. *Journal of Rehabilitation Research and Development, 42*(2), 9–24.

Gordon-Salant, S., & Fitzgibbons, P. J. (1995a). Comparing recognition of distorted speech using an equivalent signal-to-noise ratio index. *Journal of Speech and Hearing Research, 38*(3), 706–713.

Gordon-Salant, S., & Fitzgibbons, P. J. (1995b). Recognition of multiply degraded speech by young and elderly listeners. *Journal of Speech and Hearing Research, 38*(5), 1150–1156.

Gordon-Salant, S., & Fitzgibbons, P. J. (1997). Selected cognitive factors and speech recognition performance among young and elderly listeners. *Journal of Speech, Language, and Hearing Research, 40*(2), 423–431.

Gorges, M., Markewitz, B. A., & Westenskow, D. R. (2009). Improving alarm performance in the medical intensive care unit using delays and clinical context. *Anesthesia and Analgesia, 108*(5), 1546–1552.

Grady, C. L. (1998). Brain imaging and age-related changes in cognition. *Experimental Gerontology, 33*(7–8), 661–673.

Grady, C. L., Maisog, J. M., Horwitz, B., Ungerleider, L. G., Mentis, M. J., Salerno, J. A., et al. (1994). Age-related changes in cortical blood flow activation during visual processing of faces and location. *Journal of Neuroscience, 14*(3, Pt. 2), 1450–1462.

Graham, F. K. (1975). The more or less startling effects of weak prestimulation. *Psychophysiology, 12*(3), 238–248.

Graham, R. (1999). Use of auditory icons as emergency warnings: Evaluation within a vehicle collision avoidance application. *Ergonomics, 42*(9), 1233–1248.

Grant, H. M., Bredahl, L. C., Clay, J., Ferrie, J., Groves, J. E., McDorman, T. A., et al. (1998). Context-dependent memory for meaningful material: Information for students. *Applied Cognitive Psychology, 12*(6), 617–623.

Green, K. B., Pasternack, B. S., & Shore, R. E. (1982). Effects of aircraft noise on reading ability of school-age children. *Archives of Environmental Health, 37*(1), 24.

Green, P. A. (1992). *American human factors research on in-vehicle navigation systems* (Technical Report No. UMTRI-92–47). Ann Arbor, MI: University of Michigan Transportation Research Institute.

Greenwood, P. M. (2000). The frontal aging hypothesis evaluated. *Journal of the International Neuropsychological Society, 6*(6), 705–726.

Greenwood, P. M. (2007). Functional plasticity in cognitive aging: Review and hypothesis. *Neuropsychology, 21*(6), 657–673.

Greenwood, P. M., & Parasuraman, R. (in press). *Nurturing the older brain and mind.* Cambridge, MA: MIT Press.

Greenwood, P. M., Parasuraman, R., & Haxby, J. V. (1993). Changes in visuospatial attention over the adult lifespan. *Neuropsychologia, 31*(5), 471–485.

Gregory, S. W., Green, B. E., Carrothers, R. M., Dagan, K. A., & Webster, S. W. (2001). Verifying the primacy of voice fundamental frequency in social status accommodation. *Language & Communication, 21*(1), 37–60.

Grier, R. A., Warm, J. S., Dember, W. N., Matthews, G., Galinsky, T. L., Szalma, J. L., et al. (2003). The vigilance decrement reflects limitations in effortful attention, not mindlessness. *Human Factors, 45*(3), 349–359.

Griffiths, T. D., Buechel, C., Frackowiak, R. S. J., & Patterson, R. D. (1998). Analysis of temporal structure in sound by the human brain. *Nature Neuroscience, 1*(5), 422–427.

Griffiths, T. D., & Warren, J. D. (2004). What is an auditory object? *Nature Reviews Neuroscience, 5*, 887–892.

Grillon, C., Ameli, R., Charney, D. S., Krystal, J. H., & Braff, D. (1992). Startle gating deficits occur across prepulse intensities in schizophrenic patients. *Biological Psychiatry, 32*(10), 939–943.

Grodzinsky, Y., & Friederici, A. D. (2006). Neuroimaging of syntax and syntactic processing. *Current Opinion in Neurobiology, 16*(2), 240–246.

Grommes, P., & Dietrich, R. (2002). Coherence in operating room team and cockpit communication: A psycholinguistic contribution to applied linguistics. In J. E. Alatis, H. E. Hamilton, & A.-H. Tan (Eds.), *Linguistics, language, and the professions: Education, journalism, law, medicine, and technology*, 190–219. Washington, DC: Georgetown University Press.

Grosjean, F. (1980). Spoken word recognition processes and the gating paradigm. *Perception & Psychophysics, 28*(4), 267–283.

Grosjean, F. (1985). The recognition of words after their acoustic offset: Evidence and implications. *Perception & Psychophysics, 38*(4), 299–310.

Grosjean, F., & Gee, J. P. (1987). Prosodic structure and spoken word recognition. *Cognition, 25*, 135–155.

Gumenyuk, V., Korzyukov, O., Alho, K., Escera, C., & Naatanen, R. (2004). Effects of auditory distraction on electrophysiological brain activity and performance in children aged 8–13 years. *Psychophysiology, 41*(1), 30–36.

Haarmann, H. J., Just, M. A., & Carpenter, P. A. (1997). Aphasic sentence comprehension as a resource deficit: A computational approach. *Brain and Language, 59*(1), 76–120.

Haas, E., & Edworthy, J. (2006). An introduction to auditory warnings and alarms. In M. S. Wogalter (Ed.), *Handbook of warnings*, 189–198. Mahwah, NJ: Erlbaum.

Haas, E. C., & Casali, J. G. (1995). Perceived urgency of and response time to multi-tone and frequency-modulated warning signals in broadband noise. *Ergonomics, 38*(11), 2313–2326.

Haas, E. C., & Edworthy, J. (1996, August). Designing urgency into auditory warnings using pitch, speed, and loudness. *Computing and Control Engineering Journal*, 193–198.

Hagoort, P. (2003). How the brain solves the binding problem for language: A neurocomputational model of syntactic processing. *NeuroImage, 20*(Suppl. 1), S18–S29.

Hagoort, P. (2005). On Broca, brain, and binding: A new framework. *Trends in Cognitive Sciences, 9*(9), 416–423.

Haigney, D. E., Taylor, R. G., & Westerman, S. J. (2000). Concurrent mobile (cellular) phone use and driving performance: Task demand characteristics and compensatory processes. *Transportation Research Part F: Traffic Psychology and Behaviour, 3*(3), 113–121.

Haines, M. M., Stansfeld, S. A., Head, J., & Job, R. F. S. (2002). Multilevel modelling of aircraft noise on performance tests in schools around Heathrow Airport London. *Journal of Epidemiology and Community Health, 56*(2), 139–144.

Hale, T. S., Zaidel, E., McGough, J. J., Phillips, J. M., & McCracken, J. T. (2006). Atypical brain laterality in adults with ADHD during dichotic listening for emotional intonation and words. *Neuropsychologia, 44*(6), 896–904.

Hall, D. A., Hart, H. C., & Johnsrude, I. S. (2003). Relationships between human auditory cortical structure and function. *Audiology & Neuro-Otology, 8*(1), 1–18.

Hamilton, B. E. (1999). Helicopter human factors. In D. J. Garland, J. A. Wise & V. D. Hopkin (Eds.), *Handbook of aviation human factors*, 405–428. Mahwah, NJ: Lawrence Erlbaum.

Hancock, P. A., & Caird, J. K. (1993). Experimental evaluation of a model of mental workload. *Human Factors, 35*(3), 413–429.

Hancock, P. A., & Chignell, M. H. (1988). Mental workload dynamics in adaptive interface design. *IEEE Transactions on Systems, Man, and Cybernetics, 18*(4), 647–658.

Hancock, P. A., & Desmond, P. A. (2001). *Stress, workload, and fatigue.* Mahwah, NJ: Lawrence Erlbaum.

Hancock, P. A., Lesch, M., & Simmons, L. (2003). The distraction effects of phone use during a crucial driving maneuver. *Accident Analysis and Prevention, 35*(4), 501–514.

Hancock, P. A., Parasuraman, R., & Byrne, E. A. (1996). Driver-centered issues in advanced automation for motor vehicles. In R. Parasuraman & M. Mouloua (Eds.) *Automation and human performance: Theories and applications*, 337–364. Mahwah, NJ: Lawrence Erlbaum.

Harding, C., & Souleyrette, R. R. (2010). Investigating the use of 3D graphics, haptics (touch), and sound for highway location planning. *Computer-Aided Civil and Infrastructure Engineering, 25*(1), 20–38.

Hargus, S. E., & Gordon-Salant, S. (1995). Accuracy of Speech Intelligibility Index predictions for noise-masked young listeners with normal hearing and for elderly listeners with hearing impairment. *Journal of Speech and Hearing Research, 38*(1), 234–243.

Harms, L. (1986). Drivers' attentional responses to environmental variations: A dual task real traffic study. In A. G. Gale et al. (Eds.), *Vision in vehicles*, 131–138. Amsterdam: North Holland/Elsevier Science.

Harms, L. (1991). Variation in drivers' cognitive load. Effects of driving through village areas and rural junctions. *Ergonomics, 34*(2), 151–160.

Harms, L., & Patten, C. (2003). Peripheral detection as a measure of driver distraction. A study of memory-based versus system-based navigation in a built-up area. *Transportation Research Part F: Traffic Psychology and Behaviour, 6*(1), 23–36.

Hart, S. G., & Staveland, L. E. (1988). Development of NASA-TLX (Task Load Index): Results of empirical and theoretical research. In P. A. Hancock & N. Meshkati (Eds.), *Human mental workload*, 239–250. Amsterdam: North Holland Press.

Hartley, A. A., Speer, N. K., Jonides, J., Reuter-Lorenz, P. A., & Smith, E. E. (2001). Is the dissociability of working memory systems for name identity, visual-object identity, and spatial location maintained in old age? *Neuropsychology, 15*(1), 3–17.

Hartmann, W. M. (1995). The physical description of signals. In B. C. J. Moore (Ed.), *Hearing* (2nd ed., 1–40). San Diego, CA: Academic Press.

Hasher, L., Quig, M. B., & May, C. P. (1997). Inhibitory control over no-longer-relevant information: Adult age differences. *Memory & Cognition, 25*(3), 286–295.

Hasher, L., & Zacks, R. T. (1979). Automatic and effortful processes in memory. *Journal of Experimental Psychology: General, 108*(3), 356–388.

Hass, E. C., & Edworthy, J. (1996, August). Designing urgency into auditory warnings using pitch, speed, and loudness. *Computing and Control Engineering Journal*, 193–198.

Hauser, M. D., Chomsky, N., & Fitch, W. T. (2002). The faculty of language: What is it, who has it, and how did it evolve? *Science, 298*(5598), 1569–1579.

Hawkins, F. H. (1987). *Human factors in flight*. Burlington, VT: Ashgate.

Hawkins, F. H., & Orlady, H. W. (1993). *Human factors in flight* (2nd ed.). Aldershot, UK: Ashgate.

Hawkins, H., & Presson, J. (1986). Auditory information processing. In K. R. Boff, L. Kaufman, & J. P. Thomas (Eds.), *Handbook of perception and human performance. Vol. 2: Cognitive processes and performance*, 26-21–26-64. New York: Wiley.

Hazeltine, E., Ruthruff, E., & Remington, R. W. (2006). The role of input and output modality pairings in dual-task performance: Evidence for content-dependent central interference. *Cognitive Psychology, 52*(4), 291–345.

Healy, A. F., & McNamara, D. S. (1996). Verbal learning and memory: Does the modal model still work? *Annual Review of Psychology, 47*, 143–172.

Helfer, K. S., & Wilber, L. A. (1990). Hearing loss, aging, and speech perception in reverberation and noise. *Journal of Speech and Hearing Research, 33*(1), 149–155.

Helleberg, J. R., & Wickens, C. D. (2003). Effects of data-link modality and display redundancy on pilot performance: An attentional perspective. *International Journal of Aviation Psychology, 13*(3), 189–210.

Hellier, E., & Edworthy, J. (1999a). The design and validation of attensons for a high workload environment. In N. A. Stanton & J. Edworthy (Eds.), *Human factors in auditory warnings*, 283–303. Aldershot, UK: Ashgate.

Hellier, E., & Edworthy, J. (1999b). On using psychophysical techniques to achieve urgency mapping in auditory warnings. *Applied Ergonomics, 30*, 167–171.

Hellier, E., Edworthy, J., Weedon, B., Walters, K., & Adams, A. (2002). The perceived urgency of speech warnings: Semantics versus acoustics. *Human Factors, 44*(1), 1–17.

Hellier, E., Wright, D. B., Edworthy, J., & Newstead, S. (2000). On the stability of the arousal strength of warning signal words. *Applied Cognitive Psychology, 14*(6), 577–592.

Hellier, E. J., Edworthy, J., & Dennis, I. (1993). Improving auditory warning design: Quantifying and predicting the effects of different warning parameters on perceived urgency. *Human Factors, 35*(4), 693–706.

Helmholtz, H. E. F. v. (1930/1863). *The sensations of tone: As a physiological basis for the theory of music.* New York: Longmans, Green.

Hendy, K. C., Hamilton, K. M., & Landry, L. N. (1993). Measuring subjective workload: When is one scale better than many? *Human Factors, 35*(4), 579–601.

Henry, L. A. (2001). How does the severity of a learning disability affect working memory performance? *Memory, 9*(4), 233–247.

Herrington, J. D., & Capella, L. M. (1996). Effects of music in service environments: A field study. *The Journal of Services Marketing, 10*(2), 26.

Hewlett, P., Smith, A., & Lucas, E. (2009). Grazing, cognitive performance and mood. *Appetite, 52*(1), 245–248.

Hill, S. G., Iavecchia, H. P., Byers, J. C., Bittner, A. C., Zaklad, A. L., & Christ, R. E. (1992). Comparison of four subjective workload rating scales. *Human Factors, 34*(4), 429–439.

Hillyard, S. A., Hink, R. F., Schwent, V. L., & Picton, T. W. (1973). Electrical signs of selective attention in the human brain. *Science, 182*(4108), 177–179.

Hiscock, M., Inch, R., & Kinsbourne, M. (1999). Allocation of attention in dichotic listening: Differential effects on the detection and localization of signals. *Neuropsychology, 13*(3), 404–414.

Hiscock, M., Lin, J., & Kinsbourne, M. (1996). Shifts in children's ear asymmetry during verbal and nonverbal auditory-visual association tasks: A "virtual stimulus" effect. *Cortex, 32*(2), 367–374.

Hitchcock, E. M., Warm, J. S., Matthews, G., Dember, W. N., Shear, P. K., Tripp, L. D., et al. (2003). Automation cueing modulates cerebral blood flow and vigilance in a simulated air traffic control task. *Theoretical Issues in Ergonomics Science, 4*(1–2), 89–112.

Ho, C., & Spence, C. (2005). Assessing the effectiveness of various auditory cues in capturing a driver's visual attention. *Journal of Experimental Psychology: Applied, 11*(3), 157–174.

Hockey, G. R. J. (1997). Compensatory control in the regulation of human performance under stress and high workload: A cognitive-energetical framework. *Biological Psychology, 45*(1–3), 73–93.

Hockey, G. R. J., & Hamilton, P. (1970). Arousal and information selection in short-term memory. *Nature, 226*(5248), 866–867.

Hoffmann, E. R., & Macdonald, W. A. (1980). Short-term retention of traffic turn restriction signs. *Human Factors, 22*(2), 241–251.

Holt, L. L., & Lotto, A. J. (2002). Behavioral examinations of the level of auditory processing of speech context effects. *Hearing Research, 167*(1–2), 156–169.

Hong, S. H., Cho, Y.-S., Chung, W.-H., Koh, S.-J., Seo, I.-s., & Woo, H.-C. (2001). Changes in external ear resonance after ventilation tube (Grommet) insertion in children with otitis media with effusion. *International Journal of Pediatric Otorhinolaryngology, 58*(2), 147–152.

Horrey, W. J., & Wickens, C. D. (2002). *Driving and side task performance: The effects of display clutter, separation, and modality* (Technical Report No. AHFD-02-13/GM-02-2). Savoy, IL: Aviation Human Factors, Division Institute of Aviation.

Horrey, W. J., & Wickens, C. D. (2003, July). *Multiple resources modeling of task interference in vehicle control, hazard awareness and in-vehicle task performance.* Paper presented at the International Driving Symposium on Human Factors in Driver Assessment, Training, and Vehicle Design, Park City, UT.

Horrey, W. J., & Wickens, C. D. (2004). Driving and side task performance: The effects of display clutter, separation, and modality. *Human Factors, 46*(4), 611–624.

Horton, A. M. J. (2000). Prediction of brain injury severity by subscales of the Alternative Impairment Index. *International Journal of Neuroscience, 105*(1), 97–100.

Hubel, D. H. (1960). Single unit activity in lateral geniculate body and optic tract of unrestrained cats. *Journal of Physiology, 150*, 91–104.

Hubel, D. H., & Wiesel, T. N. (1959). Receptive fields of single neurones in the cat's striate cortex. *Journal of Physiology, 148*, 574–591.

Hubel, D. H., & Wiesel, T. N. (1962). Receptive fields, binocular interaction, and functional architecture in the cat's visual cortex. *Journal of Physiology, 160*, 106–154.

Hudson, A. I., & Holbrook, A. (1982). Fundamental frequency characteristics of young Black adults: Spontaneous speaking and oral reading. *Journal of Speech and Hearing Research, 25*(1), 25–28.

Hugdahl, K., Bodner, T., Weiss, E., & Benke, T. (2003). Dichotic listening performance and frontal lobe function. *Cognitive Brain Research, 16*(1), 58–65.

Hulme, C., & Snowling, M. (1988). The classification of children with reading difficulties. *Developmental Medicine and Child Neurology, 30*(3), 398–402.

Humes, L. E., & Christopherson, L. (1991). Speech identification difficulties of hearing-impaired elderly persons: The contributions of auditory processing deficits. *Journal of Speech and Hearing Research, 34*(3), 686–693.

Humes, L. E., Nelson, K. J., & Pisoni, D. B. (1991). Recognition of synthetic speech by hearing-impaired elderly listeners. *Journal of Speech and Hearing Research, 34*(5), 1180–1184.

Humes, L. E., Watson, B. U., Christensen, L. A., Cokely, C. G., Halling, D. C., & Lee, L. (1994). Factors associated with individual differences in clinical measures of speech recognition among the elderly. *Journal of Speech and Hearing Research, 37*(2), 465–474.

Hunt, M. (1993). *The story of psychology.* New York: Anchor Books, Random House.

Hunt, R. R., & Ellis, H. C. (2004). *Fundamentals of cognitive psychology.* New York: McGraw-Hill.

Hyde, K. L., & Peretz, I. (2004). Brains that are out of tune but in time. *Psychological Science, 15*(5), 356–360.

Hygge, S., Evans, G. W., & Bullinger, M. (2002). A prospective study of some effects of aircraft noise on cognitive performance in schoolchildren. *Psychological Science, 13*(5), 469–474.

Hygge, S., & Knez, I. (2001). Effects of noise, heat and indoor lighting on cognitive performance and self-reported affect. *Journal of Environmental Psychology, 21*(3), 291–299.

IEC. (2006). *Medical electrical equipment — Part 1-8: General requirements for basic safety and essential performance — Collateral standard: General requirements, tests and guidance for alarm systems in medical electrical equipment and medical electrical systems* (No. IEC 60601-1-8:2006): International Electrotechnical Commission.

Imhoff, M., & Kuhls, S. (2006). Alarm algorithms in critical care monitoring. *Anesthesia and Analgesia, 102*(5), 1525–1537.

Isreal, J. B., Chesney, G. L., Wickens, C. D., & Donchin, E. (1980). P300 and tracking difficulty: Evidence for multiple resources in dual-task performance. *Psychophysiology, 17*(3), 259–273.

Isreal, J. B., Wickens, C. D., Chesney, G. L., & Donchin, E. (1980). The event-related brain potential as an index of display-monitoring workload. *Human Factors, 22*(2), 211–224.

Issac, A. R., & Ruitenberg, B. (1999). *Air traffic control: Human performance factors.* Brookfield, VT: Ashgate.

Itoh, K., Miyazaki, K. I., & Nakada, T. (2003). Ear advantage and consonance of dichotic pitch intervals in absolute-pitch possessors. *Brain and Cognition, 53*(3), 464–471.

Jackendoff, R. (1999). The representational structures of the language faculty and their inter-actions. In C. Brown & P. Hagoort (Eds.), *The neurocognition of language*, 37–79. Oxford: Oxford University Press.

Jackendoff, R. (2007). A parallel architecture perspective on language processing. *Brain Research, 1146*, 2–22.

Jacko, J. A. (1996). The identifiability of auditory icons for use in educational software for children. *Interacting with Computers, 8*(2), 121–133.

Jackson, P. G. (1998). In search of better route guidance instructions. *Ergonomics, 41*(7), 1000–1013.

Jaeger, J. J., Lockwood, A. H., Van Valin, R. D., Jr., Kemmerer, D. L., Murphy, B. W., & Wack, D. S. (1998). Sex differences in brain regions activated by grammatical and read-ing tasks. *NeuroReport: For Rapid Communication of Neuroscience Research, 9*(12), 2803–2807.

James, W. (1918). *The principles of psychology*. New York: Holt. (Originally published 1890)

Jamesdaniel, S., Salvi, R., & Coling, D. (2009). Auditory proteomics: Methods, accomplish-ments and challenges. *Brain Research, 1277*, 24–36.

Janse, E. (2004). Word perception in fast speech: Artificially time-compressed vs. naturally produced fast speech. *Speech Communication, 42*(2), 155–173.

Jefferies, L. N., Smilek, D., Eich, E., & Enns, J. T. (2008). Emotional valence and arousal interact in attentional control. *Psychological Science, 19*(3), 290–295.

Jenstad, L. M., & Souza, P. E. (2007). Temporal envelope changes of compression and speech rate: Combined effects on recognition for older adults. *Journal of Speech, Language, and Hearing Research, 50*(5), 1123–1138.

Jentzsch, I., Leuthold, H., & Ulrich, R. (2007). Decomposing sources of response slowing in the PRP paradigm. *Journal of Experimental Psychology: Human Perception and Performance, 33*(3), 610–626.

Jermakian, J. S. (2010). *Crash avoidance potential of four passenger vehicle technologies*. Arlington, VA: Insurance Institute for Highway Safety.

Joanisse, M. F., & Seidenberg, M. S. (1999). Impairments in verb morphology after brain injury: A connectionist model. *Proceedings of the National Academy of Science of the United States of America, 96*, 7592–7597.

Joanisse, M. F., & Seidenberg, M. S. (2003). Phonology and syntax in specific language impairment: Evidence from a connectionist model. *Brain and Language, 86*(1), 40–56.

Joanisse, M. F., & Seidenberg, M. S. (2005). Imaging the past: Neural activation in fron-tal and temporal regions during regular and irregular past-tense processing. *Cognitive, Affective, & Behavioral Neuroscience, 5*(3), 282–296.

Joint Planning and Development Office. (2010). *Next generation air transportation system international strategy* (No. 09-013). Washington, DC: Joint Planning and Development Office (JPDO).

Jolicoeur, P. (1999). Dual-task interference and visual encoding. *Journal of Experimental Psychology: Human Perception and Performance, 25*(3), 596–616.

Jones, D. (1999). The cognitive psychology of auditory distraction: The 1997 BPS Broadbent Lecture. *British Journal of Psychology, 90*(2), 167–187.

Jones, D. M., & Morris, N. (1992). Irrelevant speech and cognition. In D. M. Jones & A. P. Smith (Eds.), *Handbook of human performance. Vol. 1: The physical environment*, 29–53. London: Academic Press.

Jones, D. M. (1983). Loud noise and levels of control: A study of serial reaction. In G. Rossi (Ed.), *Fifth International Congress on Noise as a Public Health Problem*, 809–917. Turin: Minerva.

Jones, L. B., Rothbart, M. K., & Posner, M. I. (2003). Development of executive attention in preschool children. *Developmental Science, 6*(5), 498–504.

Jones, M. R., Johnston, H. M., & Puente, J. (2006). Effects of auditory pattern structure on anticipatory and reactive attending. *Cognitive Psychology, 53*(1), 59–96.

Jones, M. R., Moynihan, H., MacKenzie, N., & Puente, J. (2002). Temporal aspects of stimulus-driven attending in dynamic arrays. *Psychological Science, 13*(4), 313–319.

Jonides, J., Lacey, S. C., & Nee, D. E. (2005). Processes of working memory in mind and brain. *Current Directions in Psychological Science, 14*(1), 2–5.

Jonsson, J. E., & Ricks, W. R. (1995, August). *Cognitive models of pilot categorization and prioritization of flight-deck information.* Hampton, VA: National Aeronautics and Space Administration.

Ju, M., & Luce, P. A. (2006). Representational specificity of within-category phonetic variation in the long-term mental lexicon. *Journal of Experimental Psychology: Human Perception and Performance, 32*(1), 120–138.

Jusczyk, P. W. (2003). The role of speech perception capacities in early language acquisition. In M. T. Banich & M. Mack (Eds.), *Mind, brain, and language: Multidisciplinary perspectives,* 61–83. Mahwah, NJ: Erlbaum.

Jusczyk, P. W., & Luce, P. A. (2002). Speech perception. In H. Pashler & S. Yantis (Eds.), *Steven's handbook of experimental psychology* (3rd ed.), 493–536. New York: Wiley.

Just, M. A., & Carpenter, P. A. (1992). A capacity theory of comprehension: Individual differences in working memory. *Psychological Review, 99*(1), 122–149.

Just, M. A., Carpenter, P. A., & Keller, T. A. (1996). The capacity theory of comprehension: New frontiers of evidence and arguments. *Psychological Review, 103*(4), 773–780.

Just, M. A., Carpenter, P. A., Keller, T. A., Eddy, W. F., Rep, M., van Dijl, J. M., et al. (1996). Brain activation modulated by sentence comprehension. *Science, 274*(5284), 114–116.

Just, M. A., Carpenter, P. A., & Miyake, A. (2003). Neuroindices of cognitive workload: Neuroimaging, pupillometric and event-related potential studies of brain work. *Theoretical Issues in Ergonomics Science, 4*(1–2), 56–88.

Kahneman, D. (1973). *Attention and effort.* Englewood Cliffs, NJ: Prentice Hall.

Kahneman, D., Ben-Ishai, R., & Lotan, M. (1973). Relation of a test of attention to road accidents. *Journal of Applied Psychology, 58*(1), 113–115.

Kahneman, D., Tursky, B., Shapiro, D., & Crider, A. (1969). Pupillary, heart rate, and skin resistance changes during a mental task. *Journal of Experimental Psychology, 79*(1), 164–167.

Kaitaro, T. (2001). Biological and epistemological models of localization in the nineteenth century: From Gall to Charcot. *Journal of the History of the Neurosciences, 10*(3), 262–276.

Kane, M. J., Bleckley, M. K., Conway, A. R. A., & Engle, R. W. (2001). A controlled-attention view of working-memory capacity. *Journal of Experimental Psychology: General, 130*(2), 169–183.

Kane, M. J., Hasher, L., Stoltzfus, E. R., & Zacks, R. T. (1994). Inhibitory attentional mechanisms and aging. *Psychology and Aging, 9*(1), 103–112.

Karageorghis, C. I., Jones, L., & Low, D. C. (2006). Relationship between exercise heart rate and music tempo preference. *Research Quarterly for Exercise and Sport, 77*(2), 240–250.

Karageorghis, C. I., Priest, D. L., Terry, P. C., Chatzisarantis, N. L. D., & Lane, A. M. (2006). Redesign and initial validation of an instrument to assess the motivational qualities of music in exercise: The Brunel Music Rating Inventory-2. *Journal of Sports Sciences, 24*(8), 899–909.

Karageorghis, C. I., & Terry, P. C. (1997). The psychophysical effects of music in sport and exercise: A review. *Journal of Sport Behavior, 20*(1), 54–68.

Karageorghis, C. I., Terry, P. C., & Lane, A. M. (1999). Development and initial validation of an instrument to assess the motivational qualities of music in exercise and sport: The Brunel Music Rating Inventory. *Journal of Sports Sciences, 17*(9), 713–724.

Kasai, K., Yamada, H., Kamio, S., Nakagome, K., Iwanami, A., Fukuda, M., et al. (2001). Brain lateralization for mismatch response to across- and within-category change of vowels. *NeuroReport: For Rapid Communication of Neuroscience Research, 12*(11), 2467–2471.

Kato, Y., & Takeuchi, Y. (2003). Individual differences in wayfinding strategies. *Journal of Environmental Psychology, 23*(2), 171–188.

Katz, A. N., Blasko, D. G., & Kazmerski, V. A. (2004). Saying what you don't mean: Social influences on sarcastic language processing. *Current Directions in Psychological Science, 13*(5), 186–189.

Katz, J., Stecker, N., & Henderson, D. (Eds.). (1992). *Central auditory processing: A transdisciplinary view.* St. Louis, MO: Mosby-Year Book.

Kawano, T., Iwaki, S., Azuma, Y., Moriwaki, T., & Hamada, T. (2005). Degraded voices through mobile phones and their neural effects: A possible risk of using mobile phones during driving. *Transportation Research Part F: Traffic Psychology and Behaviour, 8*(4), 331–340.

Keller, P., & Stevens, C. (2004). Meaning from environmental sounds: Types of signal-referent relations and their effect on recognizing auditory icons. *Journal of Experimental Psychology: Applied, 10*(1), 3–12.

Kello, C. T., Sibley, D. E., & Plaut, D. C. (2005). Dissociations in performance on novel versus irregular items: Single-route demonstrations with input gain in localist and distributed models. *Cognitive Science, 29*(4), 627–654.

Kemper, S. (2006). Language in adulthood. In E. Bialystok & F. I. M. Craik (Eds.), *Lifespan cognition: Mechanisms of change,* 223–238. Oxford: Oxford University Press.

Kent, R. D., & Read, C. (1992). *The acoustical analysis of speech.* San Diego, CA: Singular.

Kerns, K. (1999). Human factors in air traffic control/flight deck integration: Implications of data-link simulation research. In D. J. Garland, J. A. Wise, & V. D. Hopkin (Eds.), *Handbook of aviation human factors,* 519–546. Mahwah, NJ: Erlbaum.

Kerr, W. A. (1950). Accident proneness of factory departments. *Journal of Applied Psychology, 34*(3), 167–170.

Kimura, K., Marunaka, K., & Sugiura, S. (1997). Human factors considerations for automotive navigation systems—Legibility, comprehension, and voice guidance. In Y. I. Noy (Ed.), *Ergonomics and safety of intelligent driver interfaces,* 153–167. Mahwah, NJ: Erlbaum.

King, J., & Just, M. A. (1991). Individual differences in syntactic processing: The role of working memory. *Journal of Memory and Language, 30*(5), 580–602.

Kintsch, W., & van Dijk, T. A. (1978). Toward a model of text comprehension and production. *Psychological Review, 85*(5), 363–394.

Kirkpatrick, F. H. (1943). Music in industry. *Journal of Applied Psychology, 27*(3), 268–274.

Klapp, S. T., & Netick, A. (1988). Multiple resources for processing and storage in short-term memory. *Human Factors, 30*(5), 617–632.

Kline, D. W., & Scialfa, C. T. (1996). Visual and auditory aging. In J. E. Birren & K. W. Schaie (Eds.), *Handbook of the psychology of aging* (4th ed., 181–203). San Diego, CA: Academic Press.

Kline, D. W., & Scialfa, C. T. (1997). Sensory and perceptual functioning: Basic research and human factors implications. In A. D. Fisk & W. A. Rogers (Eds.), *Handbook of human factors and the older adult,* 27–54. San Diego, CA: Academic.

Knowles, W. B. (1963). Operator loading tasks. *Human Factors, 9*(5), 155–161.

Koelega, H. S. (1992). Extraversion and vigilance performance—30 years of inconsistencies. *Psychological Bulletin, 112*(2), 239–258.

Koelsch, S., Gunter, T., Friederici, A. D., & Schroeger, E. (2000). Brain indices of music processing: "Nonmusicians" are musical. *Journal of Cognitive Neuroscience, 12*(3), 520–541.

Koelsch, S., Gunter, T. C., Schroger, E., Tervaniemi, M., Sammler, D., & Friederici, A. D. (2001). Differentiating ERAN and MMN: An ERP study. *NeuroReport: For Rapid Communication of Neuroscience Research, 12*(7), 1385–1389.

Koelsch, S., Maess, B., Grossmann, T., & Friederici, A. D. (2003). Electric brain responses reveal gender differences in music processing. *NeuroReport: For Rapid Communication of Neuroscience Research, 14*(5), 709–713.

Kohfeld, D. L. (1971). Simple reaction time as a function of stimulus intensity in decibels of light and sound. *Journal of Experimental Psychology, 88*(2), 251–257.

Korczynski, M., & Jones, K. (2006). Instrumental music? The social origins of broadcast music in British factories. *Popular Music, 25*(2), 145–164.

Koul, R. (2003). Synthetic speech perception in individuals with and without disabilities. *AAC: Augmentative and Alternative Communication, 19*(1), 49–58.

Koul, R. K., & Allen, G. D. (1993). Segmental intelligibility and speech interference thresholds of high-quality synthetic speech. *Journal of Speech and Hearing Research, 36*(4), 790.

Kramer, A. F. (1991). Physiological metrics of mental workload: A review of recent progress. In D. L. Damos (Ed.), *Multiple-task performance*, 279–328. London: Taylor & Francis.

Kramer, A. F., Cassavaugh, N., Horrey, W. J., Becic, E., & Mayhugh, J. L. (2007). Influence of age and proximity warning devices on collision avoidance in simulated driving. *Human Factors, 49*(5), 935–949.

Kramer, A. F., Sirevaag, E. J., & Braune, R. (1987). A psychophysiological assessment of operator workload during simulated flight missions. *Human Factors, 29*, 145–160.

Kramer, A. F., Trejo, L. J., & Humphrey, D. (1995). Assessment of mental workload with task-irrelevant auditory probes. *Biological Psychology, 40*(1–2), 83–100.

Kramer, G., Walker, B., Bonebright, T., Cook, P., Flowers, J., Miner, N., et al. (1999). *The sonification report: Status of the field and research agenda* (Report prepared for the National Science Foundation by members of the International Community for Auditory Display). Santa Fe, NM: International Community for Auditory Display.

Krumhansl, C. L. (1990). *Cognitive foundations of musical pitch* (Vol. 17). New York: Oxford University Press.

Kryter, K. D. (1960). *Human engineering principles for the design of speech communication systems*. Cambridge, MA: Bolt, Beranek, and Newman.

Kryter, K. D. (1972). Speech communication. In H. Van Cott & R. Kinkdale (Eds.), *Human engineering guide to equipment design*. Washington, DC: Government Printing Office.

Kryter, K. D. (1985). *The effects of noise on man*. Orlando, FL: Academic Press.

Kryter, K. D. (1994). *The handbook of hearing and the effects of noise: Physiology, psychology, and public health*. Bingley, UK: Emerald Group.

Kryter, K. D., Williams, C., & Green, D. M. (1962). Auditory acuity and the perception of speech. *Journal of the Acoustical Society of America, 34*(9), 1217–1223.

Kuhl, P. K. (1993). Infant speech perception: A window on psycholinguistic development. *International Journal of Psycholinguistics, 9*(1), 33–56.

Kuhl, P. K. (2004). Early language acquisition: Cracking the speech code. *Nature Reviews Neuroscience, 5*(11), 831–841.

Kumari, V., Aasen, I., & Sharma, T. (2004). Sex differences in prepulse inhibition deficits in chronic schizophrenia. *Schizophrenia Research, 69*(2–3), 219–235.

Kutas, M., & Hillyard, S. A. (1980). Reading senseless sentences: Brain potentials reflect semantic incongruity. *Science, 207*(4427), 203–205.

Kutas, M., & Hillyard, S. A. (1984). Brain potentials during reading reflect word expectancy and semantic association. *Nature, 307*(5947), 161–163.

Lam, L. T. (2002). Distractions and the risk of car crash injury: The effect of drivers' age. *Journal of Safety Research, 33*(3), 411–419.

la Pointe, L. B., & Engle, R. W. (1990). Simple and complex word spans as measures of working memory capacity. *Journal of Experimental Psychology: Learning, Memory, and Cognition, 16*(6), 1118–1133.

Laroche, C., Quoc, H. T., Hetu, R., & McDuff, S. (1991). "Detectsound": A computerized model for predicting the detectability of warning signals in noisy workplaces. *Applied Acoustics, 32*(3), 193–214.

Larsen, J. D., & Baddeley, A. (2003). Disruption of verbal STM by irrelevant speech, articulatory suppression, and manual tapping: Do they have a common source? *Quarterly Journal of Experimental Psychology A, 8*, 1249–1268.

Latorella, K. A. (October, 1998). *Effects of modality on interrupted flight deck performance: Implications for data link.* Paper presented at the proceedings of the Human Factors and Ergonomics Society. Chicago, IL.

Laughery, K. R., & Pinkus, A. L. (1966). Short-term memory: Effects of acoustic similarity, presentation rate and presentation mode. *Psychonomic Science, 6*(6), 285–286.

Lawton, C. A. (1994). Gender differences in way-finding strategies: Relationship to spatial ability and spatial anxiety. *Sex Roles, 30*(11–12), 765–779.

Laver, G. D., & Burke, D. M. (1993). Why do semantic priming effects increase in old age? A meta-analysis. *Psychology and Aging, 8*(1), 34–43.

Lavie, N. (1995). Perceptual load as a necessary condition for selective attention. *Journal of Experimental Psychology: Human Perception and Performance, 21*(3), 451–468.

Learmount, D. (1995). Lessons from the cockpit. *Flight International, January 11*, 11–17.

Lee, F.-S., Matthews, L. J., Dubno, J. R., & Mills, J. H. (2005). Longitudinal study of pure-tone thresholds in older persons. *Ear and Hearing, 26*, 1–11.

Lee, J. D., Caven, B., Haake, S., & Brown, T. L. (2001). Speech-based interaction with in-vehicle computers: The effect of speech-based e-mail on drivers' attention to the roadway. *Human Factors, 43*(4), 631–640.

Lee, J. D., McGehee, D. V., Brown, T. L., & Reyes, M. L. (2002). Collision warning timing, driver distraction, and driver response to imminent rear-end collisions in a high-fidelity driving simulator. *Human Factors, 44*(2), 314–334.

Lefebvre, P. P., Malgrange, B., Staecker, H., Moonen, G., & Van De Water, T. R. (1993). Retinoic acid stimulates regeneration of mammalian auditory hair cells. *Science, 260*(5108), 692–695.

Levy, J., & Pashler, H. (2001). Is dual-task slowing instruction dependent? *Journal of Experimental Psychology: Human Perception and Performance, 27*(4), 862–869.

Levy, J., & Pashler, H. (2008). Task prioritisation in multitasking during driving: Opportunity to abort a concurrent task does not insulate braking responses from dual-task slowing. *Applied Cognitive Psychology, 22*(4), 507–525.

Levy, J., Pashler, H., & Boer, E. (2006). Central interference in driving: Is there any stopping the psychological refractory period? *Psychological Science, 17*(3), 228–235.

Li, L., Daneman, M., Qi, J. G., & Schneider, B. A. (2004). Does the information content of an irrelevant source differentially affect spoken word recognition in younger and older adults? *Journal of Experimental Psychology: Human Perception and Performance, 30*(6), 1077–1091.

Liberman, A. M. (1995). The relation of speech to reading and writing. In B. D. Gelder & J. Morais (Eds.), *Speech and reading*, 17–31. East Sussex, UK: Taylor & Francis.

Liegeois-Chauvel, C., Giraud, K., Badier, J.-M., Marquis, P., & Chauvel, P., (2003). Intracerebral evoked potentials in pitch perception reveal a function of asymmetry of human auditory cortex. In I. Peretz & R. J. Zatorre (Eds.), *The cognitive neuroscience of music*, 152–167. New York: Oxford University Press.

Lindenberger, U., & Baltes, P. B. (1994). Sensory functioning and intelligence in old age: A strong connection. *Psychology and Aging, 9*(3), 339–355.

Lindenberger, U., & Baltes, P. B. (1997). Intellectual functioning in old and very old age: Cross-sectional results from the Berlin Aging Study. *Psychology and Aging, 12*(3), 410–432.

Lindenberger, U., Marsiske, M., & Baltes, P. B. (2000). Memorizing while walking: Increase in dual-task costs from young adulthood to old age. *Psychology and Aging, 15*(3), 417–436.

Lindenberger, U., Scherer, H., & Baltes, P. B. (2001). The strong connection between sensory and cognitive performance in old age: Not due to sensory acuity reductions operating during cognitive assessment. *Psychology and Aging, 16*(2), 196–205.

Liu, Y.-C. (2000). Effect of advanced traveler information system displays on younger and older drivers' performance. *Displays, 21*(4), 161–168.

Liu, Y.-C. (2001). Comparative study of the effects of auditory, visual and multimodality displays on drivers' performance in advanced traveller information systems. *Ergonomics, 44*(4), 425–442.

Lloyd, L. L., & Kaplan, H. (1978). *Audiometric interpretation: A manual of basic audiometry.* Baltimore: University Park Press.

Lobley, K. J., Baddeley, A. D., & Gathercole, S. E. (2005). Phonological similarity effects in verbal complex span. *The Quarterly Journal of Experimental Psychology A: Human Experimental Psychology, 58*(8), 1462–1478.

Loeb, M. (1986). *Noise and human efficiency.* New York: Wiley.

Loeb, R. G. (1993). A measure of intraoperative attention to monitor displays. *Anesthesia and Analgesia, 76*(2), 337–341.

Loeb, R. G. M. D., & Fitch, W. T. P. (2002). A laboratory evaluation of an auditory display designed to enhance intraoperative monitoring. *Anesthesia and Analgesia, 94*(2), 362–368.

Logan, G. D., & Burkell, J. (1986). Dependence and independence in responding to double stimulation: A comparison of stop, change, and dual-task paradigms. *Journal of Experimental Psychology: Human Perception and Performance, 12*(4), 549–563.

Logan, J. S., Greene, B. G., & Pisoni, D. B. (1989). Segmental intelligibility of synthetic speech produced by rule. *Journal of the Acoustical Society of America, 86*(2), 566–581.

Logie, R. H. (1986). Visuo-spatial processing in working memory. *The Quarterly Journal of Experimental Psychology A: Human Experimental Psychology, 38*(2), 229–247.

Logie, R. H., Venneri, A., Sala, S. D., Redpath, T. W., & Marshall, I. (2003). Brain activation and the phonological loop: The impact of rehearsal. *Brain and Cognition, 53*(2), 293–296.

Loomis, J. M., Golledge, R. D., & Klatzky, R. L. (2001). GPS-based navigation systems for the visually impaired. In W. Barfield & T. Caudell (Eds.), *Fundamentals of wearable computers and augmented reality*, 429–446. Mahwah, NJ: Erlbaum.

Loring, D. W., & Larrabee, G. J. (2006). Sensitivity of the Halstead and Wechsler test batteries to brain damage: Evidence from Reitan's original validation sample. *Clinical Neuropsychologist, 20*(2), 221–229.

Loven, F. C., & Collins, M. J. (1988). Reverberation, masking, filtering, and level effects on speech recognition performance. *Journal of Speech and Hearing Research, 31*(4), 681–695.

Lu, Z. L., Williamson, S. J., & Kaufman, L. (1992). Behavioral lifetime of human auditory sensory memory predicted by physiological measures. *Science, 258*(5088), 1668–1670.

Luce, P. A., Goldinger, S. D., Auer, E. T. J., & Vitevitch, M. S. (2000). Phonetic priming, neighborhood activation, and PARSYN. *Perception & Psychophysics, 62*(3), 615–625.

Luce, P. A., & Pisoni, D. B. (1998). Recognizing spoken words: The neighborhood activation model. *Ear and Hearing, 19*, 1–36.

Luce, P. A., Pisoni, D. B., & Goldinger, S. D. (1990). Similarity neighborhoods of spoken words. In G. T. M. Altmann (Ed.), *Cognitive models of speech processing*, 122–147. Cambridge, MA: MIT Press.

Luck, S., & Girelli, M. (1998). Electrophysiological approaches to the study of selective attention in the human brain. In R. Parasuraman (Ed.), *The attentive brain*, 71–94. Cambridge, MA: MIT Press.

Luck, S. J. (2005). Ten simple rules for designing ERP experiments. In T. Handy (Ed.), *Event-related potentials: A methods handbook*, 17–32. Cambridge, MA: MIT Press.

Luck, S. J., & Vogel, E. K. (1997). The capacity of visual working memory for features and conjunctions. *Nature, 390*(6657), 279–281.

Lunner, T. (2003). Cognitive function in relation to hearing aid use. *International Journal of Audiology, 42*(Suppl. 1), S49–S58.

Lunner, T., Rudner, M., & Ronnberg, J. (2009). Background and basic processes: Cognition and hearing aids. *Scandinavian Journal of Psychology, 50*(5), 395–403.

Lyons, T. J., Gillingham, K. K., Teas, D. C., & Ercoline, W. R. (1990). The effects of acoustic orientation cues on instrument flight performance in a flight simulator. *Aviation, Space, and Environmental Medicine, 61*(8), 699–706.

MacDonald, J. A., Balakrishnan, J. D., Orosz, M. D., & Karplus, W. J. (2002). Intelligibility of speech in a virtual 3-D environment. *Human Factors, 44*(2), 272–286.

MacDonald, J. A., Henry, P. P., & Letowski, T. R. (2006). Spatial audio through a bone conduction interface. *International Journal of Audiology, 45*(10), 595–599.

MacDonald, M. C., & Christiansen, M. H. (2002). Reassessing working memory: Comment on Just and Carpenter (1992) and Waters and Caplan (1996). *Psychological Review, 109*(1), 35–54.

MacDonald, M. C., Just, M. A., & Carpenter, P. A. (1992). Working memory constraints on the processing of syntactic ambiguity. *Cognitive Psychology, 24*(1), 56–98.

Macken, W. J., Tremblay, S., Houghton, R. J., Nicholls, A. P., & Jones, D. M. (2003). Does auditory streaming require attention? Evidence from attentional selectivity in short-term memory. *Journal of Experimental Psychology: Human Perception and Performance, 29*(1), 43–51.

Mackworth, J. F. (1959). Paced memorizing in a continuous task. *Journal of Experimental Psychology, 58*, 206–211.

Mackworth, N. H. (1948). The breakdown of vigilance during prolonged visual search. *Quarterly Journal of Experimental Psychology A, 1*, 6–21.

Mackworth, N. H. (1949). Human problems of work design. *Nature, 164*, 982–984.

MacSweeney, M. A., Calvert, G. A., Campbell, R., McGuire, P. K., David, A. S., Williams, S. C. R., et al. (2002). Speechreading circuits in people born deaf. *Neuropsychologia, 40*(7), 801–807.

Madden, D. J. (1988). Adult age differences in the effects of sentence context and stimulus degradation during visual word recognition. *Psychology and Aging, 3*(2), 167–172.

Maki, P. M., & Resnick, S. M. (2001). Effects of estrogen on patterns of brain activity at rest and during cognitive activity: A review of neuroimaging studies. *NeuroImage, 14*(4), 789–801.

Maltz, M., & Shinar, D. (2004). Imperfect in-vehicle collision avoidance warning systems can aid drivers. *Human Factors, 46*(2), 357–366.

Maltz, M., & Shinar, D. (2007). Imperfect in-vehicle collision avoidance warning systems can aid distracted drivers. *Transportation Research Part F: Traffic Psychology and Behaviour, 10*(4), 345–357.

Marin, O. S. M., & Perry, D. W. (1999). Neurological aspects of music perception and performance. In D. Deutsch (Ed.), *The psychology of music* (2nd ed., 653–724). New York: Academic Press.

Marks, L. E. (1994). "Recalibrating" the auditory system: The perception of loudness. *Journal of Experimental Psychology: Human Perception and Performance, 20*(2), 382–396.

Marshall, D. C., Lee, J. D., & Austria, P. A. (2007). Alerts for in-vehicle information systems: Annoyance, urgency, and appropriateness. *Human Factors, 49*(1), 145–157.

Marshall, N. B., Duke, L. W., & Walley, A. C. (1996). Effects of age and Alzheimer's disease on recognition of gated spoken words. *Journal of Speech and Hearing Research, 39*, 724–733.

Marslen-Wilson, W., & Welsh, A. (1978). Processing interactions and lexical access during word recognition in continuous speech. *Cognitive Psychology, 10*, 29–63.

Marslen-Wilson, W. D. (1987). Functional parallelism in spoken word-recognition. *Cognition, 25*(1–2), 71–102.

Martin, B. A., & Boothroyd, A. (2000). Cortical, auditory, evoked potentials in response to changes of spectrum and amplitude. *Journal of the Acoustical Society of America, 107*(4), 2155–2161.

Martin, C. S., Mullennix, J. W., Pisoni, D. B., & Summers, W. V. (1989). Effects of talker variability on recall of spoken word lists. *Journal of Experimental Psychology: Learning, Memory, and Cognition, 15*(4), 676–684.

Martin, R. C., Wogalter, M. S., & Forlano, J. G. (1988). Reading comprehension in the presence of unattended speech and music. *Journal of Memory and Language, 27*(4), 382–398.

Martindale, C., Moore, K., & Anderson, K. (2005). The effect of extraneous stimulation on aesthetic preference. *Empirical Studies of the Arts, 23*(2), 83–91.

Martindale, C., Moore, K., & Borkum, J. (1990). Aesthetic preference: Anomalous findings for Berlyne's psychobiological theory. *American Journal of Psychology, 103*(1), 53–80.

Mason, R. A., Just, M. A., Keller, T. A., & Carpenter, P. A. (2003). Ambiguity in the brain: What brain imaging reveals about the processing of syntactically ambiguous sentences. *Journal of Experimental Psychology: Learning, Memory, and Cognition, 29*(6), 1319–1338.

Massaro, D. W. (1972). Preperceptual images, processing time, and perceptual units in auditory perception. *Psychological Review, 79*(2), 124–145.

Massaro, D. W. (1982). Sound to representation: An information-processing analysis. In T. Myers, J. Laver, & J. Anderson (Eds.), *The cognitive representation of speech*, 181–193. Amsterdam: North Holland.

Massaro, D. W. (1987). *Speech perception by ear and eye: A paradigm for psychological inquiry*. Hillsdale, NJ: Erlbaum.

Massaro, D. W. (1995). From speech is special to talking heads: The past to the present. In R. L. Solso & D. W. Massaro (Eds.), *The science of the mind: 2001 and beyond*, 203–220. Oxford: Oxford University Press.

Massaro, D. W., & Cohen, M. M. (1991). Integration versus interactive activation: The joint influence of stimulus and context in perception. *Cognitive Psychology, 23*, 558–614.

Mathiak, K., Hertrich, I., Kincses, W. E., Riecker, A., Lutzenberger, W., & Ackermann, H. (2003). The right supratemporal plane hears the distance of objects: Neuromagnetic correlates of virtual reality. *NeuroReport: For Rapid Communication of Neuroscience Research, 14*(3), 307–311.

Matlin, M. W., & Foley, H. J. (1997). *Sensation and perception, Fourth Edition*. Boston: Allyn & Bacon.

Matthews, G., Davies, D. R., Westerman, S. J., & Stammers, R. B. (2000). *Human performance: Cognition, stress, and individual differences*. Philadelphia: Psychology Press.

Matthews, R., Legg, S., & Charlton, S. (2003). The effect of cell phone type on drivers subjective workload during concurrent driving and conversing. *Accident Analysis and Prevention, 35*(4), 451–457.

May, J. F., Baldwin, C. L., & Parasuraman, R. (October, 2006). *Prevention of rear-end crashes in drivers with task-induced fatigue through use of auditory collision avoidance warnings.* Paper presented at the Proceedings of the Human Factors and Ergonomics Society, San Francisco.

Mayfield, C., & Moss, S. (1989). Effect of music tempo on task performance. *Psychological Reports, 65*(32), 1283–1290.

McAdams, S. (1989). Psychological constraints on form-bearing dimensions in music. *Contemporary Music Review, 4*(1), 181–198.

McAdams, S. (1993). Recognition of sound sources and events. In S. McAdams & E. Bigand (Eds.), *Thinking in sound: The cognitive psychology of human audition*, 146–198. Oxford, UK: Clarendon Press.

McAnally, K. I., & Martin, R. L. (2007). Spatial audio displays improve the detection of target messages in a contiuous monitoring task. *Human Factors, 49*(4), 688–695.

McCarley, J. S., Vais, M. J., Pringle, H., Kramer, A. F., Irwin, D. E., & Strayer, D. L. (2004). Conversation disrupts change detection in complex traffic scenes. *Human Factors, 46*(3), 424–436.

McClelland, J. L., & Elman, J. L. (1986). The TRACE model of speech perception. *Cognitive Psychology, 18*, 1–86.

McClelland, J. L., Rumelhart, D. E., & Hinton, G. E. (1986). The appeal of parallel distributed processing. In D. E. Rumelhart, J. L. McClelland, & P. R. Group (Eds.), *Parallel distributed processing. Vol. 1: Foundations*, 3–44. Cambridge, MA: MIT Press.

McCoy, S. L., Tun, P. A., Cox, L., Colangelo, M., Stewart, R. A., & Wingfield, A. (2005). Hearing loss and perceptual effort: Downstream effects on older adults' memory for speech. *Quarterly Journal of Experimental Psychology: Human Experimental Psychology, 58A*(1), 22–33.

McDermott, J., & Hauser, M. (2004). Are consonant intervals music to their ears? Spontaneous acoustic preferences in a nonhuman primate. *Cognition, 94*(2), B11–B21.

McDonald, D. P., Gilson, R. D., Mouloua, M., Dorman, J., & Fouts, P. (1999). Contextual and cognitive influences on multiple alarm heuristics. In M. W. Scerbo & M. Mouloua (Eds.), *Automation technology and human performance: Current research and trends*, 301–306. Mahwah, NJ: Erlbaum.

McEvoy, S. P., Stevenson, M. R., McCartt, A. T., Woodward, M., Haworth, C., Palamara, P., et al. (2005). Role of mobile phones in motor vehicle crashes resulting in hospital attendance: A case-crossover study. *BMJ: British Medical Journal, 331*(7514), 428.

McEvoy, S. P., Stevenson, M. R., & Woodward, M. (2007). The contribution of passengers versus mobile phone use to motor vehicle crashes resulting in hospital attendance by the driver. *Accident Analysis and Prevention, 39*(6), 1170–1176.

McGann, A., Morrow, D., Rodvold, M., & Mackintosh, M.-A. (1998). Mixed-media communication on the flight deck: A comparison of voice, data link, and mixed ATC environments. *International Journal of Aviation Psychology, 8*(2), 137–156.

McGurk, H., & MacDonald, J. A. (1976). Hearing lips and seeing voices. *Nature, 264*(5588), 746–748.

McLennan, C. T., Luce, P. A., & Charles-Luce, J. (2005). Representation of lexical form: Evidence from studies of sublexical ambiguity. *Journal of Experimental Psychology: Human Perception and Performance, 31*(6), 1308–1314.

McNamara, D. S., & Shapiro, A. M. (2005). Multimedia and hypermedia solutions for promoting metacognitive engagement, coherence, and learning. *Journal of Educational Computing Research, 33*(1), 1–29.

McQueen, J. M. (2005). Spoken-word recognition and production: Regular but not insepa-rable bedfellows. In A. Cutler (Ed.), *Twenty-first century psycholinguistics: Four cornerstones*, 229–244. Mahwah, NJ: Erlbaum.

McQueen, J. M., Norris, D., & Cutler, A. (1994). Competition in spoken word recognition: Spotting words in other words. *Journal of Experimental Psychology: Learning, Memory, and Cognition May, 20*(3), 621–638.

Medvedev, Z. A. (1990). *The legacy of Chernobyl*. New York: Norton.

Melara, R. D., Rao, A., & Tong, Y. (2002). The duality of selection: Excitatory and inhibitory processes in auditory selection attention. *Journal of Experimental Psychology: Human Perception and Performance, 28*(2), 279–306.

Meshkati, N., Hancock, P. A., & Rahimi, M. (1995). Techniques in mental workload. In J. R. Wilson & E. N. Corlett (Eds.), *Evaluation of human work: A practical ergonomics methodology,* 605–627. Philadelphia, PA.

Meyer, D. E., & Kieras, D. E. (1997). A computational theory of executive cognitive processes and multiple-task performance. 2. Accounts of psychological refractory-period phenomena. *Psychological Review, 104*(4), 749–791.

Middlebrooks, J. C., Furukawa, S., Stecker, G. C., & Mickey, B. J. (2005). Distributed representation of sound-source location in the auditory cortex. In R. Konig, P. Heil, E. Budinger, & H. Scheich (Eds.), *The auditory cortex: A synthesis of human and animal research,* 225–240. Mahwah, NJ: Erlbaum.

Miller, G. A., Heise, G. A., & Lichten, W. (1951). The intelligibility of speech as a function of the context of the test materials. *Journal of Experimental Psychology, 41,* 329–335.

Miller, G. A., & Isard, S. (1963). Some perceptual consequences of linguistic rules. *Journal of Verbal Learning and Verbal Behavior, 2*(3), 217–228.

Miller, G. A., & Licklider, J. C. R. (1950). The intelligibility of interrupted speech. *Journal of the Acoustical Society of America 22,* 167–173.

Miller, J. (1982). Divided attention: Evidence for coactivation with redundant signals. *Cognitive Psychology, 14*(2), 247–279.

Miller, J. (1991). Channel interaction and the redundant-targets effect in bimodal divided attention. *Journal of Experimental Psychology: Human Perception and Performance, 17*(1), 160–169.

Miller, J. L., Grosjean, F., & Lomanto, C. (1984). Articulation rate and its variability in spontaneous speech: A reanalysis and some implications. *Phonetica, 41,* 215–225.

Milliman, R. E. (1982). Using background music to affect the behavior of supermarket shoppers. *The Journal of Marketing, 46*(3), 86–91.

Mirabella, A., & Goldstein, D. A. (1967). The effects of ambient noise upon signal detection. *Human Factors, 9*(3), 277–284.

Mitchell, R. L. C., Elliott, R., Barry, M., Cruttenden, A., & Woodruff, P. W. R. (2003). The neural response to emotional prosody, as revealed by functional magnetic resonance imaging. *Neuropsychologia, 41*(10), 1410–1421.

Mitsutomi, M., & O'Brien, K. (2004). Fundamental aviation language issues addressed by new proficiency requirements. *ICAO Journal, 59*(1), 7–9, 26–27.

Miyake, A., Carpenter, P. A., & Just, M. A. (1995). Reduced resources and specific impairments in normal and aphasic sentence comprehension. *Cognitive Neuropsychology, 12*(6), 651–679.

Miyake, A., Just, M. A., & Carpenter, P. A. (1994). Working memory constraints on the resolution of lexical ambiguity: Maintaining multiple interpretations in neutral contexts. *Journal of Memory and Language, 33*(2), 175–202.

Molholm, S., Ritter, W., Javitt, D. C., & Foxe, J. J. (2004). Multisensory visual-auditory object recognition in humans: A high-density electrical mapping study. *Cerebral Cortex, 14*(4), 452–465.

Moller, A. R. (1999). Review of the roles of temporal and place coding of frequency in speech discrimination. *Acta Oto-Laryngologica, 119*(4), 424–430.

Momtahan, K. L. (1990). *Mapping of psychoacoustic parameters to the perceived urgency of auditory warning signals.* Unpublished master's thesis, Carleton University, Ottawa, Ontario.

Mondor, T. A., & Finley, G. A. (2003). The perceived urgency of auditory warning alarms used in the hospital operating room is inappropriate: [Caractere inapproprie de l'urgence percue des alarmes sonores utilisees dans les salles d'operation d'hopitaux]. *Canadian Journal of Anesthesia, 50*(3), 221–228.

Moore, B. C. J. E. (1995). *Hearing.* San Diego, CA: Academic Press.

Moray, N. (1959). Attention in dichotic listening: Affective cues and the influence of instructions. *Quarterly Journal of Experimental Psychology A, 11,* 56–60.

Moray, N. (1967). Where is capacity limited? A survey and a model. *Acta Psychologica, 27,* 84–92.

Moray, N. (1969). *Listening and attention.* Baltimore: Penguin.

Moray, N. (1982). Subjective mental workload. *Human Factors, 24*(1), 25–40.

Moray, N., Bates, A., & Barnett, T. (1965). Experiments on the four-eared man. *Journal of the Acoustical Society of America, 38*(2), 196–201.

Moray, N., Dessouky, M. I., Kijowski, B. A., & Adapathya, R. (1991). Strategic behavior, workload, and performance in task scheduling. *Human Factors, 33*(6), 607–629.

Morley, S., Petrie, H., O'Neill, A.-M., & McNally, P. (1999). Auditory navigation in hyperspace: Design and evaluation of a non-visual hypermedia system for blind users. *Behaviour and Information Technology, 18*(1), 18–26.

Morria, N., & Sarll, P. (2001). Drinking glucose improves listening span in students who miss breakfast. *Educational Research, 43*(2), 201–207.

Morrow, D., Wickens, C., Rantanen, E., Chang, D., & Marcus, J. (2008). Designing external aids that support older pilots' communication. *International Journal of Aviation Psychology, 18*(2), 167–182.

Mowbray, G. H. (1953). Simultaneous vision and audition: The comprehension of prose passages with varying levels of difficulty. *Journal of Experimental Psychology, 46*(5), 365–372.

Mullennix, J. W., Pisoni, D. B., & Martin, C. S. (1989). Some effects of talker variability on spoken word recognition. *Journal of the Acoustical Society of America, 85*(1), 365–378.

Munte, T. F., Kohlmetz, C., Nager, W., & Altenmuller, E. (2001). Superior auditory spatial tuning in conductors. *Nature, 409*(6820), 580.

Murdock, B. B., Jr. (1962). The serial position effect of free recall. *Journal of Experimental Psychology, 64*(5), 482–488.

Murray, D. J. (1965). Vocalization-at-presentation and immediate recall, with varying presentation-rates. *Quarterly Journal of Experimental Psychology, 17*(1), 47–56.

Murray, D. J. (1966). Vocalization at presentation and the recall of lists varying in association value. *Journal of Verbal Learning and Verbal Behavior, 5*(5), 488–491.

Murray, M. M., Camen, C., Andino, S. L. G., Clarke, S., & Bovet, P. (2006). Rapid brain discrimination of sounds of objects. *Journal of Neuroscience, 26*(4), 1293–1302.

Muthayya, S., Thomas, T., Srinivasan, K., Rao, K., Kurpad, A. V., van Klinken, J.-W., et al. (2007). Consumption of a mid-morning snack improves memory but not attention in school children. *Physiology and Behavior, 90*(1), 142–150.

Mynatt, E. D. (1997). Transforming graphical interfaces into auditory interfaces for blind users. *Human-Computer Interaction, 12*(1–2), 7–45.

Naatanen, R. (1992). *Attention and brain function.* Hillsdale, NJ: Erlbaum.

Naatanen, R., & Alho, K. (2004). Mechanisms of attention in audition as revealed by the event-related potentials of the brain. In M. I. Posner (Ed.), *Cognitive neuroscience of attention,* 194–206. New York: Guilford Press.

Naatanen, R., Pakarinen, S., Rinne, T., & Takegata, R. (2004). The mismatch negativity (MMN): Towards the optimal paradigm. *Clinical Neurophysiology, 115*(1), 140–144.

Nagel, D. C. (1988). Human error in aviation operations. In E. L. Wiener & D. C. Nagel (Eds.), *Human factors in aviation. Academic Press series in cognition and perception,* 263–303. San Diego, CA: Academic Press.

Nager, W., Kohlmetz, C., Altenmuller, E., Rodriguez-Fornells, A., & Munte, T. F. (2003). The fate of sounds in conductors' brains: An ERP study. *Cognitive Brain Research, 17*(1), 83–93.

Nantais, K. M., & Schellenberg, E. G. (1999). The Mozart effect: An artifact of preference. *Psychological Science, 10*(4), 370–373.

Navarro, J., Mars, F., Forzy, J.-F., El-Jaafari, M., & Hoc, J.-M. (2010). Objective and subjective evaluation of motor priming and warning systems applied to lateral control assistance. *Accident Analysis and Prevention, 42*(3), 904–912.

Navarro, J., Mars, F., & Hoc, J.-M. (2007). Lateral control assistance for car drivers: A comparison of motor priming and warning systems. *Human Factors, 49*(5), 950–960.

Naveh-Benjamin, M., Craik, F. I. M., Guez, J., & Dori, H. (1998). Effects of divided attention on encoding and retrieval processes in human memory: Further support for an asymmetry. *Journal of Experimental Psychology: Learning, Memory, and Cognition, 24*(5), 1091–1104.

Navon, D. (1984). Resources—A theoretical soup stone? *Psychological Review, 91*(2), 216–234.

Navon, D., & Miller, J. (2002). Queuing or sharing? A critical evaluation of the single-bottleneck notion. *Cognitive Psychology, 44*(3), 193–251.

Nelson, N. W., & McRoskey, R. L. (1978). Comprehension of Standard English at varied speaking rates by children whose major dialect is Black English. *Journal of Communication Disorders, 11*(1), 37–50.

Newman, S. D., Just, M. A., Keller, T. A., Roth, J., & Carpenter, P. A. (2003). Differential effects of syntactic and semantic processing on the subregions of Broca's area. *Cognitive Brain Research, 16*(2), 297–307.

Nicholls, A. P., & Jones, D. M. (2002). Capturing the suffix: Cognitive streaming in immediate serial recall. *Journal of Experimental Psychology: Learning, Memory, and Cognition, 28*(1), 12–28.

Nicholls, M. E. R., Searle, D. A., & Bradshaw, J. L. (2004). Read my lips: Asymmetries in the visual expression and perception of speech revealed through the McGurk effect. *Psychological Science, 15*(2), 138–141.

Nickerson, R. S. (1973). Intersensory facilitation of reaction time: Energy summation or preparation enhancement? *Psychological Review, 80*(6), 489–509.

Niemi, P., & Naatanen, R. (1981). Foreperiod and simple reaction time. *Psychological Bulletin, 89*(1), 133–162.

Ninio, A., & Kahneman, D. (1974). Reaction time in focused and in divided attention. *Journal of Experimental Psychology, 103*(3), 394–399.

Norman, D. A. (1968). Toward a theory of memory and attention. *Psychological Review, 75*(6), 522–536.

Norman, D. A. (1969). Memory while shadowing. *Quarterly Journal of Experimental Psychology A, 21*(1), 85–93.

Norman, D. A. (1976). *Memory and attention* (2nd ed.). New York: Wiley.

Norman, D. A., & Bobrow, D. G. (1975). On data-limited and resource-limited processes. *Cognitive Psychology, 7*, 44–64.

Norris, D. (1990). A Dynamic-Net model of human speech recognition. In G. T. M. Altmann (Ed.), *Cognitive models of speech processing: Psycholinguistic and computational perspectives*, 87–104. Cambridge, MA: MIT Press.

Norris, D. (1994). Shortlist: A connectionist model of continuous speech recognition. *Cognition, 52*(3), 189–234.

Norris, D. (2006). The Bayesian reader: Explaining word recognition as an optimal Bayesian decision process. *Psychological Review, 113*(2), 327–357.

Norris, D., McQueen, J. M., & Cutler, A. (1995). Competition and segmentation in spoken-word recognition: Spotting words in other words. *Journal of Experimental Psychology: Learning, Memory, and Cognition September, 21*(5), 1209–1228.

Norris, D., McQueen, J. M., & Cutler, A. (2000). Merging information in speech recognition: Feedback is never necessary. *Behavioral and Brain Sciences, 23*(3), 299–370.

North, A. C., & Hargreaves, D. J. (1996). Responses to music in a dining area. *Journal of Applied Social Psychology, 26*(6), 491–501.

North, A. C., & Hargreaves, D. J. (1999). Music and driving game performance. *Scandinavian Journal of Psychology, 40*(4), 285–292.

Noweir, M. H. (1984). Noise exposure as related to productivity, disciplinary actions, absentee-ism, and accidents among textile workers. *Journal of Safety Research, 15*(4), 163–174.

Nowicka, A., & Fersten, E. (2001). Sex-related differences in interhemispheric transmission time in the human brain. *NeuroReport: For Rapid Communication of Neuroscience Research, 12*(18), 4171–4175.

Noy, Y. I. (Ed.). (1997). *Ergonomics and Safety of Intelligent Driver Interfaces.* Mahwah, NJ: Lawrence Erlbaum.

Noyes, J. M., Cresswell, A. F., & Rankin, J. A. (1999). Designing aircraft warning systems: A case study. In N. A. Stanton & J. Edworthy (Eds.), *Human factors in auditory warnings,* 265–281. Aldershot, UK: Ashgate.

Noyes, J. M., & Starr, A. F. (2000). Civil aircraft warning systems: Future directions in infor-mation management and presentation. *International Journal of Aviation Psychology, 10*(2), 169–188.

Noyes, J. M., Starr, A. F., Frankish, C. R., & Rankin, J. A. (1995). Aircraft warning systems: Application of model-based reasoning techniques. *Ergonomics, 38*(11), 2432–2445.

Nygaard, L. C., & Pisoni, D. B. (1995). Speech perception: New directions in research and theory. In J. L. Miller & P. D. Eimas (Eds.), *Speech, language, and communication* (2nd ed.), 63–96. San Diego, CA: Academic.

Obata, J. (1934). The Effects of Noise upon Human Efficiency. *Journal of the Acoustical Society of America, 5*(4), 255.

Obleser, J., Elbert, T., Lahiri, A., & Eulitz, C. (2003). Cortical representation of vowels reflects acoustic dissimilarity determined by formant frequencies. *Cognitive Brain Research, 15*(3), 207–213.

O'Donnell, R. D., & Eggemeier, F. T. (1986). Workload assessment methodology. In K. Boff, L. Kauffman & J. P. Thomas (Eds.), *Handbook of perception and human per-formance. Vol. 2, Cognitive processes and performance,* 42.41–42.49. New York: Wiley.

Ogden, G. D., Levine, J. M., & Eisner, E. J. (1979). Measurement of workload by secondary tasks. *Human Factors, 21*(5), 529–548.

O'Hanlon, L., Kemper, S., & Wilcox, K. A. (2005). Aging, encoding, and word retrieval: Distinguishing phonological and memory processes. *Experimental Aging Research, 31*(2), 149–171.

Oldham, G. R., Cummings, A., Mischel, L. J., Schmidtke, J. M., & Zhou, J. (1995). Listen while you work? Quasi-experimental relations between personal-stereo headset use and employee work responses. *Journal of Applied Psychology, 80*(5), 547–564.

Olsho, L. W., Harkins, S. W., & Lenhardt, M. L. (1985). Aging and the auditory system. In J. E. Birren & K. W. Schaie (Eds.), *Handbook of the psychology of aging* (2nd ed.). *The handbooks of aging,* 332–377. New York: Van Nostrand Reinhold.

O'Rourke, T. B., & Holcomb, P. J. (2002). Electrophysiological evidence for the efficiency of spoken word processing. *Biological Psychology, 60*(2–3), 121–150.

Oving, A. B., Veltman, J. A., & Bronkhorst, A. W. (2004). Effectiveness of 3-D audio for warnings in the cockpit. *International Journal of Aviation Psychology, 14*(3), 257–276.

Owen, A. M., McMillan, K. M., Laird, A. R., & Bullmore, E. (2005). N-back working memory paradigm: A meta-analysis of normative functional neuroimaging studies. *Human Brain Mapping, 25*(1), 46–59.

Paivio, A. (1969). Mental imagery in associative learning and memory. *Psychological Review, 76*(3), 241–263.

Paivio, A., & Csapo, K. (1971). Short-term sequential memory for pictures and words. *Psychonomic Science, 24*(2), 50–51.

Palmer, C. (2005). Sequence memory in music performance. *Current Directions in Psychological Science, 14*(5), 247–250.

Pantev, C., Oostenveld, R., Engelien, A., Ross, B., Roberts, L. E., & Hoke, M. (1998). Increased auditory cortical representation in musicians. *Nature, 392,* 811–114.

Parasuraman, R. (1978). Auditory evoked potentials and divided attention. *Psychophysiology, 15*(5), 460–465.

Parasuraman, R. (1979). Memory load and event rate control sensitivity decrements in sustained attention. *Science, 205*(4409), 924–927.

Parasuraman, R. (1980). Effects of information processing demands on slow negative shift latencies and N100 amplitude in selective and divided attention. *Biological Psychology, 11*(3, Suppl. 4), 217–233.

Parasuraman, R. (1986). Vigilance, monitoring, and search. In K. R. Boff, L. Kaufman, & J. P. Thomas (Eds.), *Handbook of perception and human performance. Vol. 2: Cognitive processes and performance,* 43: 1–39. New York: Wiley.

Parasuraman, R. (1998). The attentive brain. Cambridge, MA: MIT Press.

Parasuraman, R. (2003). Neuroergonomics: Research and practice. *Theoretical Issues in Ergonomics Science, 4,* 5–20.

Parasuraman, R., & Caggiano, D. (2005). Neural and genetic assays of mental workload. In D. McBride & D. Schmorrow (Eds.), *Quantifying human information processing,* 123–155. Lanham, MD: Rowman and Littlefield.

Parasuraman, R., & Davies, D. R. (1984). *Varieties of attention.* San Diego, CA: Academic Press.

Parasuraman, R., & Hancock, P. A. (2001). Adaptive control of mental workload. In P. A. Hancock & P. A. Desmond (Eds.), *Stress, workload, and fatigue,* 305–333. Mahwah, NJ: Erlbaum.

Parasuraman, R., & Haxby, J. V. (1993). Attention and brain function in Alzheimer's disease: A review. *Neuropsychology, 7*(3), 242–272.

Parasuraman, R., & Rizzo, M. (2007). *Neuroergonomics: The brain at work.* Oxford: Oxford University Press.

Parasuraman, R., Sheridan, T. B., & Wickens, C. D. (2000). A model for types and levels of human interaction with automation. *IEEE Transactions on Systems, Man, and Cybernetics-Part A: Systems and Humans, 30*(3), 286–297.

Parasuraman, R., Warm, J. S., & See, J. E. (1998). Brain systems of vigilance. In R. Parasuraman (Ed.), *The attentive brain,* 221–256. Cambridge, MA: MIT Press.

Pardo, P. J., Makela, J. P., & Sams, M. (1999). Hemispheric differences in processing tone frequency and amplitude modulations. *NeuroReport: For Rapid Communication of Neuroscience Research, 10*(14), 3081–3086.

Park, D. C., Lautenschlager, G., Hedden, T., Davidson, N. S., Smith, A. D., & Smith, P. K. (2002). Models of visuospatial and verbal memory across the adult life span. *Psychology and Aging, 17*(2), 299–320.

Park, D. C., & Payer, D. (2006). Working memory across the adult lifespan. In E. Bialystok & F. I. M. Craik (Eds.), *Lifespan cognition: Mechanisms of change,* 128–142. Oxford: Oxford University Press.

Parker, S. P. A., Smith, S. E., Stephan, K. L., Martin, R. L., & McAnally, K. I. (2004). Effects of supplementing head-down displays with 3-D audio during visual target acquisition. *International Journal of Aviation Psychology, 14*(3), 277–295.

Pascual-Leone, A. (2001). The brain that plays music and is changed by it. *Annals of the New York Academy of Sciences, 930*(1), 315–329.

Pashler, H. (1994). Dual-task interference in simple tasks: Data and theory. *Psychological Bulletin, 116*(2), 220–244.

Pashler, H. (1998a). *Attention.* London: Psychology Press/Taylor & Francis.

Pashler, H. (1998b). *The psychology of attention.* Cambridge, MA: MIT Press.

Pashler, H., Harris, C. R., & Nuechterlein, K. H. (2008). Does the central bottleneck encompass voluntary selection of hedonically based choices? *Experimental Psychology, 55*(5), 313–321.

Passolunghi, M. C., & Siegel, L. S. (2001). Short-term memory, working memory, and inhibitory control in children with difficulties in arithmetic problem solving. *Journal of Experimental Child Psychology, 80,* 44–67.

Patel, A. D. (2003). Language, music, syntax and the brain. *Nature Neuroscience, 6*(7), 674–681.

Patel, A. D., Iversen, J. R., Bregman, M. R., & Schulz, I. (2009). Experimental evidence for synchronization to a musical beat in a nonhuman animal. *Current Biology, 19*(10), 880–880.

Patston, L. L. M., Corballis, M. C., Hogg, S. L., & Tippett, L. J. (2006). The neglect of musicians: Line bisection reveals an opposite bias. *Psychological Science, 17*(12), 1029–1031.

Patten, C. J. D., Kircher, A., Ostlund, J., & Nilsson, L. (2004). Using mobile telephones: Cognitive workload and attention resource allocation. *Accident Analysis and Prevention, 36*(3), 341–350.

Patterson, R. D. (1982). *Guidelines for auditory warning systems* (No. CA Paper 82017). London: Civil Aviation Authority.

Patterson, R. D. (1990a). Auditory warning sounds in the work environment. *Philosophical Transactions of the Royal Society of London. Series B, Biological Sciences, 327*(1241), 485–492.

Patterson, R. D. (1990b). Auditory warning sounds in the work environment. In D. E. Broadbent, J. T. Reason, & A. D. Baddeley (Eds.), *Human factors in hazardous situations,* 37–44. Oxford: Clarendon Press/Oxford University Press.

Patterson, R. D., & Datta, A. J. (1999). Extending the domain of auditory warning sounds: Creative use of high frequencies and temporal asymmetry. In N. A. Stanton & J. Edworthy (Eds.), *Human factors in auditory warnings,* 73–88. Aldershot, UK: Ashgate.

Paulesu, E., Frith, C. D., & Frackowiak, R. S. (1993). The neural correlates of the verbal component of working memory. *Nature, 362*(6418), 342–345.

Pauletto, S., & Hunt, A. (2009). Interactive sonification of complex data. *International Journal of Human-Computer Studies, 67*(11), 923–933.

Pavlovic, C. V. (1993). Problems in the prediction of speech recognition performance of normal-hearing and hearing-impaired individuals. In G. A. Studebaker & I. Hochberg (Eds.), (2nd ed.). Acoustical factors affecting hearing aid performance, 221–234. Boston: Allyn & Bacon.

Payne, D. G., Peters, L. J., Birkmire, D. P., & Bonto, M. A. (1994). Effects of speech intelligibility level on concurrent visual task performance. *Human Factors, 36*(3), 441–475.

Pearson, J. D., Morrell, C. H., Gordon-Salant, S., Brant, L. J., Metter, E. J., Klein, L. L., et al. (1995). Gender differences in a longitudinal study of age-associated hearing loss. *Journal of the Acoustical Society of America, 97*(2), 1196–1205.

Peelle, J. E., McMillan, C., Moore, P., Grossman, M., & Wingfield, A. (2004). Dissociable patterns of brain activity during comprehension of rapid and syntactically complex speech: Evidence from fMRI. *Brain and Language, 91*(3), 315–325.

Penny, C. G. (1989). Modality effects and the structure of short-term verbal memory. *Memory and Cognition, 17*(4), 398–422.

Peretz, I. (2003). Brain specialization for music: New evidence from congenital amusia. In I. Peretz & R. J. Zatorre (Eds.), *The cognitive neuroscience of music,* 192–203. Oxford: Oxford University Press.

Peretz, I. (2006). The nature of music from a biological perspective. *Cognition, 100*(1), 1–32.

Perfect, T. J., & Maylor, E. A. (Eds.). (2000). *Models of cognitive aging.* Oxford: Oxford University Press.

Perrott, D. R., Cisneros, J., McKinley, R. L., & D'Angelo, W. R. (1996). Aurally aided visual search under virtual and free-field listening conditions. *Human Factors, 38*(4), 702–715.

Perry, N. C., Stevens, C. J., Wiggins, M. W., & Howell, C. E. (2007). Cough once for danger: Icons versus abstract warnings as informative alerts in civil aviation. *Human Factors, 49*(6), 1061–1071.

Perry, W., Geyer, M. A., & Braff, D. L. (1999). Sensorimotor gating and thought disturbance measured in close temporal proximity in schizophrenic patients. *Archives of General Psychiatry, 56*(3), 277–281.

Peterson, G. E., & Lehiste, I. (1960). Duration of syllable nuclei in English. *Journal of the Acoustical Society of America, 32*(6), 693–703.

Petersson, K. M., Reis, A., Askeloef, S., Castro-Caldas, A., & Ingvar, M. (2000). Language processing modulated by literacy: A network analysis of verbal repetition in literate and illiterate subjects. *Journal of Cognitive Neuroscience, 12*(3), 364–382.

Petocz, A., Keller, P. E., & Stevens, C. J. (2008). Auditory warnings, signal-referent relations, and natural indicators: Re-thinking theory and application. *Journal of Experimental Psychology-Applied, 14*(2), 165–178.

Phillips, J. L., Shiffrin, R. M., & Atkinson, R. C. (1967). Effects of list length on short-term memory. *Journal of Verbal Learning and Verbal Behavior, 6*(3), 303–311.

Picard, M., & Bradley, J. S. (2001). Revisiting speech interference in classrooms. *Audiology, 40*(5), 221–244.

Picard, M., Girard, S. A., Simard, M., Larocque, R., Leroux, T., & Turcotte, F. (2008). Association of work-related accidents with noise exposure in the workplace and noise-induced hearing loss based on the experience of some 240,000 person-years of observation. *Accident Analysis and Prevention, 40*(5), 1644–1652.

Pichora-Fuller, M. K. (2008). Use of supportive context by younger and older adult listeners: Balancing bottom-up and top-down information processing. *International Journal of Audiology, 47*(S2), S72–S82.

Pichora-Fuller, M. K., & Carson, A. J. (2001). Hearing health and the listening experiences of older communicators. In M. L. Hummert & J. F. Nussbaum (Eds.), *Aging, communication, and health: Linking research and practice for successful aging*. Mahwah, NJ: Erlbaum.

Pichora-Fuller, M. K., Scheider, B. A., & Daneman, M. (1995). How young and old adults listen to and remember speech in noise. *Journal of the Acoustical Society of America, 97*(1), 593–608.

Pickett, J. M. (1956). Effects of vocal force on the intelligibility of speech sounds. *The Journal of the Acoustical Society of America, 28*(5), 902–905.

Pierce, J. R. (1999). The nature of musical sound. In D. Deutsch (Ed.), *The psychology of music* (2nd ed., 1–23). New York: Academic Press.

Pinker, S. (1994). *The language instinct: How the mind creates language*. New York: HarperCollins.

Pinker, S., & Ullman, M. T. (2002). The past and future of the past tense. *Trends in Cognitive Sciences, 6*(11), 456–463.

Pizzighello, S., & Bressan, P. (2008). Auditory attention causes visual inattentional blindness. *Perception, 37*(6), 859–866.

Plaut, D. C. (2003). Connectionist modeling of language: Examples and implications. In M. T. Banich & M. Mack (Eds.), *Mind, brain, and language: Multidisciplinary perspectives*, 143–167. Mahwah, NJ: Erlbaum.

Plomp, R. (2002). *The intelligent ear: On the nature of sound perception*. Mahwah, NJ: Erlbaum.

Plomp, R., & Mimpen, A. M. (1979). Speech reception threshold for sentences as a function of age and noise level. *Journal of the Acoustical Society of America, 66*, 1333–1342.

Polich, J., Ellerson, P. C., & Cohen, J. (1996). P300, stimulus intensity, modality, and probability. *International Journal of Psychophysiology, 23*(1–2), 55–62.

Pollitt, E. (1995). Does breakfast make a difference in school? *Journal of the American Dietetic Association, 95*(10), 1134–1139.

Populin, L. C., Tollin, D. J., & Yin, T. C. T. (2004). Effect of eye position on saccades and neuronal responses to acoustic stimuli in the superior colliculus of the behaving cat. *Journal of Neurophysiology, 92*(4), 2151–2167.

Poreh, A. (2005). Analysis of mean learning of normal participants on the Rey Auditory-Verbal Learning Test. *Psychological Assessment, 17*(2), 191–199.

Posner, M. I., & DiGirolamo, G. J. (2000). Cognitive neuroscience: Origins and promise. *Psychological Bulletin, 126*(6), 873–889.

Posner, M. I., Petersen, S. E., Fox, P. T., & Raichle, M. E. (1988). Localization of cognitive operations in the human brain. *Science, 240*(4859), 1627–1631.

Poulton, E. C. (1972). The environment at work. *Applied Ergonomics, 3*(1), 24–29.

Poulton, E. C. (1976). Arousing environmental stresses can improve performance, whatever people say. *Aviation Space and Environmental Medicine, 47*, 1193–1204.

Poulton, E. C. (1977). Continuous intense noise masks auditory feedback and inner speech. *Psychological Bulletin, 84*(5), 977–1001.

Poulton, E. C. (1978). A new look at the effects of noise upon performance. *British Journal of Psychology, 69*(4), 435–437.

Powell, H. W. R., Parker, G. J. M., Alexander, D. C., Symms, M. R., Boulby, P. A., Wheeler-Kingshott, C. A. M., et al. (2006). Hemispheric asymmetries in language-related pathways: A combined functional MRI and tractography study. *NeuroImage, 32*(1), 388–399.

Pritchard, W. S., & Hendrickson, R. (1985). The structure of human attention: Evidence for separate spatial and verbal resource pools. *Bulletin of the Psychonomic Society, 23*(3), 177–180.

Proctor, R. W., & Vu, K.-P. L. (2006). *Stimulus-response compatibility principles: Data, theory, and application.* Boca Raton, FL: Taylor & Francis/CRC Press.

Pujol, J., Lopez-Sala, A., Deus, J., Cardoner, N., Sebastian-Galles, N., Conesa, G., et al. (2002). The lateral asymmetry of the human brain studied by volumetric magnetic resonance imaging. *NeuroImage, 17*(2), 670–679.

Pylyshyn, Z. W. (1973). What the mind's eye tells the mind's brain: A critique of mental imagery. *Psychological Bulletin, 80*(1), 1–24.

Qin, L., Sakai, M., Chimoto, S., & Sato, Y. (2005). Interaction of excitatory and inhibitory frequency-receptive fields in determining fundamental frequency sensitivity of primary auditory cortex neurons in awake cats. *Cerebral Cortex, 15*(9), 1371–1383.

Quednow, B. B., Kuhn, K.-U., Hoenig, K., Maier, W., & Wagner, M. (2004). Prepulse inhibition and habituation of acoustic startle response in male MDMA ("Ecstasy") users, cannabis users, and healthy controls. *Neuropsychopharmacology, 29*(5), 982–990.

Rabbitt, P. M. (1968). Channel-capacity, intelligibility and immediate memory. *The Quarterly Journal of Experimental Psychology, 20*(3), 241–248.

Rabbitt, P. M. A. (1991). Mild hearing loss can cause apparent memory failures which increase with age and reduce IQ. *Acta Otolaryngolica, 476*(Suppl.), 167–176.

Rabbitt, P. M. A. (1991). Management of the working population. *Ergonomics, 34*(6), 775–790.

Rahman, Q., Cockburn, A., & Govier, E. (2008). A comparative analysis of functional cerebral asymmetry in lesbian women, heterosexual women, and heterosexual men. *Archives of Sexual Behavior, 37*(4), 566–571.

Rahman, Q., Kumari, V., & Wilson, G. D. (2003). Sexual orientation-related differences in prepulse inhibition of the human startle response. *Behavioral Neuroscience, 117*(5), 1096–1102.

Rakauskas, M. E., Gugerty, L. J., & Ward, N. J. (2004). Effects of naturalistic cell phone conversations on driving performance. *Journal of Safety Research, 35*(4), 453–464.

Ralston, J. V., Pisoni, D. B., Lively, S. E., Greene, B. G., & Mullennix, J. W. (1991). Comprehension of synthetic speech produced by rule: Word monitoring and sentence-by-sentence listening times. *Human Factors, 33*(4), 471–491.

Ranney, T., Watson, G. S., Mazzae, E. N., Papelis, Y. E., Ahmad, O., & Wightman, J. R. (2005). *Examination of the distraction effects of wireless phone interfaces using the National Advanced Driving Simulator—Final report on a freeway study* (No. DOT HS 809 787). Washington, DC: National Highway Traffic Safety Administration.

Rauschecker, J. P. (1998). Cortical processing of complex sounds. *Current Opinion in Neurobiology, 8*(4), 516–521.

Rauschecker, J. P., & Tian, B. (2000). Mechanisms and streams for processing of "what" and "where" in the auditory cortex. *Proceedings of the National Academy of Science of the United States of America, 97*(22), 11800–11806.

Rauschecker, J. P., Tian, B., & Hauser, M. (1995). Processing of complex sounds in the macaque nonprimary auditory cortex. *Science, 268*(5207), 111–115.

Rauscher, F. H., & Hinton, S. C. (2006). The Mozart effect: Music listening is not music instruction. *Educational Psychologist, 41*(4), 233–238.

Rauscher, F. H., Shaw, G. L., & Ky, K. N. (1993). Music and spatial task performance. *Nature, 365*, 611.

Raz, A., Deouell, L. Y., & Bentin, S. (2001). Is pre-attentive processing compromised by prolonged wakefulness? Effects of total sleep deprivation on the mismatch negativity. *Psychophysiology, 38*(5), 787–795.

Read, H. L., Winer, J. A., & Schreiner, C. E. (2002). Functional architecture of auditory cortex. *Current Opinion in Neurobiology, 12*(4), 433–440.

Readinger, W. O., Chatziastros, A., Cunningham, D. W., Bulthoff, H. H., & Cutting, J. E. (2002). Gaze-eccentricity effects on road position and steering. *Journal of Experimental Psychology: Applied, 8*(4), 247–258.

Reagan, I., & Baldwin, C. L. (2006). Facilitating route memory with auditory route guidance systems. *Journal of Environmental Psychology, 26*(2), 146–155.

Recanzone, G. H., Schreiner, C. E., & Merzenich, M. M. (1993). Plasticity in the frequency representation of primary auditory cortex following discrimination training in adult owl monkeys. *The Journal of Neuroscience, 13*(1), 87–103.

Recarte, M. A., & Nunes, L. (2002). Mental load and loss of control over speed in real driving: Towards a theory of attentional speed control. *Transportation Research Part F: Traffic Psychology and Behaviour, 5*(2), 111–122.

Redelmeier, D. A., & Tibshirani, R. J. (1997). Association between cellular-telephone calls and motor vehicle collisions. *New England Journal of Medicine, 336*(7), 453–458.

Reed, M. P., & Green, P. A. (1999). Comparison of driving performance on-road and in a low-cost simulator using a concurrent telephone dialling task. *Ergonomics, 42*(8), 1015–1037.

Reitan, R. M., & Wolfson, D. (2000). The neuropsychological similarities of mild and more severe head injury. *Archives of Clinical Neuropsychology, 15*(5), 433–442.

Reitan, R. M., & Wolfson, D. (2005). The effect of conation in determining the differential variance among brain-damaged and nonbrain-damaged persons across a broad range of neuropsychological tests. *Archives of Clinical Neuropsychology, 20*(8), 957–966.

Reuter-Lorenz, P. A., & Cappell, K. A. (2008). Neurocognitive aging and the compensation hypothesis. *Current Directions in Psychological Science, 17*(3), 177–182.

Reuter-Lorenz, P. A., & Sylvester, C.-Y. C. (2005). The cognitive neuroscience of working memory and aging. In R. Cabeza, L. Nyberg, & D. C. Park (Eds.), *Cognitive neuroscience of aging*, 186–217. New York: Oxford University Press.

Revill, K. P., Tanenhaus, M. K., & Aslin, R. N. (2008). Context and spoken word recognition in a novel lexicon. *Journal of Experimental Psychology: Learning, Memory, and Cognition, 34*(5), 1207–1223.

Risser, M. R., McNamara, D. S., Baldwin, C. L., Scerbo, M. W., & Barshi, I. (September-October, 2002). *Interference effects on the recall of words heard and read: Considerations for ATC communication.* Proceedings of the Human Factors and Ergonomics Society 46th Annual Meeting. Human Factors and Ergonomics Society. Baltimore, MD.

Risser, M. R., Scerbo, M. W., Baldwin, C. L., & McNamara, D. S. (April, 2003). *ATC commands executed in speech and text formats: Effects of task interference.* Paper presented at the Proceedings of the 12th Biennial International Symposium on Aviation Psychology. Dayton, OH.

Risser, M. R., Scerbo, M. W., Baldwin, C. L., & McNamara, D. S. (March, 2004). *Implementing voice and datalink commands under task interference during simulated flight.* Paper presented at the Proceedings of the 5th HPSAA II Conference, Human Performance, Situation Awareness and Automation Technology, Daytona Beach, FL.

Risser, M. R., Scerbo, M. W., Baldwin, C. L., & McNamara, D. S. (2006). *Interference timing and acknowledgement response with voice and datalink ATC commands.* Paper presented at the Human Factors and Ergonomics Society, San Francisco.

Robinson, C. P., & Eberts, R. E. (1987). Comparison of speech and pictorial displays in a cockpit environment. *Human Factors, 29*(1), 31–44.

Roediger, H. L., & Crowder, R. G. (1976). Recall instructions and the suffix effect. *American Journal of Psychology, 89*(1), 115–125.

Rogers, C. L., Lister, J. J., Febo, D. M., Besing, J. M., & Abrams, H. B. (2006). Effects of bilingualism, noise, and reverberation on speech perception by listeners with normal hearing. *Applied Psycholinguistics, 27*(3), 465–485.

Rogers, W. A., Lamson, N., & Rousseau, G. K. (2000). Warning research: An integrative perspective. *Human Factors, 42*(1), 102–139.

Rohrer, J. D., Ridgway, G. R., Crutch, S. J., Hailstone, J., Goll, J. C., Clarkson, M. J., et al. (2009). Progressive logopenic/phonological aphasia: Erosion of the language network. *NeuroImage, 49,* 984–993.

Romanski, L. M., Tian, B., Fritz, J., Mishkin, M., Goldman-Rakic, P. S., & Rauschecker, J. P. (1999). Dual streams of auditory afferents target multiple domains in the primate prefrontal cortex. *Nature Neuroscience, 2*(12), 1131–1136.

Romanski, L. M., Tian, B., Fritz, J. B., Mishkin, M., Goldman-Rakic, P. S., & Rauschecker, J. P. (2000). "What," "where" and "how" in auditory cortex. Reply. *Nature Neuroscience, 3*(10), 966.

Ronkainen, S. (2001, July 29–August 1). *Earcons in motion—Defining language for an intelligent mobile device.* Paper presented at the Proceedings of the 2001 International Conference on Auditory Display, Espoo, Finland.

Roring, R. W., Hines, F. G., & Charness, N. (2007). Age differences in identifying words in synthetic speech. *Human Factors: The Journal of the Human Factors and Ergonomics Society, 49*(1), 25–31.

Rosenblum, L. D. (2004). Perceiving articulatory events: Lessons for an ecological psychoacoustics. In J. Neuhoff (Ed.), *Ecological psychoacoustics,* 219–248. San Francisco: Elsevier.

Roske-Hofstrand, R. J., & Murphy, R. D. (1998). Human information processing in air traffic control. In M. W. Smolensky & E. S. Stein (Eds.), *Human factors in air traffic control,* 65–114. New York: Academic Press.

Ross, L. A., Saint-Amour, D., Leavitt, V. M., Javitt, D. C., & Foxe, J. J. (2007). Do you see what I am saying? Exploring visual enhancement of speech comprehension in noisy environments. *Cerebral Cortex, 17*(5), 1147–1153.

Routh, D. A. (1971). Independence of the modality effect and amount of silent rehearsal in immediate serial recall. *Journal of Verbal Learning and Verbal Behavior, 10*(2), 213–218.

Routh, D. A. (1976). An across-the-board modality effect in immediate serial recall. *The Quarterly Journal of Experimental Psychology, 28*(2), 285–304.

Rubio, S., Diaz, E., Martin, J., & Puente, J. M. (2004). Evaluation of subjective mental workload: A comparison of SWAT, NASA-TLX, and workload profile methods. *Applied Psychology: An International Review, 53*(1), 61–86.

Rumelhart, D. E., McClelland, J. L., & Group, P. R. (Eds.). (1986). *Parallel distributed processing, Vol. 1: Foundations.* Cambridge, MA: MIT Press.

Sacks, H., Schegloff, E. A., & Jefferson, G. (1974). A simplest systematics for the organization of turn-taking for conversation. *Language, 50*(4), 696–735.

Salame, P., & Baddeley, A. (1990). The effects of irrelevant speech on immediate free recall. *Bulletin of the Psychonomic Society, 28*(6), 540–542.

Salame, P., & Baddeley, A. D. (1982). Disruption of short-term memory by unattended speech: Implications for the structure of working memory. *Journal of Verbal Learning and Verbal Behavior, 21*(2), 150–164.

Salame, P., & Baddeley, A. D. (1987). Noise, unattended speech and short-term memory. *Ergonomics, 30*(8), 1185–1194.

Salame, P., & Baddeley, A. D. (1989). Effects of background music on phonological short-term memory. *Quarterly Journal of Experimental Psychology: Human Experimental Psychology, 41*(1-A), 107–122.

Salthouse, T. A. (1994). The nature of the influence of speed on adult age differences in cognition. *Developmental Psychology, 30*(2), 240–259.

Salthouse, T. A. (1999). Theories of cognition. In V. L. Bengtson & K. W. Schaie (Eds.), *Handbook of theories of aging,* 196–208. New York: Springer.

Salthouse, T. A. (2003). Memory aging from 18 to 80. *Alzheimer Disease and Associated Disorders, 17*(3), 162–167.

Sams, M., Kaukoranta, E., Hamalainen, M., & Naatanen, R. (1991). Cortical activity elicited by changes in auditory stimuli: Different sources for the magnetic N100m and mismatch responses. *Psychophysiology, 28*(1), 21–29.

Samuel, A. (1996a). Phoneme restoration. *Language and Cognitive Processes, 11*(6), 647–653.

Samuel, A. G. (1996b). Does lexical information influence the perceptual restoration of phonemes? *Journal of Experimental Psychology: General, 125*(1), 28–51.

Sanders, M. S., & McCormick, E. J. (1993). *Human factors in engineering and design* (7th ed.). New York: McGraw-Hill.

Sanders, T. J. M., & Gernsbacher, M. A. (2004). Accessibility in text and discourse processing. *Discourse Processes, 37*(2), 79–89.

Sanderson, P. (2006). The multimodal world of medical monitoring displays. *Applied Ergonomics, 37*(4), 501–512.

Santa, J. L. (1977). Spatial transformations of words and pictures. *Journal of Experimental Psychology: Human Learning and Memory, 3*(4), 418–427.

Sarno, K. J., & Wickens, C. D. (1995). Role of multiple resources in predicting time-sharing efficiency: Evaluation of three workload models in a multiple-task setting. *International Journal of Aviation Psychology, 5*(1), 107–130.

Scerbo, M. W., Risser, M. R., Baldwin, C. L., & McNamara, D. S. (October, 2003). *The effects of task interference and message length on implementing speech and simulated data link commands.* Paper presented at the Proceedings of the Human Factors and Ergonomics Society 47th Annual Meeting. Denver, CO.

Schall, U., & Ward, P. B. (1996). "Prepulse inhibition" facilitates a liberal response bias in an auditory discrimination task. *NeuroReport: For Rapid Communication of Neuroscience Research, 7*(2), 652–656.

Scharf, B. (1971). Fundamentals of auditory masking. *Audiology, 10*(1), 30–40.

Schellenberg, E. G. (2003). Does exposure to music have beneficial side effects? In I. Peretz & R. J. Zatorre (Eds.), *The cognitive neuroscience of music,* 430–448. Oxford: Oxford University Press.

Schellenberg, E. G. (2005). Music and cognitive abilities. *Current Directions in Psychological Science, 14*(6), 317–320.

Schieber, F., & Baldwin, C. L. (1996). Vision, audition, and aging research. In F. Blanchard-Fields & T. M. Hess (Eds.), *Perspectives on cognitive change in adulthood and aging*, 122–162. New York: McGraw-Hill.

Schirmer, A., Kotz, S. A., & Friederici, A. D. (2002). Sex differentiates the role of emotional prosody during word processing. *Cognitive Brain Research, 14*(2), 228–233.

Schneider, B. A., Daneman, M., Murphy, D. R., & See, S. K. (2000). Listening to discourse in distracting settings: The effects of aging. *Psychology and Aging, 15*(1), 110–125.

Schneider, B. A., Daneman, M., & Pichora-Fuller, M. K. (2002). Listening in aging adults: From discourse comprehension to psychoacoustics. *Canadian Journal of Experimental Psychology, 56*(3), 139–152.

Schneider, B. A., Pichora-Fuller, M., & Daneman, M. (2010). Effects of senescent changes in audition and cognition and spoken language comprehension. In S. Gordon-Salant, R. D. Frisina, A. N. Popper, & R. R. Fay (Eds.), *The aging auditory system*, 167–210. New York: Springer.

Schneider, B. A., & Pichora-Fuller, M. K. (2000). Implications of perceptual deterioration for cognitive aging research. In F. I. M. Craik & T. A. Salthouse (Eds.), *The handbook of aging and cognition* (2nd ed.), 155–219. Mahwah, NJ: Erlbaum.

Schneider, B. A., & Pichora-Fuller, M. K. (2001). Age-related changes in temporal processing: Implications for speech perception. *Seminars in Hearing, 22*(3), 227–239.

Schneider, W., & Shiffrin, R. M. (1977). Controlled and automatic human information processing: I. Detection, search, and attention. *Psychological Review, 84*(1), 1–66.

Schubert, T., Fischer, R., & Stelzel, C. (2008). Response activation in overlapping tasks and the response-selection bottleneck. *Journal of Experimental Psychology: Human Perception and Performance, 34*(2), 376–397.

Schultheis, H., & Jameson, A. (2004). Assessing cognitive load in adaptive hypermedia systems: Physiological and behavioral methods. In W. Nejdl & P. De Bra (Eds.), *Adaptive hypermedia and adaptive web-based systems*, 18–24. Eindhoven, Netherlands: Springer.

Schumacher, E. H., Seymour, T. L., Glass, J. M., Fencsik, D. E., Lauber, E. J., Kieras, D. E., et al. (2001). Virtually perfect time sharing in dual-task performance: Uncorking the central cognitive bottleneck. *Psychological Science, 12*(2), 101–108.

Schutte, P. C., & Trujillo, A. C. (September, 1996). *Flight crew task management in non-normal situations.* Paper presented at the 40th Annual Meeting of the Human Factors and Ergonomics Society, Philadelphia, PA.

Scialfa, C. T. (2002). The role of sensory factors in cognitive aging research. *Canadian Journal of Experimental Psychology/Revue canadienne de psychologie expérimentale, 56*(3), 153–163.

Scott, S. K., Young, A. W., Calder, A. J., Hellawell, D. J., Aggleton, J. P., & Johnson, M. (1997). Impaired auditory recognition of fear and anger following bilateral amygdala lesions. *Nature, 385*(6613), 254–257.

See, J. E., Howe, S. R., Warm, J. S., & Dember, W. N. (1995). Meta-analysis of the sensitivity decrement in vigilance. *Psychological Bulletin, 117*(2), 230–249.

Seth-Smith, M., Ashton, R., & McFarland, K. (1989). A dual-task study of sex differences in language reception and production. *Cortex, 25*(3), 425–431.

Shah, P., & Miyake, A. (1996). The separability of working memory resources for spatial thinking and language processing: An individual differences approach. *Journal of Experimental Psychology: General, 125*(1), 4–27.

Sharps, M. J., Price, J. L., & Bence, V. M. (1996). Visual and auditory information as determinants of primacy effects. *Journal of General Psychology, 123*(2), 123–136.

Shaw, T. H., Warm, J. S., Finomore, V., Tripp, L., Matthews, G., Weiler, E., et al. (2009). Effects of sensory modality on cerebral blood flow velocity during vigilance. *Neuroscience Letters, 461*(3), 207–211.

Shepard, R. N., & Metzler, J. (1971). Mental rotation of three-dimensional objects. *Science, 171*(3972), 701–703.

Sherer, M., Parsons, O. A., Nixon, S. J., & Adams, R. L. (1991). Clinical validity of the Speech-Sounds Perception test and the Seashore Rhythm test. *Journal of Clinical and Experimental Neuropsychology, 13*(5), 741–751.

Sherwin, B. B. (2007). Does estrogen protect against cognitive aging in women? *Current Directions in Psychological Science, 16*, 275–279.

Sherwin, B. B. (2009). Estrogen therapy: Is time of initiation critical for neuroprotection? *Nature Reviews Endocrinology, 5*(11), 620–627.

Shestakova, A., Brattico, E., Soloviev, A., Klucharev, V., & Huotilainen, M. (2004). Orderly cortical representation of vowel categories presented by multiple exemplars. *Cognitive Brain Research, 21*(3), 342–350.

Shinar, D. (2008). Looks are (almost) everything: Where drivers look to get information. *Human Factors, 50*(3), 380–384.

Shinar, D., & Schechtman, E. (2002). Headway feedback improves intervehicular distance: A field study. *Human Factors, 44*(3), 474–481.

Shinn, J. B., Baran, J. A., Moncrieff, D. W., & Musiek, F. E. (2005). Differential attention effects on dichotic listening. *Journal of the American Academy of Audiology, 16*(4), 205–218.

Shinohara, T., Bredberg, G., Ulfendahl, M., Pyykko, I., Olivius, N. P., Kaksonen, R., et al. (2002). Neurotrophic factor intervention restores auditory function in deafened animals. *Proceedings of the National Academy of Sciences of the United States of America, 99*(3), 1657–1660.

Shtyrov, Y., Kujala, T., Lyytinen, H., Ilmoniemi, R. J., & Naatanen, R. (2000). Auditory cortex evoked magnetic fields and lateralization of speech processing. *NeuroReport: For Rapid Communication of Neuroscience Research, 11*(13), 2893–2896.

Siemens, J., Lillo, C., Dumont, R. A., Reynolds, A., Williams, D. S., Gillespie, P. G., et al. (2004). Cadherin 23 is a component of the tip link in hair-cell sterocilia. *Nature, 428*(6986), 950–954.

Silman, S., & Silverman, C. A. (1991). *Auditory diagnosis: Principles and applications.* New York: Academic Press.

Silvia, P. J., & Brown, E. M. (2007). Anger, disgust, and the negative aesthetic emotions: Expanding an appraisal model of aesthetic experience. *Psychology of Aesthetics, Creativity, and the Arts, 1*(2), 100–106.

Sim, T.-C., & Martinez, C. (2005). Emotion words are remembered better in the left ear. *Laterality: Asymmetries of Body, Brain and Cognition, 10*(2), 149–159.

Simpson, B., D., Brungart, D., S., Dallman, R. C., Yasky, R. J., & Romigh, G. D. (2008, September 22–26). *Flying by ear: Blind flight with a music-based artificial horizon.* Paper presented at the Human Factors and Ergonomics Society, New York.

Simpson, C. A., & Marchionda-Frost, K. (1984). Synthesized speech rate and pitch effects on intelligibility of warning messages for pilots. *Human Factors, 26*(5), 509–517.

Simpson, C. A., & Williams, D. H. (1980). Response time effects of alerting tone and semantic context for synthesized voice cockpit warnings. *Human Factors, 22*(3), 319–330.

Sinnett, S., Soto-Faraco, S., & Spence, C. (2008). The co-occurrence of multisensory competition and facilitation. *Acta Psychologica, 128*(1), 153–161.

Siuru, B. (2001). Major field test of collision avoidance system underway in Pittsburgh. *Mass Transit Magazine* Retrieved April 26, 2002, from http://www.masstransitmag.com

Sjogren, P., Christrup, L. L., Petersen, M. A., & Hojsted, J. (2005). Neuropsychological assessment of chronic non-malignant pain patients treated in a multidisciplinary pain centre. *European Journal of Pain, 9*(4), 453–462.

Slattery, T. J. (2009). Word misperception, the neighbor frequency effect, and the role of sentence context: Evidence from eye movements. *Journal of Experimental Psychology: Human Perception and Performance, 35*(6), 1969–1975.

Slawinski, E. B., & MacNeil, J. F. (2002). Age, music, and driving performance: Detection of external warning sounds in vehicles. *Psychomusicology, 18,* 123–131.

Smith, A. (1989). A review of the effects of noise on human performance. *Scandinavian Journal of Psychology, 30*(3), 185–206.

Smith, A. P. (1982). The effects of noise and task priority on recall of order and location. *Acta Psychologica, 51*(3), 245–255.

Smith, A. P., & Jones, D. M. (1992). Noise and performance. In A. P. Smith & D. M. Jones (Eds.), *Handbook of human performance. Vol. 1: The physical environment,* 1–28. London: Harcourt Brace Jovanovich.

Smith, A. P., Jones, D. M., & Broadbent, D. E. (1981). The effects of noise on recall of categorized lists. *British Journal of Psychology, 72*(3), 299–316.

Smith, E. E., & Jonides, J. (1997). Working memory: A view from neuroimaging. *Cognitive Psychology, 33*(1), 5–42.

Smith, E. E., & Jonides, J. (1999). Storage and executive processes in the frontal lobes. *Science, 283*(5408), 1657–1661.

Smits, C., & Houtgast, T. (2007). Recognition of digits in different types of noise by normal-hearing and hearing-impaired listeners. *International Journal of Audiology, 46*(3), 134–144.

Sodnik, J., Dicke, C., Tomazic, S., & Billinghurst, M. (2008). A user study of auditory versus visual interfaces for use while driving. *International Journal of Human-Computer Studies, 66*(5), 318–332.

Sommers, M. S., Nygaard, L. C., & Pisoni, D. B. (1994). Stimulus variability and spoken word recognition: I. Effects of variability in speaking rate and overall amplitude. *Journal of the Acoustical Society of America, 96*(3), 1314–1324.

Speaks, C., Karmen, J. L., & Benitez, L. (1967). Effect of a competing message on synthetic sentence identification. *Journal of Speech and Hearing Research, 10*(2), 390–396.

Spence, C., & Driver, J. (Eds.). (2004). *Crossmodal space and crossmodal attention.* Oxford: Oxford University Press.

Spence, C., Nicholls, M. E. R., Gillespie, N., & Driver, J. (1998). Cross-modal links in exogenous covert spatial orienting between touch, audition, and vision. *Perception & Psychophysics, 60*(4), 544–557.

Spence, C., & Read, L. (2003). Speech shadowing while driving: On the difficulty of splitting attention between eye and ear. *Psychological Science, 14*(3), 251–256.

Sperandio, J.-C. (1978). The regulation of working methods as a function of work-load among air traffic controllers. *Ergonomics, 21*(3), 195–202.

Speranza, F., Daneman, M., & Schneider, B. A. (2000). How aging affects the reading of words in noisy backgrounds. *Psychology and Aging, 15*(2), 253–258.

Sperling, G. (1960). The information available in brief visual presentation. *Psychological Monographs, 74*(11), 29.

Sperling, G. (1967). Successive approximations to a model for short term memory. *Acta Psychologica, 27,* 285–292.

Squire, P. N., Barrow, J. H., Durkee, K. T., Smith, C. M., Moore, J. C., & Parasuraman, R. (2010). RIMDAS: A proposed system for reducing runway incursions. *Ergonomics in Design, 18,* 10–17.

Squire, L. R. (1986). Mechanisms of memory. *Science, 232*(4758), 1612–1619.

Srinivasan, R., & Jovanis, P. P. (1997a). Effect of in-vehicle route guidance systems on driver workload and choice of vehicle speed: Findings from a driving simulator experiment. In Y. I. Noy (Ed.), *Ergonomics and safety of intelligent driver interfaces,* 97–114. Mahwah, NJ: Lawrence Erlbaum.

Srinivasan, R., & Jovanis, P. P. (1997b). Effect of selected in-vehicle route guidance systems on driver reaction times. *Human Factors, 39*(2), 200–215.

Stansfeld, S. A., Berglund, B., Clark, C., Lopez-Barrio, I., Fischer, P., Ohrstrom, E., et al. (2005). Aircraft and road traffic noise and children's cognition and health: A cross-national study. *Lancet, 365*(9475), 1942–1949.

Stanton, N. A., & Baber, C. (1999). Speech-based alarm displays. In N. A. Stanton & J. Edworthy (Eds.), *Human factors in auditory warnings*, 243–261. Aldershot, UK: Ashgate.

Stanton, N. A., & Edworthy, J. (1999a). Auditory warnings and displays: An overview. In N. A. Stanton & J. Edworthy (Eds.), *Human factors in auditory warnings*, 3–30. Aldershot, UK: Ashgate.

Stanton, N. A., & Edworthy, J. (Eds.). (1999b). *Human factors in auditory warnings*. Aldershot, UK: Ashgate.

Steele, K. M., Bass, K. E., & Crook, M. D. (1999). The mystery of the Mozart effect: Failure to replicate. *Psychological Science, 10*(4), 366–369.

Steenari, M.-R., Vuontela, V., Paavonen, E. J., Carlson, S., Fjallberg, M., & Aronen, E. T. (2003). Working memory and sleep in 6- to 13-year-old schoolchildren. *Journal of the American Academy of Child and Adolescent Psychiatry, 42*(1), 85–92.

Stelzel, C., Schumacher, E. H., Schubert, T., & D'Esposito, M. (2006). The neural effect of stimulus-response modality compatibility on dual-task performance: An fMRI study. *Psychological Research/Psychologische Forschung, 70*(6), 514–525.

Stevens, R. D., Edwards, A. D. N., & Harling, P. A. (1997). Access to mathematics for visually disabled students through multimodal interaction. *Human-Computer Interaction, 12*(1), 47–92.

Stevens, S. S. (1972). Stability of human performance under intense noise. *Journal of Sound and Vibration, 21*(1), 35–36, IN1, 37–56.

Stine, E. A. L., Soederberg, L. M., & Morrow, D. G. (1996). Language and discourse processing through adulthood. In F. Blanchard-Fields & T. M. Hess (Eds.), *Perspectives on cognitive change in adulthood and aging*, 250–290. New York: McGraw-Hill.

Stine, E. L., Wingfield, A., & Poon, L. W. (1986). How much and how fast: Rapid processing of spoken language in later adulthood. *Psychology and Aging, 1*(4), 303–311.

Strange, W., Verbrugge, R. R., Shankweiler, D. P., & Edman, T. R. (1976). Consonant environment specifies vowel identity. *Journal of the Acoustical Society of America, 60*(1), 213–224.

Strayer, D. L., & Drews, F. A. (2004). Profiles in driver distraction: Effects of cell phone conversations on younger and older drivers. *Human Factors, 46*(4), 640–649.

Strayer, D. L., & Drews, F. A. (2007). Cell-phone-induced driver distraction. *Current Directions in Psychological Science, 16*(3), 128–131.

Strayer, D. L., Drews, F. A., & Johnston, W. A. (2003). Cell phone-induced failures of visual attention during simulated driving. *Journal of Experimental Psychology: Applied, 9*(1), 23–32.

Strayer, D. L., & Johnston, W. A. (2001). Driven to distraction: Dual-task studies of simulated driving and conversing on a cellular telephone. *Psychological Science, 12*(6), 462–466.

Streeter, L. A., Vitello, D., & Wonsiewicz, S. A. (1985). How to tell people where to go: Comparing navigational aids. *International Journal of Man Machine Studies, 22*(5), 549–562.

Sturm, J. A., & Seery, C. H. (2007). Speech and articulatory rates of school-age children in conversation and narrative contexts. *Language, Speech, and Hearing Services in Schools, 38*(1), 47–59.

Stutts, J. C., Reinfurt, D. W., Staplin, L., & Rodgman, E. A. (2001). *The role of driver distraction in traffic crashes* (Technical Report). Washington, DC: AAA Foundation.

Suied, C., Susini, P., & McAdams, S. (2008). Evaluating warning sound urgency with reaction times. *Journal of Experimental Psychology-Applied, 14*(3), 201–212.

Sumby, W. H., & Irwin, P. (1954). Visual contribution to speech intelligibility in noise. *The Journal of the Acoustical Society of America, 26*(2), 212–215.

Swanson, H., & Howell, M. (2001). Working memory, short-term memory, and speech rate as predictors of children's reading performance at different ages. *Journal of Educational Psychology, 93*(4), 720–734.

Swanson, S. J., & Dengerink, H. A. (1988). Changes in pure-tone thresholds and temporary threshold shifts as a function of menstrual-cycle and oral-contraceptives. *Journal of Speech and Hearing Research, 31*(4), 569–574.

Sweeney, J. E. (1999). Raw, demographically altered, and composite Halstead-Reitan Battery data in the evaluation of adult victims of nonimpact acceleration forces in motor vehicle accidents. *Applied Neuropsychology, 6*(2), 79–87.

Swick, D. (2005). ERPs in neuropsychological populations. In T. C. Handy (Ed.), *Event-related potentials: A methods handbook*, 299–321. Cambridge, MA: MIT Press.

Syka, J., Popelar, J., Kvasnak, E., & Suta, D. (1998). Processing of vocalization signals in neurons of the inferior colliculus and medial geniculate body. In P. W. F. Poon & J. F. Brugge (Eds.), *Central auditory processing and neural modeling*, 1–11. New York: Plenum Press.

Szalma, J. L., Warm, J. S., Matthews, G., Dember, W. N., Weiler, E. M., Meier, A., et al. (2004). Effects of sensory modality and task duration on performance, workload, and stress in sustained attention. *Human Factors, 46*(2), 219–233.

Takegata, R., Huotilainen, M., Rinne, T., Naatanen, R., & Winkler, I. (2001). Changes in acoustic features and their conjunctions are processed by separate neuronal populations. *NeuroReport: For Rapid Communication of Neuroscience Research, 12*(3), 525–529.

Tannen, R. S., Nelson, W. T., Bolia, R. S., Warm, J. S., & Dember, W. N. (2004). Evaluating adaptive multisensory displays for target localization in a flight task. *International Journal of Aviation Psychology, 14*(3), 297–312.

Taylor, J. L., Yesavage, J. A., Morrow, D. G., & Dolhert, N. (1994). The effects of information load and speech rate on younger and older aircraft pilots' ability to execute simulated air-traffic controller instructions. *Journals of Gerontology, 49*(5), P191–P200.

Taylor, P. (2009). *Text-to-speech synthesis*. Cambridge: Cambridge University Press.

Taylor-Cooke, P. A., & Fastenau, P. S. (2004). Effects of test order and modality on sustained attention in children with epilepsy. *Child Neuropsychology, 10*(3), 212–221.

Temperley, D. (2001). *The cognition of basic musical structures*. Cambridge, MA: MIT Press.

Tempest, W. (Ed.). (1985). *The noise handbook*. London: Academic Press.

Tervaniemi, M., & Hugdahl, K. (2003). Lateralization of auditory-cortex functions. *Brain Research Reviews, 43*(3), 231–246.

Thompson, L. A., Malloy, D. M., & LeBlanc, K. L. (2009). Lateralization of visuospatial attention across face regions varies with emotional prosody. *Brain and Cognition, 69*(1), 108–115.

Thompson, W. F., Schellenberg, E. G., & Husain, G. (2001). Arousal, mood, and the Mozart effect. *Psychological Science, 12*(3), 248–251.

Tiitinen, H., May, P., Reinikainen, K., & Naatanen, R. (1994). Attentive novelty detection in humans is governed by pre-attentive sensory memory. *Nature, 372*(6501), 90–92.

Tijerina, L., Johnston, S., Parmer, E., & Winterbottom, M. D. (2000). *Driver distraction with wireless telecommunications and route guidance information* (No. DOT HS No. 809-069): National Highway Traffic Safety Administration.

Till, R., Mross, E. F., & Kintsch, W. (1988). Time course of priming for associate and inference words in discourse context. *Memory and Cognition, 16*, 283–298.

Tombaugh, T. N. (2006). A comprehensive review of the Paced Auditory Serial Addition Test (PASAT). *Archives of Clinical Neuropsychology, 21*(1), 53–76.

Tombu, M., & Jolicoeur, P. (2004). Virtually no evidence for virtually perfect time-sharing. *Journal of Experimental Psychology: Human Perception and Performance, 30*(5), 795–810.

Towse, J. N., Hitch, G., Hamilton, Z., Peacock, K., & Hutton, U. M. Z. (2005). Working memory period: The endurance of mental representations. *Quarterly Journal of Experimental Psychology, 58*(3), 547–571.

Trainor, L. J., & Trehub, S. E. (1993). Musical context effects in infants and adults: Key distance. *Journal of Experimental Psychology: Human Perception and Performance, 19*(3), 615–626.

Tramo, M. J., Cariani, P. A., Koh, C. K., Makris, N., Braida, L. D. (2005). Neurophysiology and neuroanatomy of pitch perception: Auditory cortex. *Annals of the New York Academy of Sciences, 1060*, 148–174.

Tran, T. V., Letowski, T., & Abouchacra, K. S. (2000). Evaluation of acoustic beacon characteristics for navigation tasks. *Ergonomics, 43*(6), 807–827.

Trehub, S. E. (2001). Musical predispositions in infancy. In R. J. Zatorre & I. Peretz (Eds.), *The biological foundations of music* (Vol. 930, 1–16). New York: New York Academy of Sciences.

Trehub, S. E., & Hannon, E. E. (2006). Infant music perception: Domain-general or domain-specific mechanisms? *Cognition, 100*(1), 73–99.

Treisman, A. M. (1960). Contextual cues in selective listening. *Quarterly Journal of Experimental Psychology A, 12*, 242–248.

Treisman, A. (1964a). Verbal cues, language, and meaning in selective attention. *The American Journal of Psychology, 77*, 207–219.

Treisman, A. M. (1964b). Effect of irrelevant material on the efficiency of selective listening. *American Journal of Psychology, 77*(4), 533–546.

Treisman, A. M. (1964c). Selective attention in man. *British Medical Bulletin, 20*(12–16).

Treisman, A. M., & Gelade, G. (1980). A feature-integration theory of attention. *Cognitive Psychology, 12*(1), 97–136.

Tremblay, K. L., Piskosz, M., & Souza, P. (2002). Aging alters the neural representation of speech cues. *NeuroReport: For Rapid Communication of Neuroscience Research, 13*(15), 1865–1870.

Tsang, P. S., & Velazquez, V. L. (1996). Diagnosticity and multidimensional subjective workload ratings. *Ergonomics, 39*(3), 358–381.

Tsang, P. S., Velazquez, V. L., & Vidulich, M. A. (1996). Viability of resource theories in explaining time-sharing performance. *Acta Psychologica, 91*(2), 175–206.

Tschopp, K., Beckenbauer, T., & Harris, F. P. (1991a). Objective measures of sentence level with respect to loudness. *Audiology, 30*(2), 113–122.

Tschopp, K., Beckenbauer, T., & Harris, F. (1991b). A proposal for the measurement of the level of fluctuating background noise used in audiology. *Scandinavian Audiology, 20*(3), 197–202.

Tsien, C. L. M. S., & Fackler, J. C. M. D. (1997). Poor prognosis for existing monitors in the intensive care unit. *Critical Care Medicine, 25*(4), 614–619.

Tun, P. A. (1998). Fast noisy speech: Age differences in processing rapid speech with background noise. *Psychology and Aging, 13*(3), 424–434.

Tun, P. A., McCoy, S., & Wingfield, A. (2009). Aging, hearing acuity, and the attentional costs of effortful listening. *Psychology and Aging, 24*(3), 761–766.

Tun, P. A., O'Kane, G., & Wingfield, A. (2002). Distraction by competing speech in young and older adult listeners. *Psychology and Aging, 17*(3), 453–467.

Tun, P. A., Wingfield, A., & Stine, E. A. (1991). Speech-processing capacity in young and older adults: A dual-task study. *Psychology and Aging, 6*(1), 3–9.

Tun, P. A., Wingfield, A., Stine, E. A., & Mecsas, C. (1992). Rapid speech processing and divided attention: Processing rate versus processing resources as an explanation of age effects. *Psychology and Aging, 7*(4), 546–550.

Turner, M. L., Fernandez, J. E., & Nelson, K. (1996). The effect of music amplitude on the reaction to unexpected visual events. *Journal of General Psychology, 123*(1), 51–62.

Ullman, M. T. (2001a). The declarative/procedural model of lexicon and grammar. *Journal of Psycholinguistic Research, 30*(1), 37–69.

Ullman, M. T. (2001b). The neural basis of lexicon and grammar in first and second language: The declarative/procedural model. *Bilingualism: Language and Cognition, 4*(1), 105–122.

Ullman, M. T. (2001c). A neurocognitive perspective on language: The declarative/procedural model. *Nature Reviews Neuroscience, 2*, 717–726.

Ullman, M. T. (2004). Contributions of memory circuits to language: The declarative/procedural model. *Cognition, 92*(1–2), 231–270.

Ullman, M. T., Corkin, S., Coppola, M., Hickok, G., Growdon, J. H., Koroshetz, W. J., et al. (1997). A neural dissociation within language: Evidence that the mental dictionary is part of declarative memory, and that grammatical rules are processed by the procedural system. *Journal of Cognitive Neuroscience, 9*(2), 266–276.

Ullsperger, P., Freude, G., & Erdmann, U. (2001). Auditory probe sensitivity to mental workload changes—An event-related potential study. *International Journal of Psychophysiology, 40*(3), 201–209.

Ungerleider, L. G., & Haxby, J. V. (1994). "What" and "where" in the human brain. *Current Opinion in Neurobiology, 4*(2), 157–165.

Ungerleider, L. G., & Mishkin, M. (1982). Two cortical systems. In D. J. Ingle, M. A. Goodale & R. J. W. Mansfield (Eds.), *Analysis of visual behavior,* 549–586. Cambridge, MA: MIT Press.

Unknown. (1943). Absence from work; prevention of fatigue. *Conditions for Industrial Health and Efficiency (London),* (1), 1–20.

Unsworth, N., & Engle, R. W. (2006). Simple and complex memory spans and their relation to fluid abilities: Evidence from list-length effects. *Journal of Memory and Language, 54*(1), 68–80.

Unsworth, N., Heitz, R. P., Schrock, J. C., & Engle, R. W. (2005). An automated version of the operation span task. *Behavior Research Methods, 37*(3), 498–505.

Urquhart, R. L. (2003). *The effects of noise on speech intelligibility and complex cognitive performance.* Unpublished dissertation, Virginia Polytechnic Institute and State University, Blacksburg.

Valentijn, S. A. M., van Boxtel, M. P. J., van Hooren, S. A. H., Bosma, H., Beckers, H. J. M., Ponds, R. W. H. M., et al. (2005). Change in sensory functioning predicts change in cognitive functioning? Results from a 6-year follow-up in the Maastricht Aging Study. *Journal of the American Geriatrics Society, 53*(3), 374–380.

Vallar, G., & Baddeley, A. D. (1984). Fractionation of working memory: Neuropsychological evidence for a phonological short-term store. *Journal of Verbal Learning and Verbal Behavior, 23*(2), 151–161.

van Boxtel, M. P. J., van Beijsterveldt, C. E. M., Houx, P. J., Anteunis, L. J. C., Metsemakers, J. F. M., & Jolles, J. (2000). Mild hearing impairment can reduce verbal memory performance in a healthy adult population. *Journal of Clinical and Experimental Neuropsychology, 22*(1), 147–154.

Veltman, H. J. A. (2003, March). *Mental workload: Lessons learned from subjective and physiological measures.* Paper presented at the International Symposium of Aviation Psychology, Dayton, OH.

Veltman, J. A., Oving, A. B., & Bronkhorst, A. W. (2004). 3-D audio in the fighter cockpit improves task performance. *International Journal of Aviation Psychology, 14*(3), 239–256.

Venkatagiri, H. S. (1999). Clinical measurement of rate of reading and discourse in young adults. *Journal of Fluency Disorders, 24*(3), 209–226.

Verwey, W. B. (2000). On-line driver workload estimation. Effects of road situation and age on secondary task measures. *Ergonomics, 43*(2), 187–209.

Verwey, W. B., & Veltman, H. A. (1996). Detecting short periods of elevated workload: A comparison of nine workload assessment techniques. *Journal of Experimental Psychology: Applied, 2*(3), 270–285.

Vidulich, M. A. (1988). The cognitive psychology of subjective mental workload. In P. A. Hancock & N. Meshkati (Eds.), *Human mental workload: Advances in psychology, 52,* 219–229. Oxford, UK: North-Holland.

Vidulich, M. A., & Tsang, P. S. (1986). Techniques of subjective workload assessment: A comparison of SWAT and the NASA-Bipolar methods. *Ergonomics, 29*(11), 1385–1398.

Vidulich, M. A., & Wickens, C. D. (1986). Causes of dissociation between subjective workload measures and performance: Caveats for the use of subjective assessments. *Applied Ergonomics, 17*(4), 291–296.

Villaume, W. A., Brown, M. H., & Darling, R. (1994). Presbycusis, communication and older adults. In M. L. Hummert, J. M. Wiemann & J. F. Nussbaum (Eds.), *Interpersonal communication in older adulthood. Interdisciplinary theory and research,* 83–106. London: Sage.

Vingerhoets, G., Berckmoes, C., & Stroobant, N. (2003). Cerebral hemodynamics during discrimination of prosodic and semantic emotion in speech studied by transcranial Doppler ultrasonography. *Neuropsychology, 17*(1), 93–99.

Von Berg, S., Panorska, A., Uken, D., & Qeadan, F. (2009). DECtalk and VeriVox: Intelligibility, likeability, and rate preference differences for four listener groups. *Augmentative and Alternative Communication, 25*(1), 7–18.

Voyer, D., & Boudreau, V. G. (2003). Cross-modal correlation of auditory and visual language laterality tasks: A serendipitous finding. *Brain and Cognition, 53*(2), 393–397.

Vuontela, V., Steenari, M.-R., Carlson, S., Koivisto, J., Fjallberg, M., & Aronen, E. T. (2003). Audiospatial and visuospatial working memory in 6–13 year old school children. *Learning and Memory, 10*(1), 74–81.

Waasaf, M. A. L. N. (2007). Intonation and the structural organisation of texts in simultaneous interpreting. *Interpreting, 9*(2), 177–198.

Walker, B. N. (2002). Magnitude estimation of conceptual data dimensions for use in sonification. *Journal of Experimental Psychology: Applied, 8*(4), 211–221.

Walker, B. N., & Kogan, A. (2009). Spearcon performance and preference for auditory menus on a mobile phone. In C. Stephanidis (Ed.), *Universal access in HCI, Part II,* 445–454. Berlin: Springer-Verlag.

Walker, B. N., & Kramer, G. (2004). Ecological psychoacoustics and auditory displays. In J. G. Neuhoff (Ed.), *Ecological psychoacoustics,* 149–174. Amsterdam: Elsevier.

Walker, B. N., & Kramer, G. (2005). Mappings and metaphors in auditory displays: An experimental assessment. *ACM Transactions on Applied Perception, 2*(4), 407–412.

Walker, B. N., & Lindsay, J. (2006). Navigation performance with a virtual auditory display: Effects of beacon sound, capture radius, and practice. *Human Factors, 48*(2), 265–278.

Walker, B. N., Nance, A., & Lindsay, J. (2006, June 20–23). *Spearcons: Speech-based earcons improve navigation performance in auditory menus.* Paper presented at the International Conference on Auditory Display (ICAD2006), London.

Walker, J., Alicandri, E., Sedney, C., & Roberts, K. (1990). *In-vehicle navigation devices: Effects on the safety of driver performance* (Technical Report: FHWA-RD-90-053). Washington, DC: Federal Highway Administration.

Wallace, M. S., Ashman, M. N., & Matjasko, M. J. (1994). Hearing acuity of anesthesiologists and alarm detection. *Anesthesiology, 81*(1), 13–28.

Warm, J., Matthews, G., & Finomore, V. (2008). Vigilance, workload, and stress. In P. A. Hancock & J. L. Szalma (Eds.), *Performance under stress,* 115–141. Burlington, VT: Ashgate.

Warm, J. S., & Alluisi, E. A. (1971). Influence of temporal uncertainty and sensory modality of signals on watchkeeping performance. *Journal of Experimental Psychology, 87*(3), 303–308.

Warm, J. S., Parasuraman, R., & Matthews, G. (2008). Vigilance requires hard mental work and is stressful. *Human Factors, 50*(3), 433–441.

Warren, R. M., & Warren, R. P. (1970). Auditory illusions and confusions. *Scientific American, 223*, 30–36.

Warren, W. H., & Verbrugge, R. R. (1984). Auditory perception of breaking and bouncing events: A case study in ecological acoustics. *Journal of Experimental Psychology: Human Perception and Performance, 10*(5), 704–712.

Waters, G. S., & Caplan, D. (1996). The capacity theory of sentence comprehension: Critique of Just and Carpenter (1992). *Psychological Review October, 103*(4), 761–772.

Waters, G. S., & Caplan, D. (2001). Age, working memory, and on-line syntactic processing in sentence comprehension. *Psychology and Aging, 16*(1), 128–144.

Waters, G. S., Caplan, D., & Rochon, E. (1995). Processing capacity and sentence comprehension in patients with Alzheimer's disease. *Cognitive Neuropsychology, 12*, 1–30.

Watson, M., & Sanderson, P. (2004). Sonification supports eyes-free respiratory monitoring and task time-sharing. *Human Factors, 46*(3), 497–517.

Watson, M., Sanderson, P., & Russell, W. J. (2004). Tailoring reveals information requirements: The case of anaesthesia alarms. *Interacting with Computers, 16*(2), 271–293.

Waugh, N. C., & Norman, D. A. (1965). Primary memory. *Psychological Review, 72*(2), 89–104.

Wee, A. N., & Sanderson, P. M. (2008). Are melodic medical equipment alarms easily learned? *Anesthesia and Analgesia, 106*(2), 501–508.

Weinstein, N. D. (1974). Effect of noise on intellectual performance. *Journal of Applied Psychology, 59*(5), 548–554.

Weinstein, N. D. (1977). Noise and intellectual performance: A confirmation and extension. *Journal of Applied Psychology, 62*(1), 104–107.

Welford, A. T. (1952). The 'psychological refractory period' and the timing of high-speed performance—A review and a theory. *British Journal of Psychology, 43*, 2–19.

Wenzel, E. M., Wightman, F. L., & Foster, S. H. (October, 1988). *A virtual display system for conveying three-dimensional acoustic information.* Paper presented at the Proceedings of the Human Factors Society, Anaheim, CA.

Wessinger, C. M., VanMeter, J., Tian, B., Van Lare, J., Pekar, J., & Rauschecker, J. P. (2001). Hierarchical organization of the human auditory cortex revealed by functional magnetic resonance imaging. *Journal of Cognitive Neuroscience, 13*(1), 1.

WGSUA. (1988). Speech understanding and aging. Report of the Working Group on Speech Understanding and Aging. *Journal of the Acoustical Society of America, 83*, 859–895.

White, J. N., Hutchens, T. A., & Lubar, J. F. (2005). Quantitative EEG assessment during neuropsychological task performance in adults with attention deficit hyperactivity disorder. *Journal of Adult Development, 12*(2), 113–121.

Wickens, C., & Colcombe, A. (2007). Dual-task performance consequences of imperfect alerting associated with a cockpit display of traffic information. *Human Factors, 49*(5), 839–850.

Wickens, C. D. (1980). The structure of attentional resources. In R. S. Nickerson (Ed.), *Attention and performance VIII*, 239–257. Hillsdale, NJ: Erlbaum.

Wickens, C. D. (1984). Processing resources in attention. In R. Parasuraman & R. Davies (Eds.), *Varieties of attention*, 63–101. Orlando, FL: Academic Press.

Wickens, C. D. (1990). Applications of event-related potential research to problems in human factors. In J. W. Rohrbaugh, R. Parasuraman, & R. Johnson, Jr. (Eds.), *Event-related brain potentials: Basic issues and applications*, 301–309. London: Oxford University Press.

Wickens, C. D. (1991). Processing resources and attention. In D. L. Damos (Ed.), *Multiple-task performance*, 3–34. London: Taylor & Francis.

Wickens, C. D. (1992). *Engineering psychology and human performance* (2nd ed.). New York: HarperCollins.

Wickens, C. D. (2002). Multiple resources and performance prediction. *Theoretical Issues in Ergonomics Science, 3*(2), 159–177.

Wickens, C. D., & Hollands, J. G. (2000). *Engineering psychology and human performance* (3rd ed.). Upper Saddle River, NJ: Prentice Hall.

Wickens, C. D., Kramer, A. F., & Donchin, E. (1984). The event-related potential as an index of the processing demands of a complex target acquisition task. *Annals of the New York Academy of Sciences, 425,* 295–299.

Wickens, C. D., & Liu, Y. (1988). Codes and modalities in multiple resources: A success and a qualification. *Human Factors, 30*(5), 599–616.

Wierwille, W. W. (1979). Physiological measures of aircrew mental workload. *Human Factors, 21*(5), 575–593.

Wierwille, W. W., & Connor, S. A. (1983). Evaluation of 20 workload measures using a psychomotor task in a moving-base aircraft simulator. *Human Factors, 25*(1), 1–16.

Wierwille, W. W., & Eggemeier, F. (1993). Recommendations for mental workload measurement in a test and evaluation environment. *Human Factors, 35*(2), 263–281.

Wierwille, W. W., Rahimi, M., & Casali, J. G. (1985). Evaluation of 16 measures of mental workload using a simulated flight task emphasizing mediational activity. *Human Factors, 27*(5), 489–502.

Wiese, E. E., & Lee, J. D. (2004). Auditory alerts for in-vehicle information systems: The effects of temporal conflict and sound parameters on driver attitudes and performance. *Ergonomics, 47*(9), 965–986.

Wiley, T. L., Cruickshanks, K. J., Nondahl, D. M., Tweed, T. S., Klien, R., & Klien, B. E. K. (1998). Aging and word recognition in competing message. *Journal of the American Academy of Audiology, 9*(3), 191–198.

Williges, R. C., & Wierwille, W. W. (1979). Behavioral measures of aircrew mental workload. *Human Factors, 21*(5), 549–574.

Wilson, G. F., & Eggemeier, F. T. (1991). Psychophysiological assessment of workload in multi-task environments. In D. L. Damos (Ed.), *Multiple-task performance*, 329–360. London: Taylor & Francis.

Wingfield, A., & Grossman, M. (2006). Language and the aging brain: Patterns of neural compensation revealed by functional brain imaging. *Journal of Neurophysiology, 96*(6), 2830–2839.

Wingfield, A., & Tun, P. A. (1999). Working memory and spoken language comprehension: The case for age stability in conceptual short-term memory. In S. Kemper & R. Kliegl (Eds.), *Constraints on language: Aging, grammar, and memory*, 29–51. Boston: Kluwer Academic.

Wingfield, A., Tun, P. A., & McCoy, S. L. (2005). Hearing loss in older adulthood. What it is and how it interacts with cognitive performance. *Current Directions in Psychological Science, 14*(3), 144–148.

Wingfield, A., Tun, P. A., O'Kane, G., & Peelle, J. E. (2005). Language comprehension in complex environments: Distraction by competing speech in young and older adult listeners. In S. P. Shohov (Ed.), *Advances in psychology research* (Vol. 33, 3–38). Hauppauge, NY: Nova Science.

Winkler, I., Haden, G. P., Ladinig, O., Sziller, I., & Honing, H. (2009). Newborn infants detect the beat in music. *Proceedings of the National Academy of Sciences of the United States of America, 106*(7), 2468–2471.

Winkler, I., Paavilainen, P., & Naatanen, R. (1992). Can echoic memory store two traces simultaneously? A study of event-related brain potentials. *Psychophysiology, 29*(3), 337–349.

Withington, D. J. (1999). Localisable alarms. In N. A. Stanton & J. Edworthy (Eds.), *Human factors in auditory warnings*, 33–40. Aldershot: UK: Ashgate.

Wogalter, M. S., Kalsher, M. J., Frederick, L. J., Magurno, A. B., & Brewster, B. M. (1998). Hazard level perceptions of warning components and configurations. *International Journal of Cognitive Ergonomics, 2*(1–2), 123–143.

Wong, P. C. M., Jin, J. X., Gunasekera, G. M., Abel, R., Lee, E. R., & Dhar, S. (2009). Aging and cortical mechanisms of speech perception in noise. *Neuropsychologia, 47*(3), 693–703.

Xun, X., Guo, S., & Zhang, K. (1998). The effect of inputting modality of secondary task on tracking performance and mental workload. *Acta Psychologica Sinica, 30*(3), 343–347.

Yeh, Y.-Y., & Wickens, C. D. (1988). Dissociation of performance and subjective measures of workload. *Human Factors, 30,* 111–120.

Yik, W. F. (1978). The effect of visual and acoustic similarity on short-term memory for Chinese words. *Quarterly Journal of Experimental Psychology A, 30*(3), 487–494.

Yost, W. A. (2006). *Fundamentals of hearing* (5th ed.). Bingley, UK: Emerald Group.

Young, H. H., & Berry, G. L. (1979). The impact of environment on the productivity attitudes of intellectually challenged office workers. *Human Factors, 21*(4), 399–407.

Young, M. S., & Stanton, N. A. (2002). Malleable attentional resources theory: A new explanation for the effects of mental underload on performance. *Human Factors, 44*(3), 365–375.

Zatorre, R. (1998). How do our brains analyze temporal structure in sound? *Nature Neuroscience, 1*(5), 343–345.

Zatorre, R. J. (2005). Neuroscience: Finding the missing fundamental. *Nature, 436*(7054), 1093–1094.

Zatorre, R. J., Belin, P., & Penhume, V. B. (2002). Structure and function of auditory cortex: Music and speech. *Trends in Cognitive Science, 6*(1), 37–46.

Zatorre, R. J., Bouffard, M., Ahad, P., & Belin, P. (2002). Where is "where" in the human auditory cortex? *Nature Neuroscience, 5*(9), 905–909.

Zatorre, R. J., Evans, A. C., Meyer, E., & Gjedde, A. (1992). Lateralization of phonetic and pitch discrimination in speech processing. *Science, 256*(5058), 846–849.

Zatorre, R. J., & Peretz, I. (2003). *The cognitive neuroscience of music.* Oxford: Oxford University Press.

Zeitlin, L. R. (1995). Estimates of driver mental workload: A long-term field trial of two subsidiary tasks. *Human Factors, 37*(3), 611–621.

Zentner, M. R., & Kagan, J. (1998). Infants' perception of consonance and dissonance in music. *Infant Behavior and Development, 21*(3), 483–492.

Zimmer, H. D. (1998). Spatial information with pictures and words in visual short-term memory. *Psychological Research/Psychologische Forschung, 61*(4), 277–284.

Index

3-D audio cues, 238–239

A

A-weighted network, 35
access, word recognition and, 170
acoustic advantage, 188–189
acoustic beacons, 29
acoustic confusions, 66, 188
acoustic cues, 157
  invariant, 159
acoustic environment, effects of on speech
    processing, 178
acoustic factors, influence of on mental workload
    of speech processing, 172–179
acoustic information, lexical interpretation of,
    153
acoustic parameters, perceived urgency and,
    243–244
acoustic regularity, 116
acoustic signal, variability in, 158
acoustic strategies, reliance on by older adults,
    222
acoustic-sensory level, registration of sounds at,
    165
acoustical information
  binding of to lexical information, 161
  matching of semantic content with, 154
acoustics, speech processing and changes
    in, 154
adverse listening conditions, effect of on speech
    processing of older adults, 213–218
affective prosody, speech processing and,
    176–177
age-related change in speech processing,
    210–220
age-related changes in auditory processing,
    201–203
age-related cognitive changes, 209–210
age-related design, 223–224
age-related hearing loss, 203–204
  lexical selection and, 179
aging
  auditory tasks in cognitive research of,
    112–113
  human-machine interfaces and, 7
air traffic control (ATC)
  auditory communications in, 74
  communication of with pilots, 22
  communications, 181

effect of speech rate on communication errors
    in, 175
  selective listening tasks in, 98
  text vs. speech displays, 186
  use of auditory displays in, 230–231, 252–254
  verbal information processing in, 190
aircraft noise, exposure of children to, 149–150
alarms
  nonverbal, 17–18
  reliability of, 239–240
allophones, 159
allophonic cues, 157
alterants, 74–75
amplitude, 33–34
amusia, 41–42
ANN. See Artificial neural networks
anticipatory attending, 97
aphasias, 96, 189
  language processing and, 166–167
  working memory resource capacity and, 79
arousal
  performance and, 138–139
  theory, 144
articulation rate, mental workload and, 174–175
articulatory rehearsal, 66
articulatory suppression, 68
artificial neural networks, 69–70
asynchronous music, physical performance and,
    128
ATC terminology, 22
attensons, 243
attention, 53–54
  cognitive neuroscience of, 58–59
  early research in, 54–59
  vigilance and performance decrements,
    150–151
  working memory resource system, 80
attentional capacity, resource theories of, 73–75
attentional control, 97
attentional processing, 4–5
  serial vs. parallel, 68–69
  speech processing and, 156
attentional resources, limited capacity of, 72–73
attenuation theory, 56–57
audibility threshold, 35
audio displays
  alarms in medical environments, 28–29
  infotainment systems, 28
audiovisual (AV) speech, 190–192
  encoding of, 184